UAV Photogrammetry and Remote Sensing

UAV Photogrammetry and Remote Sensing

Editors

Fernando Carvajal-Ramírez
Francisco Agüera-Vega
Patricio Martínez-Carricondo

MDPI • Basel • Beijing • Wuhan • Barcelona • Belgrade • Manchester • Tokyo • Cluj • Tianjin

Editors
Fernando Carvajal-Ramírez
University of Almería
Spain

Francisco Agüera-Vega
University of Almería
Spain

Patricio Martínez-Carricondo
University of Almería
Spain

Editorial Office
MDPI
St. Alban-Anlage 66
4052 Basel, Switzerland

This is a reprint of articles from the Special Issue published online in the open access journal *Remote Sensing* (ISSN 2072-4292) (available at: https://www.mdpi.com/journal/remotesensing/special_issues/UAV_Photogrammetry_RS).

For citation purposes, cite each article independently as indicated on the article page online and as indicated below:

LastName, A.A.; LastName, B.B.; LastName, C.C. Article Title. *Journal Name* **Year**, *Volume Number*, Page Range.

ISBN 978-3-0365-1454-3 (Hbk)
ISBN 978-3-0365-1453-6 (PDF)

Contents

ditorial

Editorial for Special Issue "UAV Photogrammetry and Remote Sensing"

ernando Carvajal-Ramírez [1,*], Francisco Agüera-Vega [1] and Patricio Martínez-Carricondo [1,2]

1 Mediterranean Research Center of Economics and Sustainable Development (CIMEDES), Department of Engineering, University of Almería (Agrifood Campus of International Excellence, ceiA3), La Cañada de San Urbano, 04120 Almería, Spain; faguera@ual.es (F.A.-V.); pmc824@ual.es (P.M.-C.)
2 Peripheral Service of Research and Development Based on Drones, University of Almeria, La Cañada de San Urbano, 04120 Almería, Spain
* Correspondence: carvajal@ual.es; Tel.: +34-950-015-950

itation: Carvajal-Ramírez, F.; güera-Vega, F.; Martínez-Carricondo, Editorial for Special Issue "UAV hotogrammetry and Remote ensing". *Remote Sens.* **2021**, *13*, 2327. ttps://doi.org/10.3390/rs13122327

eceived: 4 June 2021
ccepted: 10 June 2021
ublished: 13 June 2021

1. Introduction

The concept of Remote Sensing as a way of capturing information from an object without making contact with it has, until recently, been exclusively focused on the use of earth observation satellites.

The emergence of unmanned aerial vehicles (UAV) with Global Navigation Satellite Systems (GNSS) controlled navigation and sensor-carrying capabilities has increased the number of publications related to new remote sensing from much closer distances. Previous knowledge about the behavior of the Earth's surface under the incidence of energy of different wavelengths has been successfully applied to a large amount of data recorded from UAVs, thereby increasing the special and temporal resolution of the products obtained.

More specifically, the ability of UAVs to be positioned in the air at pre-programmed coordinate points, to track flight paths, and in any case, to record the coordinates of the sensor position at the time of the shot and pitch, yaw, and roll angles have opened an interesting field of applications for low-altitude aerial photogrammetry, known as UAV Photogrammetry. In addition, photogrammetric data processing has been improved thanks to the combination of new algorithms, e.g., structure from motion (SfM), which solve the collinearity equations without the need for any control point, producing a cloud of points referenced to an arbitrary coordinate system and a full camera calibration, and multi-view stereopsis (MVS) algorithm that applies an expanding procedure of a sparse set of matched keypoints in order to obtain a dense point cloud. The set of technical advances described above allows geometric modeling of terrain surfaces with high accuracy, minimizing the need for topographic campaigns for the georeferencing of such products.

This special issue aims to compile some applications realized thanks to the synergies established between the new remote sensing from close distances and UAV Photogrammetry. The contributions are briefly described below in alphabetical order of the first author.

2. Overview of Contributions

In the paper [1], the authors carried out an interesting combination of UAV Photogrammetry and Large-Scale Airborne Lidar Data to monitor snow masses in a forested region in central Arizona, United States. They observed that in low dense forest conditions, both sources of data deliver similar snow depth maps while in high dense forest, lidar maps are more accurate. In the other hand, UAV Photogrammetry terrain model can be used to basin-scale snowpack estimation with a multi-temporal information with a lower cost that airborne lidar campaigns.

In [2], the authors stablish the optimal distribution and number of Ground Control Points (GCPs) to use in corridor maps applied to linear projects obtained in southeast Spain. They used UAV Photogrammetry based on SfM and SMV algorithms and concluded that

9–11 GCP distributed alternatively on both sides of the road, with a pair of GCPs at each end of the road yielded optimal results regarding fieldwork cost.

The paper [3] presents a valuable fusion of digital surface model (DSM) in an extremely challenging urban environment with high level detail, and UAV orthomosaic. The authors integrated three models: adaptive hierarchical image segmentation optimization, multilevel feature selection, and multiscale supervised machine learning. They concluded that the applied methodology showed an excellent potential for the mapping the selected urban landscape in Malaysia.

The quality assessment of UAV Photogrammetric products was the main concern of [4]. In this work, the geolocation procedure of UAV orthomosaics time series was optimized, obtaining a reproducibility of 99% in a grassland located in Germany and 75% in a forest area in the Spanish Pyreneess.

UAV Photogrammetry can model terrain surfaces with extreme or quasi-vertical morphologies [5]. In this work, several combinations of number of GCP, distribution and image orientation were tested in a dam belonging to Spain's hydraulic heritage located in the Almería province, obtaining similar results than terrestrial laser scanner TLS. The authors advised that the results ostensibly improve including oblique images and break lines.

Vegetation used to be an obstacle for accurate DSM. The authors of [6] applied Deep Learning and Terrain Correction models in Chinese Loess Plateau to solve the restriction of UAV Photogrammetry in a vegetation-dense area with a complex terrain due to reduced ground visibility and lack of robust filtering algorithms. They detected the vegetation with overall accuracy of 95% and the mean square error of final DTM was 0.024 m.

In other cases, the target cover to be detected is precisely vegetation that frequently is modelled through standard vegetation indexes. In [7], the authors studied useful correlations between certain parameters of chemical analysis carried out in agriculture crops and vegetation indexes obtained from UAV-Photogrammetry assessments.

In the design process of a UAV photogrammetric project, the resolution of the images to be captured is established according to the minimum size of the smallest target to be detected. The user wonders how much information is lost when the imagery resolution decreases. In [8], the authors apply the deep convolutional neural network approach, based on a single image super-resolution, on low-resolution UAV imagery for spatial resolution enhancement. Using these high-resolution images in a SfM Photogrammetric process, they observed that the number of points in dense point cloud is about 17 times more than those extracted from a low-resolution image set.

In the paper [9], the authors carried out an interesting practical application for the 3D reconstruction of Power Lines based on UAV Photogrammetry. They stablished a 3D corridor around power lines and detected some objects inside this volume as obstacles that could threat safety of the infrastructure. They compared UAV Photogrammetry models with total station survey and terrestrial laser scanning, concluding that the accuracy is consistent.

UAV Photogrammetry can be used as a valuable source of data in architectural design process [10]. They proposed a virtual integration of UAV Photrogrammetry products and architectural design using building information modeling (BIM) technology, observing error reductions, and significate time and cost saving.

The implementation of GNSS-based UAV navigation capability in real time kinematic (RTK) mode has reduced or even eliminated the need for topographic campaigns to obtain GCPs. However, this technology implies systematic errors in elevation. In [11], the authors observed a linear relationship between these errors and the deviation in the focal length adjustment and proposed the combination of two flights with different image axis angles for their elimination.

One of the difficulties that large UAV projects present is the management of high quantity of images. In [12], the authors proposed an intelligent method for image selections in UAV-Photogrammetry projects that can be used to avoid the time-consuming manual image selection process, maintaining overlaps needed for point cloud extraction and avoiding reductions of products accuracy.

3. Conclusions

The set of contributions to this special issue points out that the possible scientific applications in the field of UAV Photogrammetry and Remote Sensing from close distances is very wide.

UAVs not only take advantage of the previous knowledge in Remote Sensing acquired over the years, but they also improve and expand its possibilities thanks to the control that users have over the sensors.

UAV photogrammetry has been shown as a previous process in all those remote sensing applications whose observed targets are spatially distributed along the terrain surface, obtaining orthomosaics and digital surface models with high spatial and temporal resolution.

Author Contributions: Conceptualization and investigation, F.C.-R., F.A.-V. and P.M.-C.; writing—original draft preparation, F.C.-R.; writing—review and editing, F.C.-R.; visualization and supervision, F.A.-V. and P.M.-C. All authors have read and agreed to the published version of the manuscript.

Funding: This research received no external funding.

Institutional Review Board Statement: Not applicable.

Informed Consent Statement: Not applicable.

Data Availability Statement: Not applicable.

Acknowledgments: The guest editors of the special issue "UAV Photogrammetry and Remote Sensing" would first like to sincerely thank the authors who have contributed with their innovate work. We would like to acknowledge the valuable work of the reviewers who have improved the quality of the published papers and thank the editorial staff for their support and kind invitation to be guest editors of a special issue of the prestigious journal "Remote Sensing".

Conflicts of Interest: The authors declare no conflict of interest.

References

1. Broxton, P.D.; van Leeuwen, W.J.D. Structure from motion of multi-angle RPAS imagery complements larger-scale airborne lidar data for cost-effective snow monitoring in mountain forests. *Remote Sens.* **2020**, *12*, 2311. [CrossRef]
2. Ferrer-González, E.; Agüera-Vega, F.; Carvajal-Ramírez, F.; Martínez-Carricondo, P. UAV Photogrammetry Accuracy Assessment for Corridor Mapping Based on the Number and Distribution of Ground Control Points. *Remote Sens.* **2020**, *12*, 2447. [CrossRef]
3. Gibril, M.B.A.; Kalantar, B.; Al-Ruzouq, R.; Ueda, N.; Saeidi, V.; Shanableh, A.; Mansor, S.; Shafri, H.Z.M. Mapping heterogeneous urban landscapes from the fusion of digital surface model and unmanned aerial vehicle-based images using adaptive multiscale image segmentation and classification. *Remote Sens.* **2020**, *12*, 1081. [CrossRef]
4. Ludwig, M.; Runge, C.M.; Friess, N.; Koch, T.L.; Richter, S.; Seyfried, S.; Wraase, L.; Lobo, A.; Sebastià, M.T.; Reudenbach, C.; et al. Quality assessment of photogrammetric methods—A workflow for reproducible UAS orthomosaics. *Remote Sens.* **2020**, *12*, 3831. [CrossRef]
5. Martínez-Carricondo, P.; Agüera-Vega, F.; Carvajal-Ramírez, F. Use of UAV-Photogrammetry for Quasi-Vertical Wall Surveying. *Remote Sens.* **2020**, *12*, 2221. [CrossRef]
6. Na, J.; Xue, K.; Xiong, L.; Tang, G.; Ding, H.; Strobl, J.; Pfeifer, N. UAV-based terrain modeling under vegetation in the chinese loess plateau: A deep learning and terrain correction ensemble framework. *Remote Sens.* **2020**, *12*, 3318. [CrossRef]
7. Janoušek, J.; Jambor, V.; Marcoň, P.; Dohnal, P.; Synková, H.; Fiala, P. Using UAV-Based Photogrammetry to Obtain Correlation between the Vegetation Indices and Chemical Analysis of Agricultural Crops. *Remote Sens.* **2021**, *13*, 1878. [CrossRef]
8. Pashaei, M.; Starek, M.J.; Kamangir, H.; Berryhill, J. Deep learning-based single image super-resolution: An investigation for dense scene reconstruction with UAS photogrammetry. *Remote Sens.* **2020**, *12*, 1757. [CrossRef]
9. Pastucha, E.; Puniach, E.; Ścisłowicz, A.; Ćwiakała, P.; Niewiem, W.; Wiącek, P. 3D reconstruction of power lines using uav images to monitor corridor clearance. *Remote Sens.* **2020**, *12*, 3698. [CrossRef]
10. Rizo-Maestre, C.; González-Avilés, Á.; Galiano-Garrigós, A.; Andújar-Montoya, M.D.; Puchol-García, J.A. UAV + BIM: Incorporation of photogrammetric techniques in architectural projects with building information modeling versus classical work processes. *Remote Sens.* **2020**, *12*, 2329. [CrossRef]

11. Štroner, M.; Urban, R.; Seidl, J.; Reindl, T.; Brouček, J. Photogrammetry Using UAV-Mounted GNSS RTK: Georeferencing Strategies without GCPs. *Remote Sens.* **2021**, *13*, 1336. [CrossRef]
12. Lim, P.; Rhee, S.; Seo, J.; Kim, J.; Chi, J.; Lee, S.; Kim, T. An Optimal Image—Selection Algorithm for Large-Scale Stereoscopic Mapping of UAV Images. *Remote Sens.* **2021**, *11*, 2118. [CrossRef]

rticle

Jsing UAV-Based Photogrammetry to Obtain Correlation between the Vegetation Indices and Chemical Analysis of Agricultural Crops

iří Janoušek [1,*], Václav Jambor [2], Petr Marcoň [1], Přemysl Dohnal [1], Hana Synková [2] and Pavel Fiala [1]

[1] Faculty of Electrical Engineering and Communication, Brno University of Technology, 61600 Brno, Czech Republic; marcon@vutbr.cz (P.M.); dohnalp@feec.vutbr.cz (P.D.); fialap@vutbr.cz (P.F.)
[2] NutriVet s.r.o., Vídeňská 1023, 69123 Pohořelice, Czech Republic; jambor.vaclav@nutrivet.cz (V.J.); nutrivet@nutrivet.cz (H.S.)
* Correspondence: xjanou09@vutbr.cz

Abstract: The optimum corn harvest time differs between individual harvest scenarios, depending on the intended use of the crop and on the technical equipment of the actual farm. It is therefore economically significant to specify the period as precisely as possible. The harvest maturity of silage corn is currently determined from the targeted sampling of plants cultivated over large areas. In this context, the paper presents an alternative, more detail-oriented approach for estimating the correct harvest time; the method focuses on the relationship between the ripeness data obtained via photogrammetry and the parameters produced by the chemical analysis of corn. The relevant imaging methodology utilizing a spectral camera-equipped unmanned aerial vehicle (UAV) allows the user to acquire the spectral reflectance values and to compute the vegetation indices. Furthermore, the authors discuss the statistical data analysis centered on both the nutritional values found in the laboratory corn samples and on the information obtained from the multispectral images. This discussion is associated with a detailed insight into the computation of correlation coefficients. Statistically significant linear relationships between the vegetation indices, the normalized difference red edge index (NDRE) and the normalized difference vegetation index (NDVI) in particular, and nutritional values such as dry matter, starch, and crude protein are evaluated to indicate different aspects of and paths toward predicting the optimum harvest time. The results are discussed in terms of the actual limitations of the method, the benefits for agricultural practice, and planned research.

Keywords: multispectral imaging; vegetation indices; nutritional analysis; correlation; photogrammetry; optimal harvest time; UAV

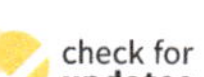

itation: Janoušek, J.; Jambor, V.; Iarcoň, P.; Dohnal, P.; Synková, H.; ala, P. Using UAV-Based hotogrammetry to Obtain orrelation between the Vegetation ıdices and Chemical Analysis of gricultural Crops. *Remote Sens.* **2021**, 3, 1878. https://doi.org/10.3390/ 13101878

cademic Editors: ernando Carvajal-Ramírez, rancisco Agüera-Vega and atricio Martínez-Carricondo

eceived: 31 March 2021
ccepted: 10 May 2021
ublished: 11 May 2021

ublisher's Note: MDPI stays neutral ith regard to jurisdictional claims in ublished maps and institutional affil- tions.

1. Introduction

Precision agriculture (or site-specific crop management) is an internationally recognized concept and term referring to land cultivation by means of nontraditional technologies that were first designed and developed at the end of the 1980s [1–3]. The aim of the concept rests in adjusting cultivating procedures to suit local conditions, the main principle being to perform the crop-growing tasks at the right place, intensity, and time [4,5].

The standard process to estimate the condition of crops during the growth phase, especially when the correct harvest time has to be defined, involves a land survey in which sample plants are manually collected and then chemically analyzed in a laboratory. Such an approach, however, is labor- and time-intensive because it relies mainly on direct human inspection inside the crop fields, which are usually inhomogeneous and thus difficult to characterize accurately through a single analysis. An effective alternative then appears to lie in remote sensing, a technique applicable in determining crop maturity degrees over large areas. The procedure yields rapid information on spatial and temporal changes in the monitored quantities [6], allowing farmers to recognize and differentiate between the specific conditions that characterize individual portions of the land; this task can also be performed via other survey methods, but only with considerable difficulty. The noninvasive evaluation of crop quality by means of multispectral imaging facilitates reform steps in agricultural management [7–10]. In the discussed field, remote sensing generally offers two functional options, namely, satellite imagery [11–15] and unmanned aerial vehicle (UAV) photogrammetry [16–20].

Advancements in unmanned aerial vehicles (UAVs) and the related developments concerning their use in remote sensing have made the technology a promising tool in recent decades [21]. Most importantly, UAVs (drones) as a remote sensing platform have shown major potential in crop-growth monitoring [22,23], where they ensure a proper balance between the image quality, sensing efficiency, and operating cost. The spectral information and vegetation indices derived from UAV-delivered multispectral or hyperspectral data have now been widely tested for this purpose [24–26].

This article describes the monitoring of a selected corn hybrid within pre-defined growth intervals, to find a relationship between the variation in the nutritional values of crops and changes in UAV-based photogrammetry images. In this context, one of the main problems investigated is the connection between the data obtained from chemical analyses of sample plants and the vegetation indices calculated from spectral reflectivities, which are monitored by using a multispectral camera. The aim of the research is to establish experimentally whether the optimum corn harvest time and quantity can be predicted with an emphasis on searching for a mathematical relationship between a variation in the content of dry matter in the sampled corn and changes in the vegetation indices. The dry matter content constitutes a significant nutritional parameter for both biogas stations and livestock production [27,28]. The optimum corn harvest time differs according to the intended use of the crop and the technical equipment and installations in the relevant works; for this reason, it is then important to specify the ideal harvest time as precisely as possible, considering the circumstances [29–31].

At present, the maturity of silage corn is normally estimated based on the targeted sampling of plants over large corn-growing farm areas (usually having an acreage of dozens of hectares). The counts of samples differ markedly, depending on the desired accuracy. The proportion of dry matter, nutritional substances, and other parameters are defined by means of a 7- to 10-day laboratory analysis. Due to the cost, the heterogeneity of the samples, and labor intensity, the actual survey can be carried out with only a limited set of samples and does not effectively cover the areal changes in the vegetation. Within the Czech Republic, by extension, the diversity of pedoclimatic conditions and the sizes of land units point to a substantial imbalance in the properties of land managed by enterprises, and embody preconditions for the successful implementation of the principles of precision agriculture.

Currently, systems are available that can assess comprehensively the quantity and quality of crops; to provide relevant examples, we can refer to John Deere's HarvestLab 3000 and the Evo NIR sensor by Dinamica Generale S.p.A. Such systems utilize NIR (near infrared) cameras mounted on the harvesters, and are operated only during the harvest period [32–34]. When seeking the optimum nutritional values, agricultural technologists and researchers employ various types of prediction; the authors of source [35], for instance exploited superspectral airborne imagery to predict corn grain yield and ear weight, and to discriminate between growth stages and irrigation treatments. The use of multispectral imaging to determine the phenophase, however, is somewhat contrasted by the fact that the authors focused solely on utilizing the normalized difference vegetation index (NDVI). Predicting the optimum harvest time is associated with diverse factors, including the volume of dry matter in the plants. The problem of dry matter in forage corn grown for silage is discussed in paper [36], with an emphasis on measuring the NDVI of the plants.

By contrast, our concept has been formulated to deliver area-specific harvest estimates that involve more input elements than merely the NDVI or chemical analyses performed with a limited number of plants from over the entire field. This article thus outlines original options for predicting the optimum harvest time, and these are based on searching for correlations between the chemical analysis of sampled corn and the images acquired with a spectral camera in the course of a UAV photogrammetry cycle. By another definition, the novelty of this article rests in that the statistical analysis is applied to reveal hitherto unexplored relationships between nutritional parameters, acquired through chemical analyses and vegetation indices yielded via processing data collected by a multispectral camera. These relationships will be utilized in estimating the optimum harvest time for the entire area of the selected cornfield.

To summarize the various views and perspectives, we can point out that this subsector still offers ample room for new approaches and interpretations.

As regards the actual structure of this article, the text is organized into three sections: Section 2 presents the chemical analysis of the samples, the remote sensing, and the data correlation methodology, Section 3 introduces the results, and Section 4 contains the discussion and conclusion.

2. Materials and Methods

The data collection and mathematical processing are characterized in the block diagram in Figure 1. The information relating to the investigated agricultural land is captured via UAV photogrammetry and manual selection.

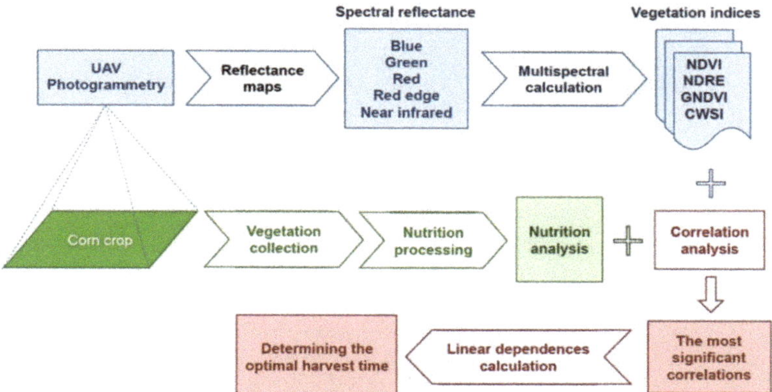

Figure 1. A block diagram defining the data collection and processing.

The multispectral images delivered by the UAV-mounted camera enable us to compile relevant reflectivity maps, which then facilitate computing the vegetation indices. To obtain the nutritional indicators in the sampled corn, we performed laboratory-based chemical analysis. The vegetation indices and the results of the analysis were then correlated at various stages of the growth phenophases; this step allowed identifying the time when the crop yield is ideal for ensuring the production of silage or methane.

2.1. Study Site

The experimental monitoring and sampling were carried out over agricultural land managed by the enterprise Bonagro Blažovice, a.s., the type of crop involved being the corn hybrid LG Apotheos FAO 500, delivered by Limagrain Central Europe S.E. The land is located in the vicinity of the village of Prace (Figure 2), in the South Moravian Region, Moravia, the Czech Republic; the coordinates of the test fields are 49.1472789N, 16.7701758E. In terms of the climate, the land, being situated within a temperate zone and at an altitude of 260 m, generally experiences warm to hot summers. During the monitored phenophase, the average air temperature and precipitation reached 14.7 °C and 12.7 mm a week, respectively. The average amounts of precipitation differed between the individual sampling cases. The concrete values equaled 17.5 mm per week in the first 3 weeks, zero (no rain) in the following 3 weeks, and 45.5 mm in the last week.

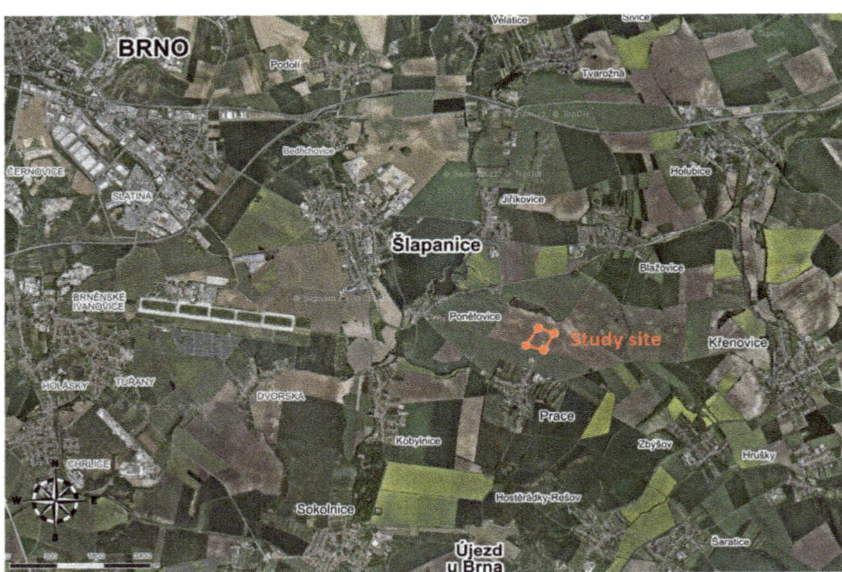

Figure 2. The location of the fields observed in the experiment.

The experiment started with the initial corn sampling on 12 August 2020, when the crop was going through the second half of the phenological stage of growth and was still earless. Procedurally, in the area of interest, we performed imaging and collected samples for chemical analysis, invariably at weekly intervals. In total, the samples were collected at eight time intervals, and the last sampling took place on 5 October 2020; by that date compared to corn not involved in the experiment, the condition of the plants had already corresponded to a later post-harvest stage. The sampling and imaging were regularly executed between 12 p.m and 2 p.m.

2.2. Imaging Methodology

The initial step consisted of acquiring a sufficient quantity of various data by using a multispectral camera mounted on a UAV (Figure 3a,b). For this purpose, we employed a MicaSense RedEdge camera on a DJI Matrice 600 Pro aerial vehicle. The camera operates in 5 narrow spectral bands, and each of the sensors has a resolution of 1280 × 940 pixels. The device ensures the narrowband recording of wavelengths in regions sensitive to the human eye, namely, the range of 400 to 700 nm in blue—B, green—G, and red—R, and also within the rim of the red sector of visible light (red edge—RE); the near-infrared (NIR) range invisible to the human eye, is recorded also. The concrete parameters of the bands are summarized in Table 1. To carry out the scanning, we preset the automatic sequential image shooting mode, based on the exact position of the aerial vehicle. The images were stored on a memory card, together with the metadata comprising the concrete GPS locations where the images were taken. Importantly, the camera contains a light sensor module to correct the exposure at varying light intensities; the sensor thus automatically adjusts the camera exposure according to the angle of the incident beams and the brightness. To compensate for the reflectivity, we calibrated the sensors before each flight by taking images of the gray calibration panel indicating the known parameters (Figure 3c). The panel ensures that the images remain stable regardless of the light conditions.

The flight path was created via the application Pix4D Capture [37]. The total area of the monitored crops exhibited a rectangular shape and dimensions of 401 m × 331 m (approximately 13.2 ha) (Figure 4). The length of the flight path equaled 4477 m, and the actual survey flight took 31 min, with an image overlap of 70%. The UAV performed the imaging at a speed and height above ground level of 8.6 km/h and 40 m, respectively. In each monitored band, we acquired 450 images with a resolution of 2.78 cm/pixel.

(a)

(b)

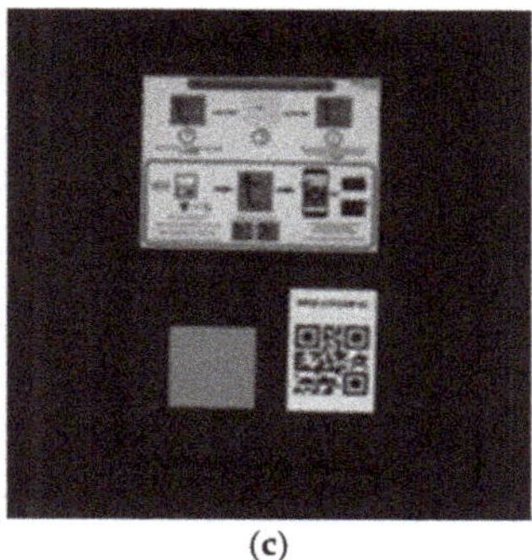
(c)

Figure 3. (**a**) The UAV in operation; (**b**) the RedEdge multispectral camera; (**c**) the calibration panel showing the known reflectivity value for each of the bands recorded.

Table 1. The parameters of the bands recorded by the RedEdge camera.

Band	Band Name	Wavelength [nm]	Bandwidth [nm]
1	Blue (B)	475	20
2	Green (G)	560	20
3	Red (R)	668	10
4	Red edge (RE)	717	10
5	Near-infrared (NIR)	840	40

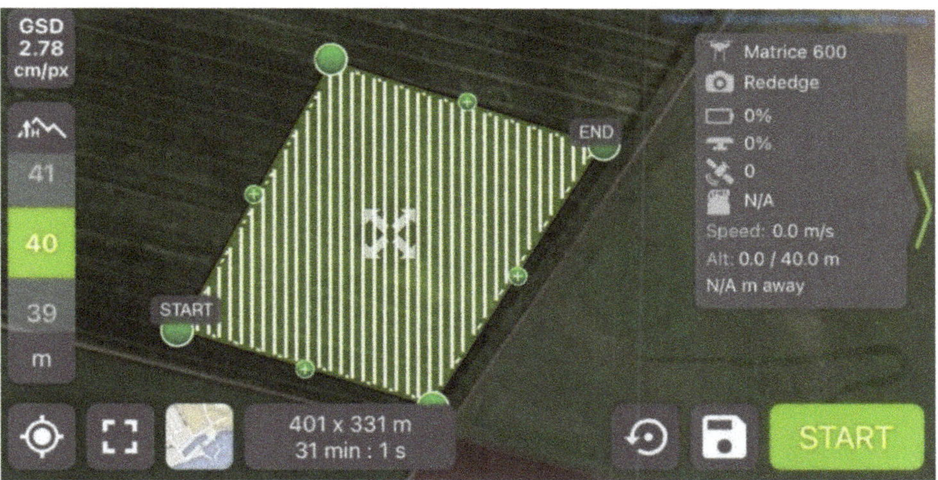

Figure 4. The area of interest visualized in Pix4D Capture.

2.3. Nutrition Value Processing

The biomass from the harvested crops comprises exclusively whole corn plants. The basic indicator relating to the phenophase of crops rests in determining the dry matter content; the term dry matter then represents the solid, waterless portion of fodder. In corn the dry matter indicates vegetation maturity [29]. During the vegetation growing season the chemistry of corn plants changes; in the course of the earless phases, the energy is stored particularly in the fibers. Fodder for dairy cows, however, requires ear starch. Thus to ensure that the silage contains both the fibers and the starch, the crops are harvested at wax maturity, when the plants' dry matter content reaches 280 to 330 g/kg. In such cases the milk line stage attains the level of 2/3 in the corn grain. Another vegetation maturity indicator lies in the corn's capability of being silaged, namely, producing the fermentation acids that conserve the silage.

The quality of the fermentation processes is fundamentally influenced by the harvest time and the total biomass quantity.

Every year, the properties of the crops and the silaging are highly variable, depending on the weather, the selected hybrid and its Food and Agriculture Organization (FAO designation (the number of vegetation days), and the quality of sowing and care. During the growth, the dry matter content increases, and the fibers lignify. Furthermore, the development of the ears causes the volume of starch to rise, while the amount of sugar decreases due to their transformation into grain starch. In this manner, an easily silageable plant becomes one that can be silaged with medium difficulty, and the chopped crop then has to be shortened to allow effective packing-down and air removal. All of these changes play a major role in the specification of the harvest time [6,30].

The sampling was invariably performed at identical time intervals, together with the multispectral imaging. To monitor the quality of the corn hybrid, we opted for sampling according to the methodology recommended by the Central Institute for Supervising and Testing in Agriculture, Brno, Moravia, the Czech Republic [38]. Before commencing the inspection, we selected 3 spots in various sectors inside the area to obtain representative data of the growth homogeneity. Each of the spots provided 10 successively neighboring plants, and these were immediately transported to the Pohořelice-based laboratories operated by the company NutriVet, s.r.o. After being separated, the samples were ground and dried at 60 °C for approximately 24 h to yield a stable content of dry matter. At the sampling and measurement stage, the plants were still earless and could thus be shredded without prior disjoining. The dried mass was homogenized by grinding in a laboratory mill with 1-mm screen openings. Subsequently, each sample was analyzed twice to supply at different stages of the procedure, information relating to the following structural, nutritional, and chemical quantities: FM—fresh matter; EW—ear weight; DM—dry matter CP—crude protein (established from the dry matter); CF—crude fiber; starch—starch content; ash—ash content; NDF—neutral detergent fiber; DNDF—digestibility (NDF and DOM—organic matter digestibility. The data obtained then facilitated computing the hectare yield indicators, namely, the yields of fresh matter (YFM) and dry matter (YDM).

All of the analyses involving the chemical quantities indicated above were executed by applying common techniques. The contents were determined via the methods specified by the Association of Official Analytical Chemists (AOAC), which are represented through numerical codes; here, each code stands for a method used with a particular substance Thus, we can provide the following list: DM (# 934.01), ash (# 942.05), crude protein (# 976.05), starch (# 920.40), NDF (# 2002.04), and DNFD (# 973.18) [39–42]. The outcomes then enabled us to compute, for each sampling phase, the average values in the monitored substances. After the fifth sampling, when the corn had already developed the ears, the analysis already involved separating the ears from the parent plants and weighing them without the leaves. The procedures were completed by establishing the dry matter and starch contents.

Usually, corn sampling to assess the condition and phenophase takes place at diverse spots. At the milk line stage, the samples began to be transported to laboratories to determine the dry matter contents in both the grain and the entire plant. Based on the level of plant development and the dry matter volume, the harvest time is preliminarily specified and differentiated according to the intended use, namely, milk or methane production (the latter in biogas stations).

2.4. Vegetation Reflectivity Preprocessing

The scanned multispectral wavelength bands for the blue, green, red, red edge, and near-infrared sectors interact with the vegetation differently, depending on the solar radiation, the absorption and reflection of which result from and show dissimilarities in the overall chemical composition and the contents of water, pigments, and nutrients. The high contrast of variations in the near-infrared band ensures broad usability when setting up vegetation indices. Furthermore, the narrow red edge band also exhibits strong reflectivity changes, from the absorption of red to the considerable reflection of near-infrared radiation. Out of the monitored bands, the near-infrared spectrum has the strongest reflectivity, and, together with the red band, is the most frequently applied option when assembling vegetation indices [20].

The data sensing was carried out at eight time intervals corresponding to specific 7-day phenophases of plants. For each of the monitored spectral bands, we formed a TIF image that embodied a reflectivity map covering the entire area. In these maps, we defined the homogeneous growth subareas that matched the sampling spots. The multispectral images of the individual scanned phases were processed using the Pix4D Mapper software. We then set up a color matrix of the vegetation pixels, assigning this matrix to each image data band. The images of the gray calibrating body indicating the known reflectivity values allowed us to acquire the mean reflectivity value in the preset section of the monitored growth area.

The patterns of the scanned spectral band values will produce a spectral reflectance curve, which represents the quantity of radiation reflected over the entire range of the wavelength bands observed. The spectral reflectance, $\rho_{(\lambda)}$, defines the energy proportion between the reflected $E_R\,(\lambda)$ and the incident $E_i\,(\lambda)$ solar radiation at a certain wavelength; utilizing the formulas employed in sources [43–46], we then have:

$$\rho_{(\lambda)} = \frac{E_R\,(\lambda)}{E_i\,(\lambda)} \cdot 100\ [\%]. \tag{1}$$

2.5. Multispectral Indices

The indices usually originate from computing at least two spectral images, selected in such a manner that the vegetation reflectivity changes become prominent. In the majority of cases, the indices are functionally equivalent, and more than 150 have been presented in the literature to date; however, only a small subset of these rest on a solid biophysical basis or were systematically tested [47–52]. Our experiment verifies possible correlations in three proportional indices, computed via a normalized proportion of surface reflectivities.

Each vegetation index focuses on certain vegetation properties and has a specific applicability. To facilitate the analysis, we used the proportional indices NDVI, NDRE, and GNDVI; all of these instruments are computed identically, the only difference being that they contain diverse spectral bands. Together, the indices then embody a comprehensive cross-section through the observed wavelengths (Figure 4).

The formula for calculating the normalized vegetation indices and the spectral band reflectivities reads:

$$\mathrm{NDVI} = \frac{\rho_{NIR} - \rho_{Red}}{\rho_{NIR} + \rho_{Red}}, \tag{2}$$

$$\mathrm{NDRE} = \frac{\rho_{NIR} - \rho_{RedEdge}}{\rho_{NIR} + \rho_{RedEdge}}, \tag{3}$$

$$\mathrm{GNDVI} = \frac{\rho_{NIR} - \rho_{Green}}{\rho_{NIR} + \rho_{Green}}. \tag{4}$$

The normalized difference vegetation index, NDVI, constitutes a numerical indicator of plant health and a source of details on vegetation changes. The index also performs the following functions of informing on the amounts of water stress and the chlorophyll in a plant, assessing the monitored vegetation surface through the proportion between the red and infrared sectors of the spectrum, and recognizing tiny vegetation differences, due to the reflectivity of the near-infrared spectrum [41].

The NDVI takes values between -1 and 1; the higher values usually represent "greener" plants having a photosynthetic capacity greater than that of the other components within the area of interest. In permanent crops, grasses, and cereals, but also in some row crops at the later stages of full growth, the chlorophyll content reaches a point where the index "saturates" close to the maximum value (NDVI 1.0). In such cases, detecting differences between plants by using the NDVI becomes problematic. At the later growth stages, the vegetation aging causes the NDVI values to decline [53–55].

The NDVI utilizes the red band, which is intensively absorbed by the upper portions of the overall plant surface. The lower levels of the plant thus do not contribute significantly to the actual measurement, worsening the correlation between the NDVI and the volumetric properties of the plant. This effect becomes more important in tall plants that carry multiple layers of leaves, especially at the later stages [53].

The normalized difference red edge index (NDRE) utilizes, similarly to the NDVI, the near-infrared band and the frequency band that is situated in the transition region between the visible and the infrared spectra, namely, the red edge band ($\rho_{RedEdge}$) [56].

In the NDRE, the computation allows us to better penetrate permanent or late crops, as the absorption by only or primarily the upper level of the plant is not as intensive as in the NDVI. Moreover, the NDRE is somewhat less sensitive to saturation in thick vegetation and therefore offers superior effectivity in the measurement of changes, when the NDVI takes near values +1.0 [53,56].

The green normalized difference vegetation index (GNDVI) exploits for the computation the wavelength of the green spectrum instead of that of the red one, with ρ_{NIR} representing the reflectivity values in the near-infrared band and ρ_{Green} denoting the values in the green band [57].

The benefit of this index lies in its high correlation with the biophysical parameters of the investigated plants and its low sensitivity to other areas monitored. At the green wavelengths, the reflectivity better responds to variation in the biomass quantity. Furthermore, the green band delivers a higher probability of capturing differences in the lack of nutrients, which then manifest themselves in the resulting production of crops. Assuming these advantages, the index has the potential to eliminate the insufficient sensitivity of the NDVI (due to the green component of the spectrum) [58].

Figure 5 below shows the maps of the vegetation indices characterizing the examined land at the fifth stage of scanning.

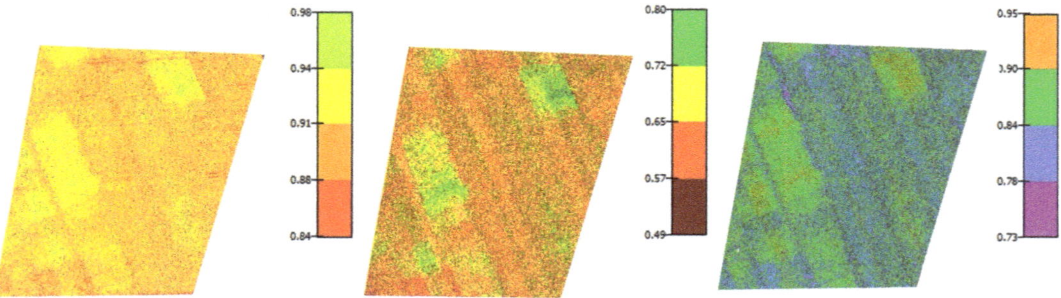

Figure 5. The maps of the investigated land at the fifth stage of scanning: (from left to right) the NDVI, NDRE, and GNDVI maps.

2.6. Correlation Analysis

To determine the relationships between the chemical analyses and the reflectivities of the spectral images, we sought the correlation coefficients. In this context, correlation does not imply causality: we only searched for a mutual linear relationship. The correlation rate was specified through the calculated correlation coefficient, which may take a value from −1 to +1. The resulting values of the correlation coefficient +1 establish a completely direct relationship, and the first variable tends to grow; by contrast, the values of the coefficient −1 establish a wholly indirect relationship, and the first variable tends to decline. If the coefficient equals zero, then no linear relationship exists between the monitored parameter and the reflectivity or the vegetation index.

To decide whether the correlation coefficients were large enough to enable us to plausibly assume a mutual relationship, we needed to calculate their statistical significance. The statistically significant value was calculated according to Student's t-distribution, with degrees of freedom $n-2$. We used:

$$t_{score} = \frac{r\sqrt{n-2}}{\sqrt{1-r^2}}, \tag{5}$$

where r is the Pearson correlation coefficient.

When searching for the statistically significant value, we selected a significance level of 2%, and the Student's critical value equaled 3.143. If the coefficient is higher than the critical value, the correlation can be considered statistically significant.

3. Results

This section outlines the results obtained from the nutritional analysis and presents the spectral curves of the reflectivities at the scanned wavelengths, acquired via processing the multispectral images and computing the vegetation indices. These aspects were completed with a description of the process of calculating the correlation coefficients associated with the relationships between the laboratory results' variation and the data from the multispectral images.

3.1. Nutrition Analysis

In each of the corn samples, on the individual sampling days, we invariably weighed the total mass of 10 plants. Subsequently, the FM (fresh matter) and EW (ear weight) rates were established in each sample; the latter rate, however, began to be determined only with the 5th sampling. The relevant chemical analysis then allowed us to establish the contents of structural, nutritional, and other substances. Out of all the sampled and analyzed values, we computed—invariably for one sampling stage—the average value of the given parameter. The values resulting from the individual sampling instances are summarized in Table 2.

3.2. Multispectral Image Processing

The spectral reflectivities acquired from the spectral maps capturing the monitored vegetation are minimal—in all the scanning phases—in the visible part of the spectrum as compared to the reflectivity changes in the near-infrared band (see Table 3).

Figure 6 displays the spectral curves of the reflectivities to define the condition of the plants with respect to that of the overall vegetation. The green spectrum, with a wavelength of 560 nm, forms the local reflectivity maximum in the visible sector of the spectrum; the higher reflectivity, compared to those of the blue (475 nm) and red (668 nm) bands, stems from a strong correlation with the chlorophyll contained in the plants. The intensive absorption exhibited by the chlorophyll in the photosynthesis within the blue and the red spectra causes low reflectivity; in these spectra, the chlorophyll absorbs approximately 90% of the incident radiation.

Table 2. The laboratory results obtained from plants collected at the different sampling stages.

Characteristics	Sample Number							
	1	2	3	4	5	6	7	8
Nutrition characteristics								
[1] DM [g/kg]	197.2	193.2	224.4	319.7	308.8	355.7	359.4	449.5
[2] CP [g/kg DM]	114.6	102.6	98.1	91.1	88.0	78.4	79.0	76.6
[3] CF [g/kg DM]	364.0	335.0	306.8	238.0	250.1	198.8	223.3	225.7
[4] Starch [g/kg DM]	6.0	21.8	153.9	273.2	319.8	344.5	349.6	398.8
[5] Ash [g/kg DM]	68.4	56.5	55.8	48.4	45.7	39.0	44.0	41.4
[6] NDF [g/kg]	684.1	639.6	598.6	403.0	452.4	410.0	432.3	448.1
[7] DNDF [%]	43.6	51.6	55.6	50.3	52.7	58.9	50.2	56.4
[8] DOM [%]	52.3	60.5	58.7	73.3	76.2	76.9	77.1	78.8
Yield characteristics								
[9] FM [kg/10 plants]	8.7	9.2	10.6	10.6	9.9	9.8	9.8	7.3
[10] EW [kg/10 plants]	0	0	0	0	3.4	3.4	3.6	2.9
[11] YFM [kg/ha]	69,360	73,947	84,560	84,667	78,960	78,693	78,347	58,320
[12] YDM [kg/ha]	13,737	14,317	18,979	27,080	24,456	28,013	28,173	26,186

[1] DM—dry matter; [2] CP—crude protein (determined from the DM); [3] CF—crude fiber; [4] starch—starch content; [5] Ash—ash content; [6] NDF—neutral detergent fiber; [7] DNDF—digestibility (NDF); [8] DOM—digestibility (organic matter); [9] FM—fresh matter; [10] EW—ear weight; [11] YFM—yield of fresh matter; and [12] YDM—yield of dry matter.

Table 3. Evaluation of the vegetation reflectivities at the individual sampling stages.

Sample Number	Spectral Band Reflectance [%]				
	Blue	Green	Red	Red Edge	NIR
1	3	7	1	16	57
2	3	6	3	14	59
3	4	7	3	15	60
4	3	8	4	17	62
5	4	11	8	29	83
6	6	14	10	39	88
7	6	14	9	30	68
8	6	13	9	26	43

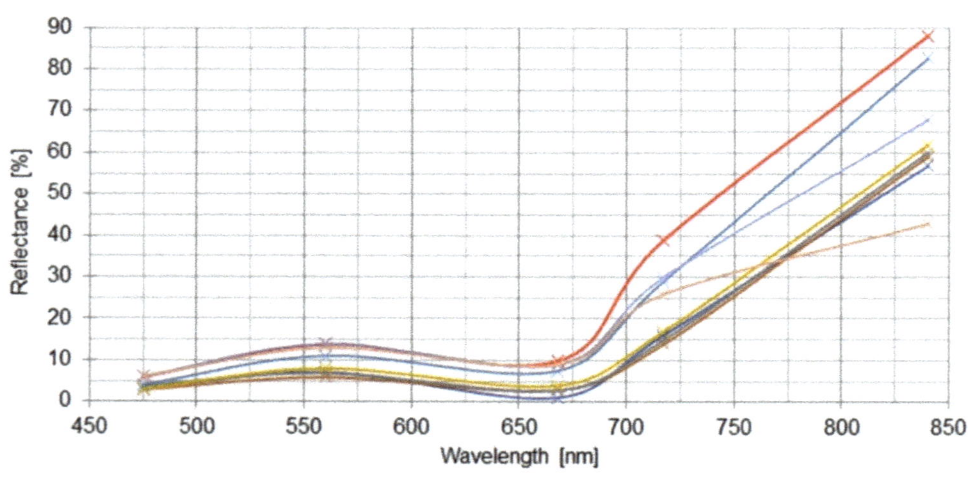

Figure 6. The reflectivity relationships at the individual wavelengths and sampling stages.

In the near-infrared band, the vegetation shows 840 nm, with 717 nm being the value for the red-edge portion; the reflectivity is thus markedly higher than in the visible spectrum. The discussed band allows us to clearly discern the growth variation between the individual scanning stages. With the progressing phenophase, the reflectivity in the near-infrared band exhibits a tendency to rise; however, a decrease begins after the maximum period has been reached and the plants have started to age.

To select suitable indicators for defining the vegetation changes in a time sequence, we first need to distinguish the differences in the reflectivities at the separate wavelengths and in the computed indices. The calculated average values of the indices from the investigated portions of land as related to the eight time intervals within the crop growth period are outlined in Table 4.

Table 4. The calculated values of the vegetation indices.

Sampling Number	Indices		
	NDVI	NDRE	GNDVI
1	0.97	0.59	0.82
2	0.91	0.65	0.84
3	0.90	0.60	0.79
4	0.89	0.57	0.77
5	0.83	0.50	0.81
6	0.83	0.46	0.76
7	0.77	0.39	0.74
8	0.72	0.36	0.64

3.3. Correlation Analysis

All the measured spectral band reflectivity averages (blue, green, red, red edge, and near-infrared) and the calculated values of the vegetation indices (the NDVI, NDRE, and GNDVI) were correlated with the average nutritional values established at the individual sampling stages via laboratory analyses performed on the sampled plants (see Table 5). The linear correlation rates from Table 5 are presented in Table 6.

Table 5. The correlation coefficients relevant for the data acquired through multispectral imaging and by means of the chemical analyses.

Characteristics	Spectral Reflectivity					Vegetation Indices		
	Blue	Green	Red	Red Edge	NIR	NDVI	NDRE	GNDVI
Nutrition characteristics								
[1] DM [g/kg]	0.804	0.861	0.866	0.691	0.012	−0.920	−0.924	−0.901
[2] CP [g/kg DM]	−0.858	−0.892	−0.953	−0.788	−0.261	0.922	0.854	0.746
[3] CF [g/kg DM]	−0.755	−0.858	−0.904	−0.804	−0.394	0.798	0.777	0.654
[4] Starch [g/kg DM]	0.783	0.875	0.905	0.767	0.267	−0.884	−0.867	−0.751
[5] Ash [g/kg DM]	−0.772	−0.838	−0.936	−0.795	−0.373	0.853	0.756	0.639
[6] NDF [g/kg]	−0.602	−0.764	−0.814	−0.711	−0.391	0.719	0.697	0.579
[7] DNDF [%]	0.607	0.480	0.640	0.524	0.260	−0.563	−0.381	−0.456
[8] DOM [%]	0.703	0.844	0.912	0.765	0.325	−0.863	−0.803	−0.651
Yield characteristics								
[9] FM [kg/ 10 plants]	−0.289	−0.217	−0.165	−0.056	0.574	0.375	0.420	0.530
[10] EW [kg/ 10 plants]	0.840	0.947	0.945	0.925	0.473	−0.822	−0.867	−0.513
[11] YFM [kg/ha]	−0.289	−0.217	−0.165	−0.056	0.574	0.375	0.420	0.530
[12] YDM [kg/ha]	0.699	0.824	0.842	0.741	0.359	−0.761	−0.769	−0.645

[1] DM—dry matter; [2] CP—crude protein (determined from the DM); [3] CF—crude fiber; [4] starch—starch content; [5] Ash—ash content; [6] NDF—neutral detergent fiber; [7] DNDF—digestibility (NDF); [8] DOM—digestibility (organic matter); [9] FM—fresh matter; [10] EW—ear weight; [11] YFM—yield of fresh matter; [12] YDM—yield of dry matter.

Table 6. The linear correlation rates.

I Value I	0 to 0.2	0.2 to 0.4	0.4 to 0.6	0.6 to 0.8	0.8 to 1
Correlation	Very weak	Weak	Medium	Strong	Very strong

In the individual correlation coefficients, we calculated the statistical significance values, and these were subsequently compared with the critical value of 3.143. The statistically significant values are highlighted in Table 7.

Table 7. The statistical significance of the correlation data acquired through multispectral imaging and by means of the chemical analyses.

Characteristics	Spectral Reflectivity					Vegetation Indices		
	Blue	Green	Red	Red Edge	NIR	NDVI	NDRE	GNDVI
Nutrition characteristics								
[1] DM [g/kg]	3.32	4.15	4.25	2.34	0.03	5.74	5.90	5.08
[2] CP [g/kg DM]	4.09	4.83	7.73	3.14	0.66	5.83	4.02	2.74
[3] CF [g/kg DM]	2.82	4.08	5.19	3.31	1.05	3.24	3.03	2.12
[4] Starch [g/kg DM]	3.08	4.43	5.21	2.93	0.68	4.62	4.26	2.79
[5] Ash [g/kg DM]	2.97	3.76	6.53	3.21	0.98	4.00	2.82	2.03
[6] NDF [g/kg]	1.85	2.90	3.43	2.48	1.04	2.54	2.38	1.74
[7] DNDF [%]	1.87	1.34	2.04	1.51	0.66	1.67	1.01	1.26
[8] DOM [%]	2.42	3.85	5.44	2.91	0.84	4.19	3.30	2.10
Yield characteristics								
[9] FM [kg/10 plants]	0.74	0.55	0.41	0.14	1.72	0.99	1.13	1.53
[10] EW [kg/10 plants]	3.80	7.26	7.06	5.98	1.31	3.53	4.26	1.46
[11] YFM [kg/ha]	0.74	0.55	0.41	0.14	1.72	0.99	1.13	1.53
[12] YDM [kg/ha]	2.40	3.57	3.83	2.71	0.94	2.87	2.95	2.07

[1] DM—dry matter; [2] CP—crude protein (determined from the DM); [3] CF—crude fiber; [4] starch—starch content; [5] Ash—ash content; [6] NDF—neutral detergent fiber; [7] DNDF—digestibility (NDF); [8] DOM—digestibility (organic matter); [9] FM—fresh matter; [10] EW—ear weight; [11] YFM—yield of fresh matter; and [12] YDM—yield of dry matter.

3.4. Classifying the Vegetation Relationships

The charts below visualize the linear relationships between the selected nutritional values and the monitored reflectivities or calculated vegetation indices (see Figure 7).

Of the established nutritional values, we selected the dry matter, nitrogen substances, and starch, due to their high statistical significance with respect to all of the imaging values, and also because they embody the most vital organic nutrients that determine the eventual quality of the corn at harvest. Using these organic nutrient values, which exhibit the most significant correlation indices, we compared the vegetation indices and major spectral bands, namely, red, green, and blue.

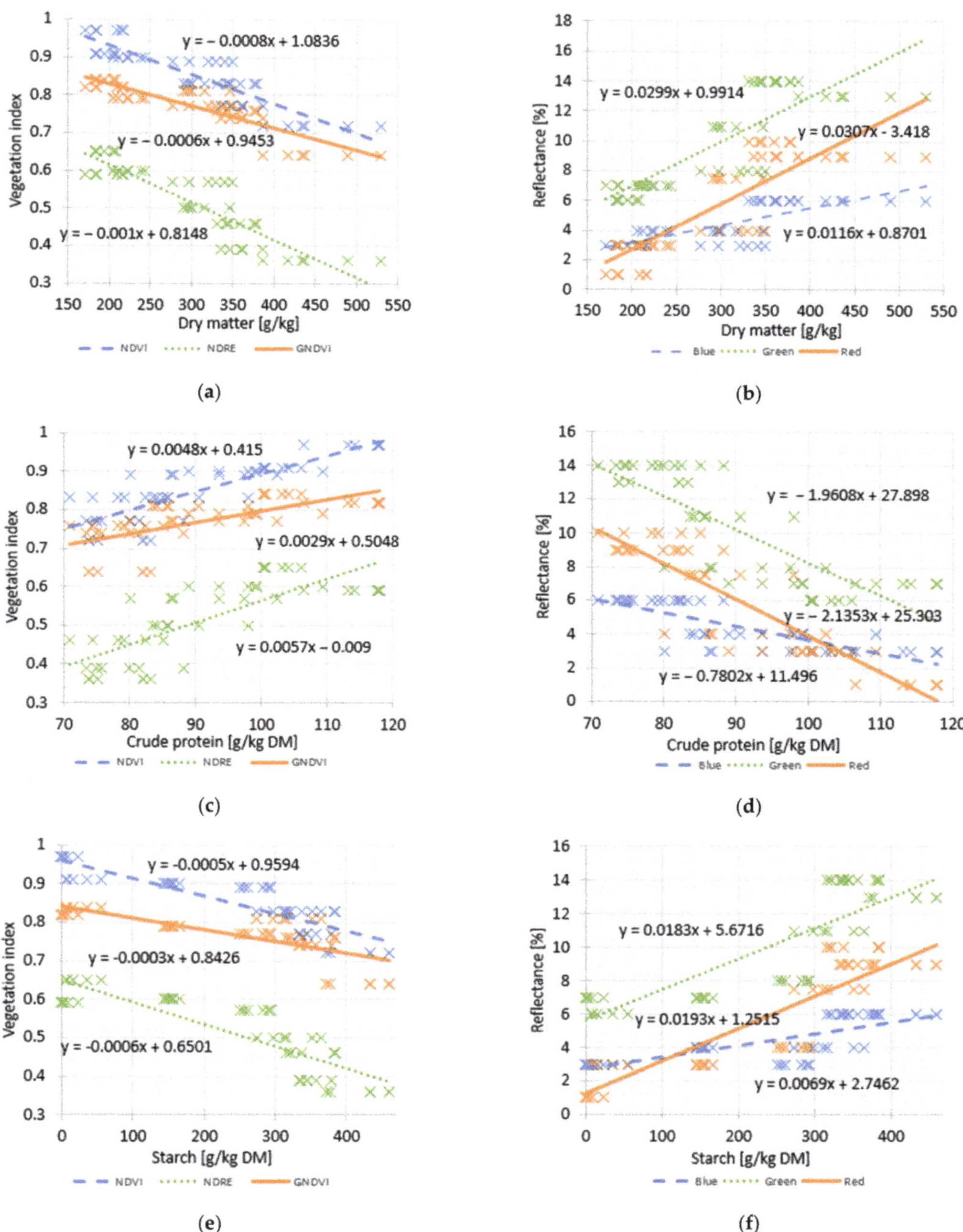

Figure 7. The laboratory-established linear relationships (left) between the vegetation indices (NDVI, NDRE, and GNDVI) (right), and the reflectivity in the red, green, and blue spectral bands, relating: (**a**) the vegetation indices to the dry matter; (**b**) the reflectance to the dry matter; (**c**) the vegetation indices to the crude protein; (**d**) the reflectance to crude protein; (**e**) the vegetation indices to the starch; (**f**) the reflectance to the starch.

4. Discussion

Analyzing multispectral images based on an exact knowledge of vegetation health is one of the procedures that support the transition from traditional agricultural methods to precision agriculture. An increase in the quality of harvested corn and a reduced fodder consumption following on from the ability to closely determine the optimum harvest time will generate novel approaches to the contactless analysis of plants at various growth stages, together with a major potential for automated and rapidly expandable applicability in most types of vegetation. In the case of corn, the optimum harvest time is established according to the content of dry matter, depending on whether the chopped crop is intended to be silaged or to produce methane in a biogas station. For an identification of the correct period, it is therefore necessary to know exactly the nutritional values of the crop on the entire land concerned.

Within the research, we achieved the preset goals, namely, defining the relationships between the nutritional parameters acquired through chemical analyses and the vegetation indices yielded via multispectral imaging of the entire area of the cornfield.

The subsections below characterize the results of the nutritional analysis and the outcomes of the multispectral image processing, including the computation of the vegetation indices. The core subsection presents the mutual correlation analysis of the relationships between the patterns of changes in the laboratory results and the data obtained from the UAV-based multispectral images. In this context, possible correlation uncertainties are also considered.

4.1. Nutritional Analysis

The evaluated nutritional indicators (Table 2) show that, in corn, the progressing phenophase is associated with an increasing content of dry matter (DM). Furthermore, the rising proportion of the grain is accompanied by a growing share of starch in the entire plant; the starch then embodies the central source of energy for the plant to be harvested. In the other parts of the organism, a decrease occurs in the nitrogen substances and the digestibility of the fiber is markedly reduced due to lignification. Interestingly, in this context, no correlation has been found to date between the fiber content and digestibility. The ideal harvest time was identified with the interval separating the 4th and 5th sampling stages. This optimum period was determined through the dry matter values, which, in the discussed phases, amount to 280–330 g/kg. More concretely, in all of the laboratory samples, the values at the 4th stage ranged from 277.9 g/kg to 349.8 g/kg, while at the 5th stage they already ranged between 291.8 g/kg and 347.4 g/kg. Another factor of importance rests in the average volumes of starch; at the 4th stage, the relevant value reached 273 g/kg/DM, and in the 5th phase it already equaled 319.8 g/kg/DM, with the ideal level of 300 g/kg/DM corresponding to 2/3 of the milk line stage. The intensive increase in the dry matter between the third and the fourth phases was induced by considerable precipitation; however, the fall of the precipitation rate down to zero then caused a sharp change in the nutritional values.

4.2. Multispectral Image Processing

The reflectivity relationships in the various portions of the spectrum confirm the expected scenario and represent the changing condition of the monitored crops over time (Figure 6). From the perspective of the reflectivity level, the spectral curves can be divided between two regions, namely, the visible part of the spectrum and the near-infrared sector. We can then observe a very low reflectivity in the visible portion of the spectrum (up to 670 nm), the relevant value being not more than 14% (the green band); thus, the radiation is mostly absorbed. The reason for this rests in the large quantity of biomass in the observed area, suggesting that the radiation is consumed through photosynthesis. The other set of monitored wavelengths gradually passes into the near-infrared region. This progressive transition is accompanied by more prominent differences (the red edge band) between the sampling stages, with the increasing reflectivity being the highest—at 39%—in the red edge band in the 6th observed phase. After the maximum, the reflectivity values decline slightly. The most conspicuous differences characterize the NIR band (840 nm), where the greatest reflectivity divergence in the monitored growth stage reaches up to 45%. Similarly to the red edge region, the NIR band attains the maximum value in the 6th phase, which, too, is followed by a decline in the values. All of the monitored spectra are important for the subsequent computation of the vegetation indices.

Considering the calculated vegetation indices in Table 4, the tendency towards a steady minor decrease allows us to assume, without prior knowledge of the nutritional values, a later phenological phase in the plants. Through the monitored period, the vegetation index NDVI ranged between 0.72 and 0.97, the average value being 0.85. This index exhibited higher values than its counterparts. In the NDRE, the range was 0.36–0.65, with an average of 0.52, and the GNDVI showed a scope of 0.64–0.84 and an average of 0.77. The GNDVI thus possesses the smallest resolving ability. By contrast, the best sensitivity is obtained from the NDRE, where differences in a broad band of reflectivities are discernible.

The values of the NDRE enable us to observe a reflectivity shift towards lower levels, compared to the other two indices; such a scenario arises from calculating the proportion between the reflectivities with the red edge spectral component, which exhibits higher reflectivity variations.

4.3. Correlation Analysis

Within the research, we established that, as regards determining the changes through the eight investigated growth phases in the selected corn hybrid, the best correlation is found between the dry matter values and the NDRE index, Table 5. To evaluate the statistical significance of the correlation coefficients, we employed Student's *t*-test, applying the significance level of 2%; this is matched by the Student's T critical value of 3.143, Table 7.

This subsection discusses the statistically significant values of the correlation coefficients. A strong correlation can be established in the GNDVI (−0.901) and NDVI (−0.920). Furthermore, statistically significant values lie also in the correlations between the CP and the NDVI (0.922), the NDRE (0.854), and the GNDVI (0.746). The previously mentioned starch content, which also exerts a major impact on the resulting quality of the harvested corn, markedly correlates with the indices NDVI (−0.884) and NDRE (−0.867). The values of the indices NDVI, NDRE, and GNDVI also correlate very well with the calculated OM values. We can then infer from these facts that the strong correlations in the indices NDVI and NDRE are usable not only for determining the convenient harvest time but also for predicting the quantity of the organic matter (OM) obtainable from the yielded crop (correlation with the NDVI at −0.924); these steps are then prominent in establishing the organic matter yield.

The evaluated correlations in the individual narrow spectral bands lead to the assumption of a strong correlation in the red band, which correlates markedly with all the determined nutritional values, the strongest correlation being that with the CP (−0.953). By contrast, the weakest values of the correlation coefficient R are found in the NIR spectral band.

4.4. Classification of the Vegetation Relationships

The diagrams capturing the vegetation relationships (see Figure 7) allow us to derive formulas that facilitate predicting the most optimum harvest values by utilizing the dry matter-, crude protein-, and starch-related data. In this context, the details outlined in the previous subsection indicate that the greatest importance for the prediction rests particularly with the indices NDRE, as related to the DM, and the NDVI, as related to starch. The obtained NDRE linear relationship ($y = -0.007\,x + 1.0836$) proposes that the ideal index values for the DM range within the NDVI value interval of 0.888–0.853. As regards the NDVI relationship ($y = -0.0045\,x + 0.9594$), the ideal value amounts to 300 g/kg DM and 0.824 (NDVI starch).

4.5. Comparing the Results with Previous Research Data

As this article contains unique, novel results, we can only refer to research papers that associate with our experiment in a merely marginal manner. However, let us note that a similar investigation was described in source [40], whose authors demonstrated that the relative data of the NDVI, PH, and a combination of both are usable when predicting the DM yield of fodder corn grown for silage. By comparison, we can point out that our study includes more types of chemical analyses; moreover, we established that the NDRE index best correlates with dry matter variations and is, besides the NDVI, therefore suitable for determining the ideal harvest time in the selected corn hybrid.

Another project that marginally resembles ours is characterized in source [24], with a focus on exploiting superspectral airborne imagery to predict corn grain yield and ear weight, and to discriminate between growth stages and irrigation treatments. Although the authors of [24] utilize multispectral imaging to determine the phenophase in plants, they concentrate solely on the NDVI index and do not state any correlations with chemical analyses, as is the case with our study.

4.6. Limitations and Future Work

The use of the methods characterized herein is accompanied by uncertainties including, for example, adverse weather conditions that may impair or destroy the entire concept of the fieldwork. For the purposes of future research, some of these uncertainties can be minimized via diversifying the land to support the experiments.

The vegetation indices are unstable due to short-term changes in the weather and in the solar radiation intensity. To minimize the error in the results of the multispectral imaging, we needed to carry out relevant calibration (Figure 3c). This task was executed for the individual crops, at various vegetation periods and in diverse weather conditions, but at identical times of the day. In plant imaging, this calibration will eliminate the inaccuracies that arise from the single-use sampling.

Another drawback to our method probably consists in that we employ data for a given hybrid, site, and year. However, as we do not follow absolute values but instead changes in the chemical composition and vegetation indices, it is possible to assume that the results will be applicable more widely to diverse corn hybrids.

The repeatability and stability of the results are certainly limited by the initial choice of a sown hybrid. The established linear relationships between the nutritional values and vegetation indices, and possibly also the reflectivities of the individual spectral bands, relate to hybrids that exhibit common plant phenophases. Different values may be revealed in *stay-green hybrids*, characterized by prolonged vitality and lower dry matter volumes; these hybrids, however, must be distinguished separately, via the criteria of nutritional values and reflectivity in the different wavelength spectra.

An alternative to UAV-based remote sensing rests in satellite imaging; this procedure may embody a more easily available and less costly option where the reflectivities have to be computed over a very large or highly particularized land area.

The planned expansion of corn growth monitoring research involves, among other steps, assigning images to already completed measurement cycles and improving the precision of the correlation curves. Importantly, we intend to compare the individual crops at the various vegetation stages, realizing that nutrition differences constitute merely one of the sources of variations in spectral behavior; other relevant factors include, for example, marked discrepancies between hybrids of the same crop.

5. Conclusions

To characterize the main outcomes of the research in general terms, we can claim that the preset aims and objectives were met. We revealed *new mathematical relationships* between the nutritional parameters acquired through chemical analyses and the vegetation indices (such as the NDRE and NDVI) established via multispectral imaging. The defined relationships then allowed us to compute the relevant nutritional values from the multispectral images of the entire monitored cornfield, without the need to perform a chemical analysis (see Figure 7). The nutritional value data corresponded to the average value of the field and may significantly help the farmers in estimating the optimum harvest time.

Considering the applied methodology and procedural options, the optimum harvest time can be predicted solely via remote sensing with a multispectral camera and by utilizing the formulas set out in Figure 7, which enable us to compute the optimum values of the multispectral index (the y variable in the formula) by applying the known concrete values of the monitored substances (the x variable in the formula). Remarkably, the NDRE and NDVI indices facilitate, based on their high statistical significance (Table 7), predicting the contents of not only the dry matter, namely, the most significant value in this context, but also the starch and crude protein.

Such innovative evaluation of the discussed factors will effectively reduce the cost of additional chemical analyses, and both farmers and researchers will be able to estimate the required quantity over the entire area of the field(s). Thus, compared to a chemical analysis of a limited number of samples, it is possible to estimate more precisely in heterogeneous vegetation the optimum harvest time with respect to the nutritional values. The authors of this paper consider determining the optimum harvest time important as regards the quantity of dry matter to generate methane and the actual production chain between the fodder and the milk.

The multispectral imaging method nevertheless features certain limitations, as described in Section 4.6. The relationships indicated in Figure 7 then can (and will) be made more precise via further experimentation.

Author Contributions: Concept, J.J. and V.J.; methodology, J.J., V.J. and P.M.; data curation, J.J., V.J. and H.S.; original draft preparation and final writing, J.J., V.J., P.M., H.S., and P.D.; funding acquisition, P.F. All authors have read and agreed to the published version of the manuscript.

Funding: This paper was funded by the general student development project at Brno University of Technology.

Institutional Review Board Statement: Not applicable.

Informed Consent Statement: Not applicable.

Data Availability Statement: Not applicable.

Acknowledgments: This paper was funded from the general student development project materialized at Brno University of Technology.

Conflicts of Interest: The authors declare no conflict of interest.

References

1. Fairchild, D.S. Soil Information System for Farming by Kind of Soil. In Proceedings of the International Interactive Workshop on Soil Resources: Their Inventory, Analysis and Interpretations for Use in the 1990's, St. Paul, MN, USA, 22–24 March 1988; pp. 159–164.
2. Hedley, C. The role of precision agriculture for improved nutrient management on farms. *J. Sci. Food Agric.* **2014**, *95*, 12–19. [CrossRef]
3. Mulla, D.J. Twenty five years of remote sensing in precision agriculture: Key advances and remaining knowledge gaps. *Biosyst. Eng.* **2013**, *114*, 358–371. [CrossRef]
4. Stafford, J.V. Implementing Precision Agriculture in the 21st Century. *J. Agric. Eng. Res.* **2000**, *76*, 267–275. [CrossRef]
5. Dwivedi, A.; Naresh, R.; Kumar, R.; Yadav, R.; Kumar, R. Precision agriculture. Promoting Agri-Hortucultural, Technological Innovatons. 2017. Available online: https://www.researchgate.net/publication/322156374_PRECISION_AGRICULTURE#fullTextFileContent (accessed on 12 March 2021).
6. Dixon, D.J.; Callow, J.N.; Duncan, J.M.; Setterfield, S.A.; Pauli, N. Satellite prediction of forest flowering phenology. *Remote Sens. Environ.* **2021**, *255*, 112197. [CrossRef]
7. Weiss, M.; Jacob, F.; Duveiller, G. Remote sensing for agricultural applications: A meta-review. *Remote Sens. Environ.* **2020**, *236*, 111402. [CrossRef]
8. Tillett, R. Image analysis for agricultural processes: A review of potential opportunities. *J. Agric. Eng. Res.* **1991**, *50*, 247–258. [CrossRef]

9. Wójtowicz, M.; Wójtowicz, A.; Piekarczyk, J. Application of Remote Sensing Methods in Agriculture. *Commun. Biometry Crop. Sc* **2016**, *11*, 31–50.
10. Shanmugapriya, P.; Rathika, S.; Ramesh, T.; Janaki, P. Applications of Remote Sensing in Agriculture—A Review. *Int. J. Cur Microbiol. Appl. Sci.* **2019**, *8*, 2270–2283. [CrossRef]
11. Řezník, T.; Pavelka, T.; Herman, L.; Lukas, V.; Širůček, P.; Leitgeb, Š.; Leitner, F. Prediction of Yield Productivity Zones from Landsat 8 and Sentinel-2A/B and Their Evaluation Using Farm Machinery Measurements. *Remote Sens.* **2020**, *12*, 1917. [CrossRef
12. Řezník, T.; Lukas, V.; Charvát, K.; Křivánek, Z.; Kepka, M.; Herman, L.; Řezníková, H. Disaster Risk Reduction in Agricultur through Geospatial (Big) Data Processing. *ISPRS Int. J. Geo-Inform.* **2017**, *6*, 238. [CrossRef]
13. Liu, S.; Chen, Y.; Ma, Y.; Kong, X.; Zhang, X.; Zhang, D. Mapping Ratoon Rice Planting Area in Central China Using Sentinel- Time Stacks and The Phenology-Based Algorithm. *Remote Sens.* **2020**, *12*, 3400. [CrossRef]
14. Numbisi, F.; Van Coillie, F. Does Sentinel-1A Backscatter Capture the Spatial Variability in Canopy Gaps of Tropical Agroforests A Proof-of-Concept in Cocoa Landscapes in Cameroon. *Remote Sens.* **2020**, *12*, 4163. [CrossRef]
15. Rahetlah, B.V.; Salgado, P.; Andrianarisoa, B.; Tillard, E.; Razafindrazaka, H.; Le Mezo, L.; Ramalanjaona, V.L. Relationshi between Normalized Difference Vegetation Index (NDVI) and Forage Biomass Yield in the Vakinankaratra Region, Madagasca *Livest. Res. Rural Dev.* **2014**, *26*, 1–11.
16. Ferrer-González, E.; Agüera-Vega, F.; Carvajal-Ramírez, F.; Martínez-Carricondo, P. UAV Photogrammetry Accuracy Assessmen for Corridor Mapping Based on the Number and Distribution of Ground Control Points. *Remote Sens.* **2020**, *12*, 2447. [CrossRef
17. Martínez-Carricondo, P.; Agüera-Vega, F.; Carvajal-Ramírez, F. Use of UAV-Photogrammetry for Quasi-Vertical Wall Surveyin *Remote Sens.* **2020**, *12*, 2221. [CrossRef]
18. Ali, A.M.; Darvishzadeh, R.; Shahi, K.R.; Skidmore, A. Validating the Predictive Power of Statistical Models in Retrieving Lea Dry Matter Content of a Coastal Wetland from a Sentinel-2 Image. *Remote Sens.* **2019**, *11*, 1936. [CrossRef]
19. Zhao, L.; Shi, Y.; Liu, B.; Hovis, C.; Duan, Y.; Shi, Z. Finer Classification of Crops by Fusing UAV Images and Sentinel-2A Dat *Remote Sens.* **2019**, *11*, 3012. [CrossRef]
20. Da Silva Júnior, M.C.; de Carvalho Pinto, F.D.; Marçal, D.; de Queiroz, E.A.; Gracia, L.M.; Gil, J.G. Correlation between Vegetatio Indices and Nitrogen Leaf Content and Dry Matter Production in Brachiaria Decumbens. *Image Anal. Agric. Prod. Process* **2006**, *6* 145–150.
21. Colomina, I.; Molina, P. Unmanned aerial systems for photogrammetry and remote sensing: A review. *ISPRS J. Photogramn Remote Sens.* **2014**, *92*, 79–97. [CrossRef]
22. Yang, G.; Liu, J.; Zhao, C.; Li, Z.; Huang, Y.; Yu, H.; Xu, B.; Yang, X.; Zhu, D.; Zhang, X.; et al. Unmanned Aerial Vehicle Remot Sensing for Field-Based Crop Phenotyping: Current Status and Perspectives. *Front. Plant. Sci.* **2017**, *8*, 1111. [CrossRef] [PubMed
23. Yao, X.; Wang, N.; Liu, Y.; Cheng, T.; Tian, Y.; Chen, Q.; Zhu, Y. Estimation of Wheat LAI at Middle to High Levels Usin Unmanned Aerial Vehicle Narrowband Multispectral Imagery. *Remote Sens.* **2017**, *9*, 1304. [CrossRef]
24. Li, S.; Yuan, F.; Ata-Ui-Karim, S.T.; Zheng, H.; Cheng, T.; Liu, X.; Tian, Y.; Zhu, Y.; Cao, W.; Cao, Q. Combining Color Indices an Textures of UAV-Based Digital Imagery for Rice LAI Estimation. *Remote Sens.* **2019**, *11*, 1763. [CrossRef]
25. Yue, J.; Feng, H.; Jin, X.; Yuan, H.; Li, Z.; Zhou, C.; Yang, G.; Tian, Q. A Comparison of Crop Parameters Estimation Using Image from UAV-Mounted Snapshot Hyperspectral Sensor and High-Definition Digital Camera. *Remote Sens.* **2018**, *10*, 1138. [CrossRef
26. Xu, X.Q.; Lu, J.S.; Zhang, N.; Yang, T.C.; He, J.Y.; Yao, X.; Cheng, T.; Zhu, Y.; Cao, W.X.; Tian, Y.C. Inversion of rice canop chlorophyll content and leaf area index based on coupling of radiative transfer and Bayesian network models. *ISPRS J. Photogramn Remote Sens.* **2019**, *150*, 185–196. [CrossRef]
27. Schittenhelm, S. Chemical composition and methane yield of maize hybrids with contrasting maturity. *Eur. J. Agron.* **2008**, *2* 72–79. [CrossRef]
28. Tilley, J.M.A.; Terry, R.A. A Two-Stage Technique for the in Vitro Digestion of Forage Crops. *Grass Forage Sci.* **1963**, *18*, 104–11 [CrossRef]
29. Berger, L.L.; Paterson, J.A.; Klopfenstein, T.J.; Britton, R.A. Effect of Harvest Date and Chemical Treatment on the Feeding Valu of Corn Stalklage2. *J. Anim. Sci.* **1979**, *49*, 1312–1316. [CrossRef]
30. Timing Is Everything for Corn Silage. Available online: https://hayandforage.com/article-permalink-3111.html (accessed on 1 March 2021).
31. Maximizing Corn Silage Quality by Monitoring Dry Matter. Available online: https://www.hubbardfeeds.com/blog maximizing-corn-silage-quality-monitoring-dry-matter (accessed on 12 March 2021).
32. Corson, D.; Waghorn, G.; Ulyatt, M.; Lee, J. NIRS: Forage analysis and livestock feeding. *Proc. N. Z. Grassl. Assoc.* **1999**, 127–13 [CrossRef]
33. Lundberg, K.; Hoffman, P.; Bauman, L.; Berzaghi, P. Prediction of Forage Energy Content by Near Infrared Reflectance Spec troscopy and Summative Equations. *Prof. Anim. Sci.* **2004**, *20*, 262–269. [CrossRef]
34. García-Sánchez, F.; Galvez-Sola, L.; Nicolás, J.J.M.; Muelas-Domingo, R.; Nieves, M. Using Near-Infrared Spectroscopy i Agricultural Systems. *Dev. Near-Infrared Spectrosc.* **2017**, *1*, 97–127. [CrossRef]
35. Herrmann, I.; Bdolach, E.; Montekyo, Y.; Rachmilevitch, S.; Townsend, P.A.; Karnieli, A. Assessment of maize yield and phenolog by drone-mounted superspectral camera. *Precis. Agric.* **2020**, *21*, 51–76. [CrossRef]
36. Islam, M.; Garcia, S. Prediction of Dry Matter Yield of Hybrid Forage Corn Grown for Silage. *Crop. Sci.* **2014**, *54*, 2362–237 [CrossRef]

37. Pix4Dcapture: Free Drone Flight Planning Mobile App. Available online: https://www.pix4d.com/product/pix4dcapture (accessed on 17 April 2021).
38. ÚKZÚZ. *Methods of Plant Variety State Tests CISTA, Pursuant to the Valid Wording from the Year 1999*; ÚKZÚZ: Brno, Czech Republic, 1999.
39. Official Methods of Analysis of AOAC International—20th Edition. 2016. Available online: https://www.techstreet.com/standards/official-methods-of-analysis-of-aoac-international-20th-edition-2016?product_id=1937367#product (accessed on 10 March 2021).
40. Licitra, G.; Hernandez, T.; Van Soest, P. Standardization of procedures for nitrogen fractionation of ruminant feeds. *Anim. Feed. Sci. Technol.* **1996**, *57*, 347–358. [CrossRef]
41. Van Soest, P.J.; Wine, R.H.; Moore, L.A. Estimation of the True Digestibility of Forages by the In Vitro Digestion of Cell Walls. Available online: https://www.cabdirect.org/cabdirect/abstract/19670700081 (accessed on 1 May 2021).
42. Brahmakshatriya, R.; Donker, J. Five Methods for Determination of Silage Dry Matter. *J. Dairy Sci.* **1971**, *54*, 1470–1474. [CrossRef]
43. Aggarwal, S. Principles of Remote Sensing. In *Satellite Remote Sensing and GIS Applications in Agricultural Meteorology*; World Meteorological Organisation: Geneva, Switzerland, 2004; Volume 23, pp. 23–38.
44. Konik, M.; Kowalczuk, P.; Zabłocka, M.; Makarewicz, A.; Meler, J.; Zdun, A.; Darecki, M. Empirical Relationships between Remote-Sensing Reflectance and Selected Inherent Optical Properties in Nordic Sea Surface Waters for the MODIS and OLCI Ocean Colour Sensors. *Remote Sens.* **2020**, *12*, 2774. [CrossRef]
45. Mobley, C.D. Estimation of the remote-sensing reflectance from above-surface measurements. *Appl. Opt.* **1999**, *38*, 7442–7455. [CrossRef]
46. Sivakumar, M.; Roy, P.; Harmsen, K.; Saha, S. *Satellite Remote Sensing and GIS Applications in Agriculture Meteorology*; World Meteorological Organisation: Geneva, Switzerland, 2003.
47. Panda, S.S.; Ames, D.P.; Panigrahi, S. Application of Vegetation Indices for Agricultural Crop Yield Prediction Using Neural Network Techniques. *Remote Sens.* **2010**, *2*, 673–696. [CrossRef]
48. Gamon, J.A.; Surfus, J.S. Assessing leaf pigment content and activity with a reflectometer. *New Phytol.* **1999**, *143*, 105–117. [CrossRef]
49. Xue, J.; Su, B. Significant Remote Sensing Vegetation Indices: A Review of Developments and Applications. *J. Sens.* **2017**, *2017*, 1353691. [CrossRef]
50. Wang, J.; Rich, P.M.; Price, K.P. Temporal responses of NDVI to precipitation and temperature in the central Great Plains, USA. *Int. J. Remote Sens.* **2003**, *24*, 2345–2364. [CrossRef]
51. Lavigne, H.; Van der Zande, D.; Ruddick, K.; Dos Santos, J.C.; Gohin, F.; Brotas, V.; Kratzer, S. Quality-control tests for OC4, OC5 and NIR-red satellite chlorophyll-a algorithms applied to coastal waters. *Remote Sens. Environ.* **2021**, *255*, 112237. [CrossRef]
52. Carneiro, F.M.; Furlani, C.E.A.; Zerbato, C.; De Menezes, P.C.; Gírio, L.A.D.S. Correlations among vegetation indices and peanut traits during different crop development stages. *Eng. Agríc.* **2019**, *39*, 33–40. [CrossRef]
53. Cammarano, D.; Fitzgerald, G.; Basso, B.; O'Leary, G.; Chen, D.; Grace, P.; Fiorentino, C. Use of the Canopy Chlorophyl Content Index (CCCI) for Remote Estimation of Wheat Nitrogen Content in Rainfed Environments. *Agron. J.* **2011**, *103*, 1597–1603. [CrossRef]
54. Buma, W.G.; Lee, S.-I. Multispectral Image-Based Estimation of Drought Patterns and Intensity around Lake Chad, Africa. *Remote Sens.* **2019**, *11*, 2534. [CrossRef]
55. Marino, S.; Alvino, A. Agronomic Traits Analysis of Ten Winter Wheat Cultivars Clustered by UAV-Derived Vegetation Indices. *Remote Sens.* **2020**, *12*, 249. [CrossRef]
56. Price, J. Leaf area index estimation from visible and near-infrared reflectance data. *Remote Sens. Environ.* **1995**, *52*, 55–65. [CrossRef]
57. Bausch, W.C.; Halvorson, A.D.; Cipra, J. Quickbird satellite and ground-based multispectral data correlations with agronomic parameters of irrigated maize grown in small plots. *Biosyst. Eng.* **2008**, *101*, 306–315. [CrossRef]
58. Gitelson, A.A.; Merzlyak, M.N. Remote sensing of chlorophyll concentration in higher plant leaves. *Adv. Space Res.* **1998**, *22*, 689–692. [CrossRef]

 *remote sensing*

MDPI

Article

Photogrammetry Using UAV-Mounted GNSS RTK: Georeferencing Strategies without GCPs

Martin Štroner *, Rudolf Urban, Jan Seidl, Tomáš Reindl and Josef Brouček

Department of Special Geodesy, Faculty of Civil Engineering, Czech Technical University in Prague, Thákurova 7, 166 29 Prague, Czech Republic; rudolf.urban@fsv.cvut.cz (R.U.); jan.seidl@fsv.cvut.cz (J.S.); tomas.reindl@fsv.cvut.cz (T.R.); josef.broucek@fsv.cvut.cz (J.B.)

* Correspondence: martin.stroner@fsv.cvut.cz

Abstract: Georeferencing using ground control points (GCPs) is the most common strategy in photogrammetry modeling using unmanned aerial vehicle (UAV)-acquired imagery. With the increased availability of UAVs with onboard global navigation satellite system–real-time kinematic (GNSS RTK), georeferencing without GCPs is becoming a promising alternative. However, systematic elevation error remains a problem with this technique. We aimed to analyze the reasons for this systematic error and propose strategies for its elimination. Multiple flights differing in the flight altitude and image acquisition axis were performed at two real-world sites. A flight height of 100 m with a vertical (nadiral) image acquisition axis was considered primary, supplemented with flight altitudes of 75 m and 125 m with a vertical image acquisition axis and two flights at 100 m with oblique image acquisition axes (30° and 15°). Each of these flights was performed twice to produce a full double grid. Models were reconstructed from individual flights and their combinations. The elevation error from individual flights or even combinations yielded systematic elevation errors of up to several decimeters. This error was linearly dependent on the deviation of the focal length from the reference value. A combination of two flights at the same altitude (with nadiral and oblique image acquisition) was capable of reducing the systematic elevation error to less than 0.03 m. This study is the first to demonstrate the linear dependence between the systematic elevation error of the models based only on the onboard GNSS RTK data and the deviation in the determined internal orientation parameters (focal length). In addition, we have shown that a combination of two flights with different image acquisition axes can eliminate this systematic error even in real-world conditions and that georeferencing without GCPs is, therefore, a feasible alternative to the use of GCPs.

Keywords: drone; GNSS RTK; UAV; photogrammetry; precision; accuracy; elevation

Citation: Štroner, M.; Urban, R.; Seidl, J.; Reindl, T.; Brouček, J. Photogrammetry Using UAV-Mounted GNSS RTK: Georeferencing Strategies without GCPs. *Remote Sens.* **2021**, *13*, 1336. https://doi.org/10.3390/rs13071336

Academic Editor: Fernando Carvajal-Ramírez

Received: 10 March 2021
Accepted: 28 March 2021
Published: 31 March 2021

1. Introduction

UAV photogrammetry combined with the structure from motion (SfM) technique is a well-established method for the mapping and creation of digital terrain models (DTMs), digital elevation models (DEMs), and/or other spatial models. This technique is often used in mining [1–3], for monitoring of various natural phenomena and geohazards [4,5], for the detection of agricultural crops/trees [6–8], dam and riverbed erosion [9], modeling topographic features [10], updating cadastral data [11], solar irradiation estimates [12], etc. SfM has become so popular that it is currently also used besides ground and UAV photogrammetry in mobile measurement systems [13]; even attempts at creating DTMs using a mobile phone have been reported [14–16]. UAV photogrammetry simplifies the work, makes it faster, and improves the quality, although the resulting model accuracy depends on many circumstances, such as the configuration and number of ground control points (GCPs) [17,18], camera pitch [19,20], image overlap [21], flight trajectory [22], camera calibration method [23,24], software used for reconstruction [25], or the quality of global navigation satellite system (GNSS) signal processing [26–28]. As well as the point cloud, other outputs, such as orthomosaics, are also widely used [29]. Photogrammetry outcomes are usually evaluated using independent laser scanning or control points (CPs), the position of which is measured by ground surveys, which facilitate the assessment of model deformations [30–32]. Correct determination of the elements of internal and external orientation, traditionally performed using GCPs, is crucial for creating an accurate photogrammetric model. At present, the use of UAVs equipped with an onboard global navigation satellite system–real-time kinematic (GNSS RTK) receiver is on the rise. This equipment used to be prohibitively expensive but, lately, it has become more affordable. DJI Phantom 4 RTK multicopter is an example of such a low-end UAV. The knowledge of the camera position during image acquisition (with centimeter accuracy) has been suggested to be able to fully substitute the presence of GCPs for georeferencing of the photogrammetric model [33]. Elimination of the need for GCPs would be a great advantage, potentially making the measurement simpler and/or cheaper; in inaccessible areas, it could be crucial for even making the measurement possible. The resulting model accuracy should correspond to that of the GNSS RTK; the standard deviation should, therefore, not exceed several centimeters. Many studies (e.g., [33]) used such an approach, with satisfactory results. Other studies, however, reported that this approach, combined with the calculation of internal orientation elements when solving bundle adjustment, i.e., without a known camera calibration, yields results systematically shifted in the elevation axis that may be, in some instances, significantly higher than the expected accuracy [34–36]. The expected accuracy of the resulting point cloud is 1–2x the ground sampling distance (GSD) [37]; nevertheless, in our previous research, the systematic error was as high as tens of centimeters despite a ground sampling distance (GSD) of 0.03 m [34]. Similar results were presented, for example, by Forlani et al. [38]. The association of such high inaccuracy with the incorrect determination of the focal length (f) was suggested in those studies. Additionally, the systematic error was shown to be more or less random as it differed even between repeated flights on the same site with the same conditions (the same UAV and processed in the same way) [34]. A small number of GCPs [39], or even a single one [34], was, however, shown to significantly reduce this problem. Several studies also suggested methods that should suppress or even remove this phenomenon. Besides the aforementioned use of a minimum number of GCPs, the use of oblique images is another promising alternative [20,36,40]. The presented study aims to propose and test strategies enabling the acquisition of a quality photogrammetric model without the high systematic elevation error while avoiding the use of GCPs. Proof that the source of the error indeed lies in the incorrect focal length determination would show the path to resolving this problem—choosing a flight configuration ensuring a sufficiently accurate calculation of internal orientation elements. We performed image acquisition at two study sites with various flight parameters to independently evaluate the results associated with different external conditions. By processing this acquired imagery in various combinations, we aimed to find a working strategy (configuration) for safe and accurate use of the GNSS RTK-based approach without the systematic elevation error.

2. Materials and Methods

2.1. Data Acquisition

To facilitate the generalization of the results of this study, the experiment was performed in two study sites. The first site—Brownfield—had only a minimum of dense undergrowth, with low buildings, and concrete and natural surfaces without monochromatic areas (Figure 1). The other site—Rural—was characterized by continuous rapeseed fields alternating with more or less dense forest and shrubs (Figure 2). These two sites differ with respect to the present surfaces and their properties. Figures 1 and 2 show the point clouds in colors indicating the reliability of the individual points acquired through SfM in Agisoft Metashape (confidence), with lower numbers indicating lower reliability (confidence; possible range 1–255). The Brownfield site can be, therefore, considered highly suitable for SfM modeling, while the Rural site can be considered problematic, which is particularly true for the part of the site that is covered by dense tree vegetation.

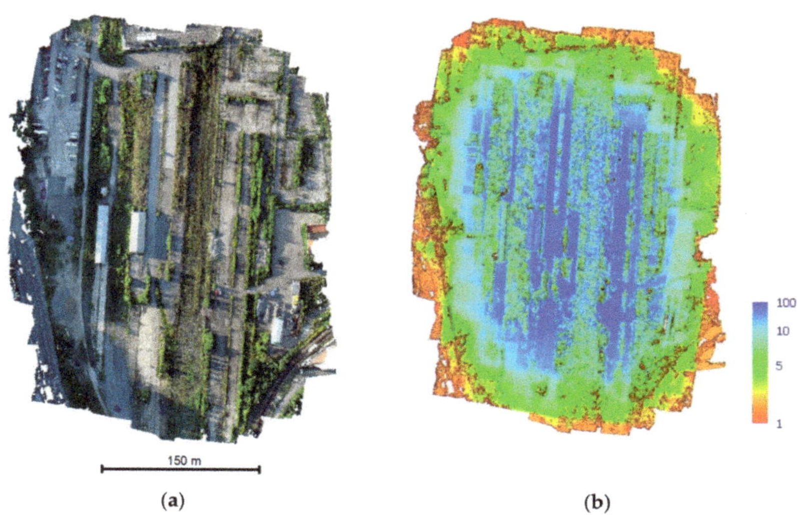

(a) (b)

Figure 1. Brownfield site ((**a**) orthomosaic, (**b**) point cloud color-coded according to confidence).

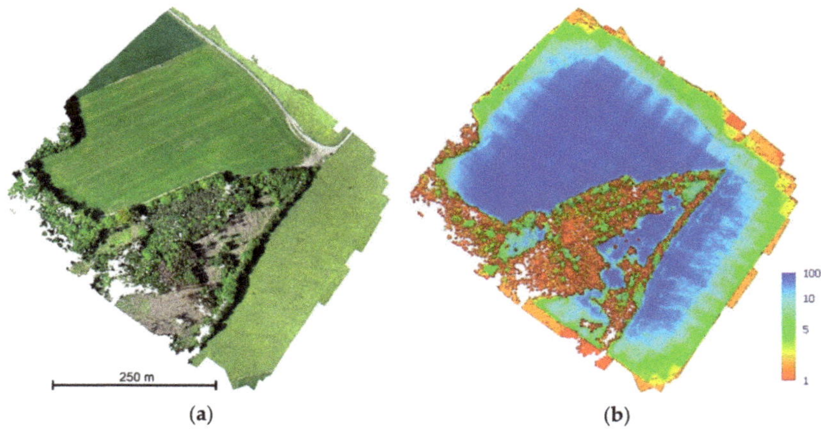

(a) (b)

Figure 2. Rural site ((**a**) orthomosaic, (**b**) point cloud color-coded according to confidence).

The DJI Phantom 4 RTK UAV mounted with a camera equipped with an FC6310R lens (f = 8.8 mm), resolution of 4864 × 3648 pixels, and with a pixel size of 2.61 × 2.61 µm (total price approx. EUR 6000), was used for image acquisition. The GNSS RTK receiver was connected to the CZEPOS permanent reference station network.

The primary flight altitude was set to 100 m above ground with a vertical image acquisition axis and ground sampling distance (GSD) of 0.03 m; at the same altitude, flights with the image acquisition axis angled by 15° and 30° from the vertical direction were performed. In addition, flights at altitudes of 75 m and 125 m above ground with a nadiral imagery acquisition axis were performed (Figure 3).

Each flight was performed with a gridded flight plan; two perpendicular flights were performed for each flight setup (forming a double grid, see Figure 4).

Altogether, 10 flights with 75% front and side overlaps were performed. Table 1 shows the parameters of individual flights with the hereinafter used designations and numbers of usable images.

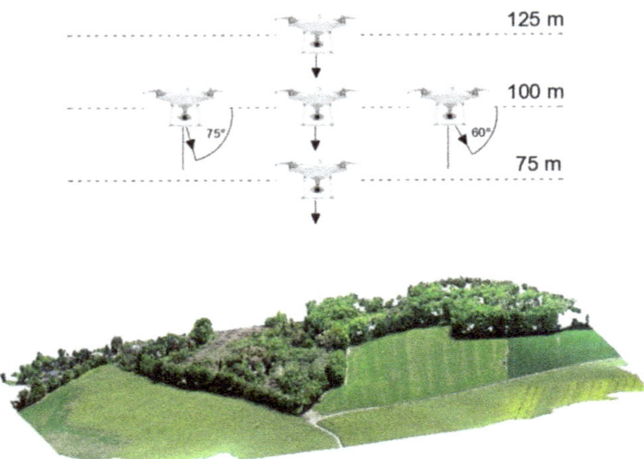

Figure 3. Flight altitudes and image acquisition directions.

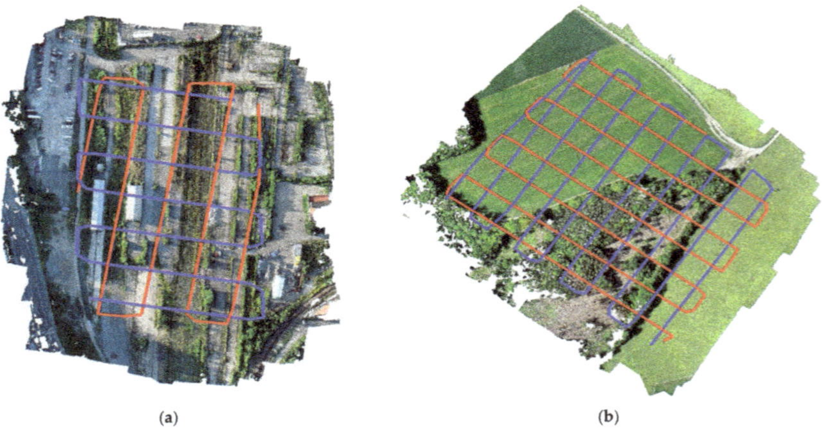

Figure 4. Flight trajectory: (a) Brownfield; (b) Rural. Flight 1 in red, Flight 2 in blue.

Points that subsequently served, depending on the calculation method, as either control points or ground control points were stabilized at each of the sites. Where possible, these were marked out as a cross painted with a contrast matte spray (Figure 5 left). Where this was not possible, especially in the undergrowth, wooden boards with black and white targets were used (Figure 5, right) and stabilized using a 10 cm long nail. The dimensions of the crosses/targets were approx. 0.40 m × 0.40 m.

(Ground) control points were distributed as evenly as possible throughout both areas. In all, 25 points were stabilized at the Brownfield site and 30 at the Rural site (Figure 6). The (ground) control points were surveyed using a GNSS RTK Trimble Geo XR receiver with a Zephyr 2 antenna connected into the CZEPOS permanent reference station network (czepos.cuzk.cz). Measurement of each control point was taken three times (before, between, and after UAV flights) for detecting potential errors or variations caused, e.g., by the change of the configuration or satellite availability. The expected nominal accuracy of each coordinate was 0.03 m.

Table 1. Image acquisition flights and their properties.

Site	Designation	Flight Altitude above the Terrain (m)	Imagery Acquisition—Deviation from the Nadiral Direction (°)	Number of Images—Flight 1	Number of Images—Flight 2
Brownfield	75 m	75	0	78	76
	100 m	100	0	50	53
	125 m	125	0	39	41
	60° (100 m)	100	30	67	80
	75° (100 m)	100	15	58	66
Rural	75 m	75	0	176	183
	100 m	100	0	112	122
	125 m	125	0	84	84
	60° (100 m)	100	30	147	160
	75° (100 m)	100	15	128	140

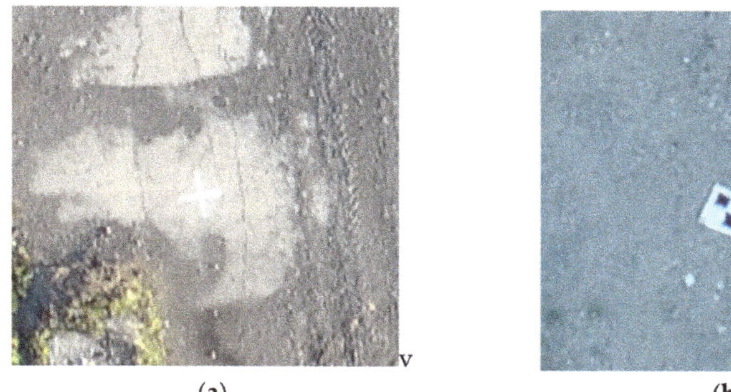

Figure 5. (Ground) control points; (**a**) marking using a spray; (**b**) using black and white target.

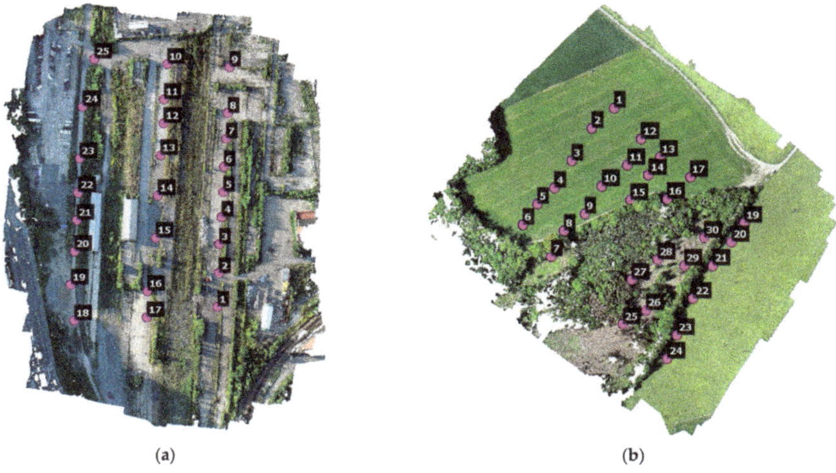

Figure 6. Distribution of the (ground) control points throughout the study sites—Brownfield (**a**), Rural (**b**).

2.2. Data Processing

Both GNSS RTK measurements (UAV onboard and ground receiver) were processed using the same methods. First, the terrestrial GNSS RTK measurements were exported from the receiver in the WGS 84 coordinate system (latitude, longitude, ellipsoidal height). Similarly, the spatial position (in the same coordinate system) was extracted from the GNSS RTK data-containing images using Exiftool. The offset between the GNSS receiver antenna reference point (ARP) and the camera center (CC) was automatically considered by the software. Subsequently, all data were converted into the Czech national coordinate positioning system (S-JTSK, System of Unified Trigonometric Cadastral Network) and the Bpv vertical datum (Balt after adjustment) using the EasyTransform software (http://adjustsolutions.cz/easytransform/, accessed on 1 March 2021) to ensure that the same algorithm was used on all data and thereby to eliminate potential systematic errors that could occur as a result of different transformation algorithms. The three measurements taken for each point by the terrestrial GNSS RTK receiver were used for the calculation of the standard deviation. Then, images were processed in Agisoft Metashape 1.6.1 using the structure from motion calculation method (SfM) with the custom settings listed in Table 2.

Table 2. Agisoft Metashape settings used for calculations.

Setting	Value
Align Photos	
-Accuracy	High
-Key point limit	40,000
-Tie point limit	4000
Optimize Camera Alignment	Fit all constants (f, cx, cy, k1–k4, p1–p4)
Build Dense Cloud	
-Quality	High
-Depth filtering	Moderate
Digital Elevation Model	
-Coordinate System	S-JTSK, Bpv
-Parameters	
-Source data	Dense Cloud
-Interpolation	Enabled
-Advanced	
-Resolution	2.8 cm/pix (implicit)
(Settings not detailed above were kept at default).	

The UAV was not equipped with a professional metric camera. Although pre-calibration is generally recommended, the stability of the parameters in non-metric UAV-mounted cameras cannot be fully ascertained and other studies have shown that laboratory calibration does not provide better results than the method used in this study [41,42]. The interior orientation parameters were, therefore, calculated in the usual way.

In all, five duplicate flights were performed at each site. The geometries of the imagery from individual flights differed (the bundles intersected at different points due to different flight altitudes and/or camera angles). Each flight (i.e., 10 flights at each site) was processed separately. In addition, joint processing of the two duplicate flights was also performed (i.e., 5 for each site), which allowed investigation of whether a simple increase in the number of images can improve the accuracy.

The primary flight was at an altitude of 100 m with nadiral image acquisition; the remaining flights were performed to acquire data for testing possible accuracy improvement strategies. Hence, all possible combinations of the 100 m altitude flight with nadiral image acquisition (the primary flight) and flights with other configurations were calculated, i.e., 100m_1 + 75m_1; 100m_1 + 125m_1; 100m_1 + 60°_1; 100m_1 + 75°_1; 100m_1 + 75m_2; 100m_1 + 125m_2; 100m_1 + 60°_2; 100m_1 + 75°_2; the same was done for the second primary flight (100m_2). Therefore, 16 such combinations were calculated for each site.

Each of the variants described above was calculated with onboard GNSS RTK data only (All_RTK) and with a combination of the onboard GNSS RTK receiver data with GCP coordinates (All_Combined).

The processing sequence was the same in all flights—alignment, sparse cloud computation, system optimization, dense cloud calculation. Dense cloud was subsequently manually cropped along the convex envelope of GCPs and filtered in CloudCompare 2.11.0 using the CSF filter to remove trees and shrubs that could potentially cause uncertainty in comparisons of the dense clouds.

Subsequently, parameters of the regression and coefficient of determination describing the relationship between the difference in the focal length f (acquired from the respective flight and from all available data from the respective site, i.e., the All_Combined calculation variant) and the (mean) systematic error in the entire model elevation were calculated for each site. Similarly, the correlation between the systematic error in model elevation and the principal point offset difference (Cx, Cy) was also calculated. The aim was to show the association between the elevation error and the calculation of internal orientation elements.

3. Results

3.1. The Accuracy of the GNSS RTK Ground Geodetic Survey

The accuracy of the GCP/checkpoints survey was evaluated through a calculation of standard deviations in individual coordinates S_X, S_Y, S_H from the replicates detailed in Table 3. The results confirm the expected accuracy of 0.03 m in each coordinate.

Table 3. Agisoft Metashape settings used for calculations.

Site	S_X (m)	S_Y (m)	S_H (m)	S in 3D Position (m)
Rural	0.0063	0.0046	0.0069	0.010
Brownfield	0.0061	0.0051	0.0067	0.010

3.2. Elevation Errors and Related Parameters

As shown, e.g., in our previous study [34], the elevation component of the resulting model is the most problematic one when using direct georeferencing based only on onboard GNSS RTK measurements. Therefore, the mean difference in elevation of the control points between the onboard GNSS RTK and the geodetic survey was calculated to obtain the systematic error (Tables 4 and 5). The residual error was then characterized by the standard deviation, the total error is described by the root mean square error (RMS). Similarly, residual systematic error between the recorded and adjusted camera elevations was calculated for the camera coordinates; the mean difference for individual flights was always equal to zero, and the standard deviation did not exceed 0.03 m (approx. 1x GSD), which confirms that the coordinates recorded during flights were correct. Tables 4 and 5 also detail other parameters, including the focal length f and the coordinates of the principal point offset Cx and Cy. The first two rows of each table represent the reference values calculated from all images using all available images together with the measured GCP coordinates (All_Combined) and using GNSS RTK coordinates only (All_RTK).

These results indicate that systematic (average) elevation error in individual flights is highly variable, with values as high as 0.85 m in some flights (Brownfield 75° (100 m)$^{-1}$). It is also apparent that even results derived from two flights at the same height and with the same image acquisition angle differ (both between sites and within the same site). The standard deviation of elevation is approximately 0.03 m, which is in accordance with the expected GNSS RTK measurement accuracy.

The All_Combined variants demonstrate a very good agreement of all measured images and coordinates, i.e., that no outliers or erroneous measurements are present. The agreement of internal orientation elements is also very good (a bit worse between sites).

It should be also noted that a higher systematic elevation error was observed in flights where a higher difference between the focal length f calculated for the particular flight and the most likely value determined from the All_Combined calculation occurred. Similar conclusions can be made with respect to the coordinates of the principal point offset.

Table 4. Results of calculation variants—individual flights—Brownfield.

Calculation Variant	Mean Difference (m)	StDev (m)	RMS (m)	f (Pixels)	Cx (Pixels)	Cy (Pixels)
All_Combined	0.0201	0.0063	0.0211	*3685.4568*	9.9062	28.3165
All_RTK	0.0254	0.0063	0.0261	*3685.3472*	9.9037	28.3546
75m_1	0.0412	0.0080	0.0419	3683.5605	9.5506	27.7429
75m_2	0.0769	0.0107	0.0776	3682.2500	9.5145	27.7860
100m_1	0.2715	0.0115	0.2717	3676.2500	9.8893	27.4269
100m_2	0.1252	0.0104	0.1256	3680.7900	9.3435	27.6945
125m_1	−0.1701	0.0225	0.1716	3691.0449	9.8862	27.5684
125m_2	0.2200	0.0153	0.2205	3679.5466	9.5863	27.7279
60° (100m)_1	−0.1398	0.0093	0.1401	3687.8589	9.6792	22.6777
60° (100m)_2	−0.0541	0.0100	0.0550	3687.0000	9.5963	25.4033
75° (100m)_1	−0.8557	0.0115	0.8558	3716.3172	12.1673	13.5000
75° (100m)_2	−0.1610	0.0113	0.1613	3693.5500	11.0213	26.6421

Table 5. Results of calculation variants—individual flights—Rural.

Calculation Variant	Mean (m)	StDev (m)	RMS (m)	f (Pixels)	Cx (Pixels)	Cy (Pixels)
All_Combined	−0.0126	0.0237	0.0265	*3685.1506*	9.0573	29.2059
All_RTK	−0.0145	0.0241	0.0278	*3685.1948*	9.0573	29.1926
75m_1	−0.1813	0.0211	0.1825	3692.6899	9.4703	27.8337
75m_2	0.1829	0.0376	0.1866	3675.7142	8.9789	30.0060
100m_1	0.0369	0.0243	0.0440	3684.3132	9.5869	30.4881
100m_2	0.3154	0.0318	0.3170	3675.1982	9.4326	30.2998
125m_1	−0.0755	0.0231	0.0789	3685.7920	8.2727	27.9129
125m_2	0.4000	0.0238	0.4007	3672.1744	8.1549	28.0104
60° (100m)_1	−0.0971	0.0297	0.1014	3687.4122	9.9673	24.2402
60° (100m)_2	0.0761	0.0274	0.0807	3684.7123	8.3047	31.9379
75° (100m)_1	−0.2793	0.0403	0.2821	3693.5554	9.6960	23.1674
75° (100m)_2	−0.0715	0.0315	0.0780	3686.4340	9.0964	29.4237

Tables 6 and 7 show the results of calculations performed using both corresponding flights. Obviously, the increase in the number of images from the two mutually perpendicular flights led to a reduction of the systematic (mean) elevation error, which is particularly true for the Brownfield site. However, it still exceeds the expected measurement accuracy.

Table 6. Results of calculation variants: joint calculation of duplicate flights—Brownfield.

Calculation Variant	Mean (m)	StDev (m)	RMS (m)	F (Pixels)	Cx (Pixels)	Cy (Pixels)
All_Combined	0.0201	0.0063	0.0211	3685.4568	9.9062	28.3165
75m_1 + 75m_2	−0.0869	0.0080	0.0873	3690.0200	9.6071	28.0833
100m_1 + 100m_2	0.0089	0.0070	0.0112	3685.3100	9.6407	27.9082
125m_1 + 125m_2	0.0217	0.0142	0.0257	3685.2400	9.7382	27.7874
60°_1 + 60°_2 (100m)	−0.0517	0.0073	0.0522	3686.6200	9.4244	25.6123
75°_1 + 75°_2 (100m)	−0.1596	0.0104	0.1600	3693.7200	11.1567	26.3864

Table 7. Results of calculation variants: joint calculation of duplicate flights—Rural.

Calculation Variant	Mean (m)	StDev (m)	RMS (m)	F (Pixels)	Cx (Pixels)	Cy (Pixels)
All_Combined	−0.0126	0.0237	0.0265	*3685.1506*	9.0573	29.2059
75m_1 + 75m_2	−0.1879	0.0266	0.1897	3692.8619	9.2088	29.2684
100m_1 + 100m_2	−0.1088	0.0253	0.1116	3689.8473	9.4996	30.5776
125m_1 + 125m_2	−0.0865	0.0215	0.0890	3686.0700	8.3529	28.1841
60°_1 + 60°_2 (100m)	0.0962	0.0235	0.0989	3684.0718	8.5973	31.4189
75°_1 + 75°_2 (100m)	−0.0343	0.0299	0.0452	3685.7482	9.0306	28.9058

Tables 8 and 9 detail the values of combined calculations. It is obvious that in the Brownfield site, basically any non-homogeneous combination of flight parameters improved the results to such a degree that the maximum mean elevation error did not exceed 0.05 m and the total error (RMS) of 0.053 m, which is still less than two GSDs. On the other hand, no such improvement was observed when data from flights performed at two different heights were combined at the Rural site; the systematic error still remained up to 0.4 m. However, the combination of the flight at 100 m and flights with oblique image acquisition improved the systematic errors; the improvement increased when increasing the image acquisition angle from the nadiral direction. The standard deviations are similar in all these cases, approximately 0.03 m.

Table 8. Results of calculation variants: non-homogenous combinations of the flight at 100 m with nadiral axis of image acquisition and other flights—Brownfield.

Calculation Variant	Mean (m)	StDev (m)	RMS (m)	F (Pixels)	Cx (Pixels)	Cy (Pixels)
All_Combined	0.0201	0.0063	0.0211	*3685.4568*	9.9062	28.3165
100m_1 + 75m_1	−0.0197	0.0073	0.0210	3686.6469	9.6673	27.8125
100m_1 + 75m_2	−0.0246	0.0084	0.0259	3687.0021	9.8057	27.7864
100m_2 + 75m_1	−0.0010	0.0074	0.0073	3685.7385	9.4378	27.8479
100m_2 + 75m_2	−0.0046	0.0086	0.0096	3686.0403	9.4761	27.9108
100m_1 + 125m_1	0.0012	0.0155	0.0152	3685.9121	9.7736	27.6163
100m_1 + 125m_2	0.0061	0.0105	0.0119	3685.9438	9.8531	27.7257
100m_2 + 125m_1	−0.0272	0.0146	0.0308	3686.5826	9.5823	27.8234
100m_2 + 125m_2	−0.0141	0.0106	0.0175	3686.1716	9.4705	27.8232
100m_1 + 60°_1	0.0181	0.0079	0.0196	3685.2030	9.8340	27.4012
100m_1 + 60°_2	0.0218	0.0093	0.0237	3685.4790	9.8563	27.5374
100m_2 + 60°_1	0.0137	0.0084	0.0159	3684.9970	9.4563	27.8556
100m_2 + 60°_2	0.0188	0.0078	0.0203	3685.2289	9.4574	27.7306
100m_1 + 75°_1	−0.0311	0.0099	0.0326	3689.4667	10.4964	27.9219
100m_1 + 75°_2	−0.0409	0.0122	0.0426	3689.7403	10.4778	28.1390
100m_2 + 75°_1	−0.0439	0.0079	0.0446	3689.0670	10.2551	28.0417
100m_2 + 75°_2	−0.0504	0.0159	0.0527	3689.8025	10.2318	28.2601

Table 9. Results of calculation variants: non-homogenous combinations of the flight at 100 m with nadiral axis of image acquisition and other flights—Rural.

Calculation Variant	Mean (m)	StDev (m)	RMS (m)	F (Pixels)	Cx (Pixels)	Cy (Pixels)
All_Combined	−0.0126	0.0237	0.0265	*3685.1506*	9.0573	29.2059
100m_1 + 75m_1	−0.1148	0.0219	0.1168	3689.6683	9.6098	29.3811
100m_1 + 75m_2	−0.1343	0.0319	0.1379	3690.3496	9.4486	30.4642
100m_2 + 75m_1	−0.2110	0.0246	0.2124	3693.8792	9.5771	29.1683
100m_2 + 75m_2	−0.2231	0.0358	0.2258	3694.4234	9.5165	30.2986
100m_1 + 125m_1	0.2818	0.0246	0.2828	3675.5925	8.8262	29.1363
100m_1 + 125m_2	0.2790	0.0238	0.2800	3675.6822	8.9231	29.2559
100m_2 + 125m_1	0.4269	0.0269	0.4277	3671.3294	8.8258	28.9144
100m_2 + 125m_2	0.4379	0.0249	0.4386	3670.9576	8.8740	29.2264
100m_1 + 60°_1	0.0127	0.0251	0.0277	3684.4032	9.5076	29.8877
100m_1 + 60°_2	−0.0036	0.0248	0.0246	3685.6338	9.2217	30.2379
100m_2 + 60°_1	0.0334	0.0261	0.0422	3684.4453	9.5665	29.8517
100m_2 + 60°_2	0.0085	0.0264	0.0273	3685.4680	9.3255	30.0001
100m_1 + 75°_1	0.0400	0.0253	0.0471	3683.1644	9.3780	29.8556
100m_1 + 75°_2	0.0010	0.0265	0.0261	3684.9419	9.4006	30.6061
100m_2 + 75°_1	0.0628	0.0266	0.0680	3683.2049	9.3918	29.8083
100m_2 + 75°_2	0.0126	0.0299	0.0320	3684.9628	9.4726	30.3318

3.3. *Analysis of the Association between the Systematic Error in Elevation and the Deviation of the Focal Length f*

The data detailed in Tables 4–9 were used for the calculation of the differences between the determined focal lengths and the reference value determined from the All-Combined calculation variant. The relationship is shown in Figures 7 and 8, along with the regression coefficients and determination coefficient (calculated in MS Excel).

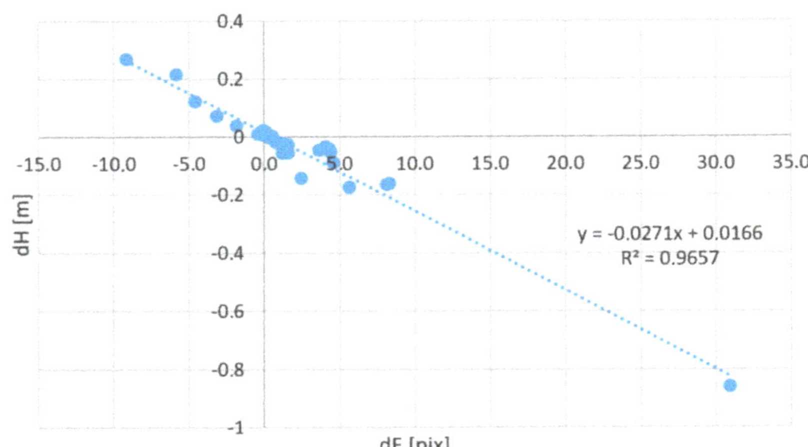

Figure 7. The systematic error of elevation as a function of the deviation of the focal length f from the reference value—Brownfields).

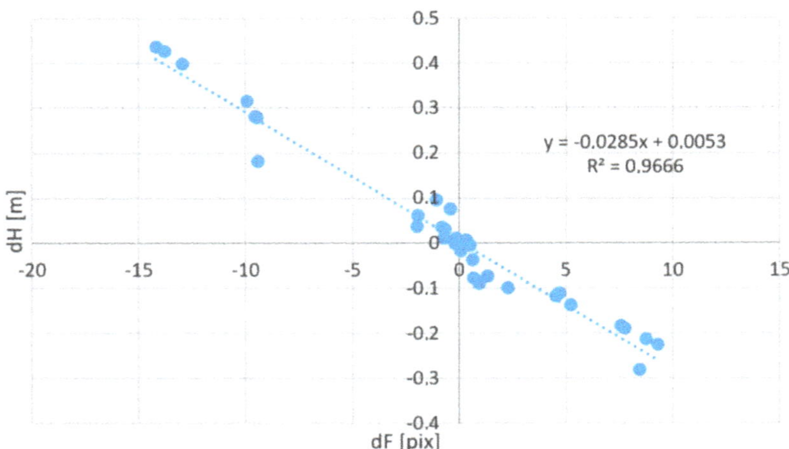

Figure 8. The systematic error of elevation as a function of the deviation of the focal length f from the reference value—Rural.

Both the graphical record and the determination coefficient R^2, which is in both cases close to 1 (0.97), prove a practically linear association of the systematic error and the focal length error, which confirms the hypothesis stated, for example, in our previous paper [34].

The linear regression parameters can be further interpreted. The constant element represents the independent constant error (mutual difference) between coordinates determined by the onboard UAV GNSS RTK receiver and the terrestrial GNSS RTK survey

The detected size of the difference (max 0.017 m) corresponds to the declared accuracy of 0.03 m. The line direction can also be interpreted; using triangle similarity, the following simple equation can be derived:

$$dH = dF \cdot \frac{h}{f}, \tag{1}$$

describing the geometric configuration of the regression line direction as h/f, where h is the flight altitude above the terrain and f is the focal length. The primary flight altitude in the performed experiments was always h = 100 m and the focal length f = 3685 pix, the ratio h/f was therefore 0.027 m/pix, which is very close to the experimentally determined h/f values (0.0271 for the Brownfield site and −0.0285 for the Rural site).

The practically linear relationship between the constant chamber deviation and the systematic error in elevation proves that the inaccuracy of the internal orientation element calculation is indeed the source of that error. The focal length f determined during model calculation can, therefore, be used as an indicator of this error.

A similar approach to the calculation of the correlation coefficient was also applied to the coordinates of the principal point offset Cx and Cy. However, in this case, no reasonable correlation was found; there is, therefore, no direct relationship between the erroneous computation of the principal point offset and the systematic elevation error.

3.4. Comparison of Dense Clouds

Thus far, all comparisons and evaluations in this paper were performed using control points (CPs) in CloudCompare software. It is, therefore, necessary to show that this acquired information also has general validity, i.e., that it is also valid for the generated dense cloud point. For this purpose, the data were cropped, filtered to remove trees and shrubs, and the resulting cloud points were compared. The data from All_Combined calculations (utilizing all available data) were, again, used as the reference. Below, results of the comparisons of the All_Combined calculations to calculations from the primary flight providing the worse result from each location are shown. The comparison of the point clouds from the 100 m^{-1} flight and All_Combined dataset for Brownfield returned a systematic elevation error of 0.268 m with a standard deviation of elevation of 0.038 m, as illustrated by Figure 9. This systematic shift corresponds very well to the value calculated from control points (see Table 4). The comparison of the All_Combined data and 100m_2 flight for the Rural site illustrated in Figure 10 revealed a systematic error of 0.305 m with a standard deviation of 0.029 m. Again, this systematic shift corresponds very well to the value obtained from the control points (0.315 m; see Table 5). The last presented comparison focused on the difference between All_Combined and All_RTK point clouds, which demonstrates the practical agreement of the resulting point clouds; it is obvious that no deformation has occurred and the elevation difference is practically constant (the systematic shift is 0.002 m and the standard deviation is 0.006 m).

The comparisons of the point cloud elevations to the All_Combined variant detailed in Figures 9–11 clearly show that the average difference value corresponds to that calculated using control points, and the same can be said about the standard deviation. The results of analyses of the remaining point clouds were in agreement with this statement, which proves that the difference is indeed caused by a systematic shift of the point cloud in the nadiral direction without deformation in the horizontal plane.

It is also obvious that the All_RTK and All_Combined calculation variants produced practically identical results, which implies that if the internal orientation elements are determined correctly, the result is correct even if only onboard RTK data are used.

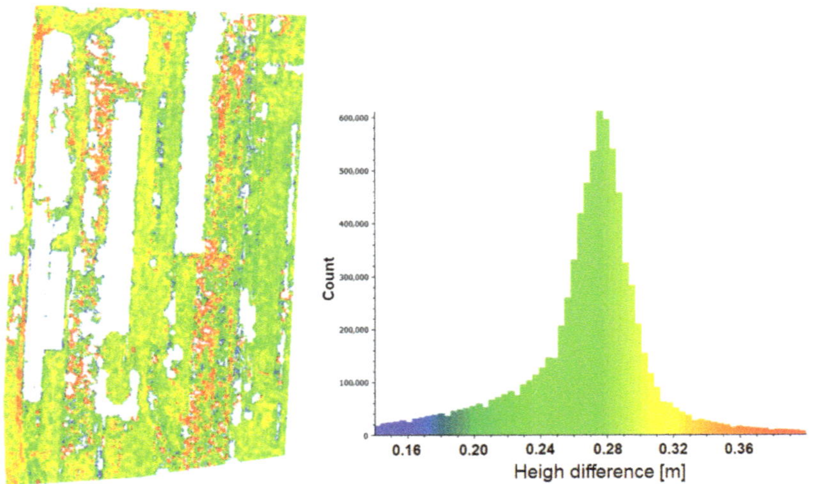

Figure 9. Height differences between 100m_1 RTK and All_Combined point clouds—Brownfield (area cropped along outer control points).

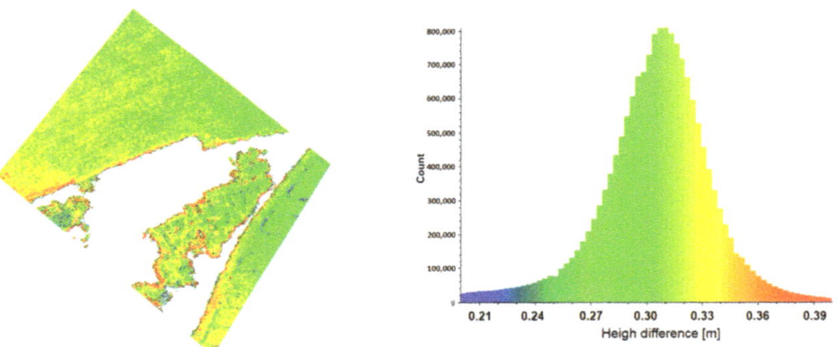

Figure 10. Comparison of All_Combined to the worst result of individual flights, i.e., the flight 100m_2—Rural (area cropped along outer control points).

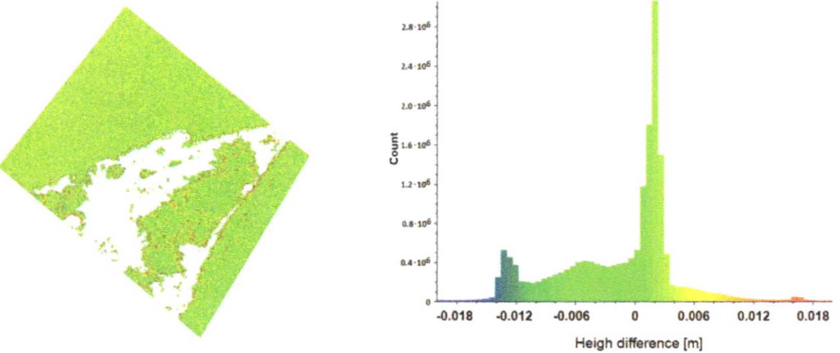

Figure 11. Comparison of All_RTK to All_Combined —Rural.

4. Discussion

Many studies have tested the use of UAV photogrammetry with onboard GNSS RTK for preparing a point cloud representing a DTM and subsequent calculations. Although, in principle, it is possible to perform the whole calculation without the use of GCPs, practical testing revealed that simultaneous calculation of the internal and external orientation elements may, in some cases, lead to a systematic elevation error, although all measurements are correct. This elevation error is random in a way, as it differed even in duplicate flights with the same configuration and the same UAV. This is in accordance with the findings by other authors (e.g., [34,36,38,39]), regardless of whether a fixed-wing or rotary-wing UAV was used.

James and Robson described in 2014 a so-called doming effect when creating a photogrammetric model solely from photographs without the use of reference points. This was not observed in our paper, which can be explained by the fact that the current algorithms used in Agisoft Metashape already consider the camera coordinates during construction of the model [43].

A detailed analysis of the results revealed that incorrect calculation of the internal orientation elements, particularly of the focal length f, is the most likely reason for this problem (e.g., [34,38,40]). Various strategies to deal with this problem have been proposed, such as the use of oblique images (e.g., [19,36,40]), including a small number of GCPs (e.g., [36,39,40,44]) or camera pre-calibration (e.g., [34,42]). Disregarding the use of GCPs (a reliable solution which, however, negates the principal goal of this study, i.e., the simplification of the measurement by avoiding the use of GCPs), it is necessary to try to propose a measurement strategy (flight configuration, processing method, other techniques) to prevent the systematic error. In this study, we aimed to prove the association between the systematic elevation error and erroneous determination of the internal orientation elements and to test various strategies for eliminating the systematic elevation error. We have shown experimentally in two sites (33 models calculated at each site) that the systematic error in elevation and deviation of the focal length are practically linearly dependent, with a coefficient of determination of more than 0.96. The regression coefficients also correspond very well to the relationship between the flight altitude and focal length (Equation (1)). This implies that it is necessary to adopt a strategy ensuring the best possible determination of the internal orientation elements. Here, we must also note that the correct camera calibration is indeed a principal problem of photogrammetry. There are methods that can be used for this purpose, but they usually rely on the use of GCPs and a very specific image acquisition configuration. Additionally, it is also well known that pre/post-calibration, i.e., calibration independent of the particular flight, brings (when using UAVs with non-metric cameras) poorer results than the calibration performed within the scope of the particular flight [45]. It is, therefore, necessary to choose approaches that are actually feasible from the perspective of the camera mounted on the UAV and that can be simply implemented in the image acquisition flight. Here, we combined the primary flight (100 m altitude, nadiral image acquisition) with imagery acquired from other flights. The results of the evaluation of such combined calculations revealed that:

1. Performing duplicate flights, even if the second flight is perpendicular to the first one (double-grid), brings only a minor improvement; however, the accuracy still remains above the expected limit of 1–2 GSD. In some cases, it may still fail with a systematic shift of up to 0.18 m.
2. Geometrically different combinations (i.e., the primary flight combined with flights at other altitudes or with different camera angles) led to a significant improvement. This was especially apparent at the Brownfield site where any of these combinations led to the expected accuracy (elevation difference below 0.05 m). Still, the error at the Rural site remained in some instances as high as 0.4 m. It is, therefore, obvious that the quality and selection of the key points for image matching affect the quality of the calibration.

3. The best results were obtained from the combinations of the primary flight and flights with oblique image acquisition. This was the only strategy that worked well at both sites in all tested combinations. The variant with the higher angle (30° from the vertical direction) provided the best results, with even the worst systematic error not exceeding 0.03 m (1 GSD).

In our experiment, only relatively small camera angles (15° and 30° from the vertical direction) were used to prevent disruption of image alignment, which could pose problems in rugged terrain (i.e., terrain with sloped surfaces). Obviously, the higher the difference in the camera angle, the greater are the differences between the appearance of the same area in the images and, therefore, the more difficult the image matching. Some studies [36,40] used a greater angle (45°) but these were performed on an "ideal" flat terrain. The possible angles and their effect on the resulting accuracy should be subject to further research.

5. Conclusions

UAVs equipped with onboard GNSS RTK receivers are becoming financially more accessible. Their usage for SfM modeling without the need for GCPs seems to be a natural path to take; however, the usual flight configuration with nadiral image acquisition and self calibration often produces results with significant systematic elevation error. In this study, we have proved that the principal cause for this is the incorrect determination of the internal orientation parameters as the resulting elevation error is practically directly proportional to the error in the determination of the focal length (coefficient of determination of 0.96). This was also confirmed by the numerical agreement with the geometrical relationship.

To be able to use an onboard GNSS RTK receiver for direct georeferencing, it is, therefore, necessary to ensure correct calibration of the internal orientation elements; where the use of GCPs and/or accurate camera calibration is not possible or feasible, the results can be improved by adjusting the flight geometry and calculation method. We have tested strategies of joint calculations of two flights that were (i) geometrically identical and (ii) geometrically different. A major difference was revealed between sites; at the site with surfaces suitable for SfM, all tested non-homogeneous flight combinations yielded satisfactory results and the systematic error was reduced to approx. 1 GSD; on the contrary, at the Rural site, covered predominantly with rapeseed, shrubs, trees, and other vegetation, combining different flight altitudes did not result in sufficient improvement. However, combinations of two flights at the same altitude with different camera acquisition axes (nadiral and oblique) performed very well. The combination with the higher difference in acquisition angle (i.e., of nadiral and 30 deg from nadiral image acquisition axes) performed best, capable of reducing the systematic elevation error to approx. 1 GSD even in the Rural area with a surface highly unsuitable for photogrammetry.

Author Contributions: Conceptualization: M.Š.; methodology: M.Š.; ground data acquisition: T.R. and J.S.; UAV data acquisition: J.B.; data processing of UAV imagery: T.R. and J.S.; analysis and interpretation of results: M.Š. and R.U.; manuscript drafting: M.Š. and R.U. All authors have read and agreed to the published version of the manuscript.

Funding: This research was funded by the Grant Agency of CTU in Prague—grant number SGS21/053/OHK1/1T/11 "Optimization of acquisition and processing of 3D data for purpose of engineering surveying, geodesy in underground spaces and 3D scanning".

Institutional Review Board Statement: Not applicable.

Informed Consent Statement: Not applicable.

Data Availability Statement: Not applicable.

Conflicts of Interest: The authors declare no conflict of interest. The funders had no role in the design of the study; in the collection, analyses, or interpretation of data; in the writing of the manuscript, or in the decision to publish the results.

References

1. Park, S.; Choi, Y. Applications of Unmanned Aerial Vehicles in Mining from Exploration to Reclamation: A Review. *Minerals* **2020**, *10*, 663. [CrossRef]
2. Kršák, B.; Blišťan, P.; Pauliková, A.; Puškárová, P.; Kovanič, L.; Palková, J.; Zelizňaková, V. Use of low-cost UAV photogrammetry to analyze the accuracy of a digital elevation model in a case study. *Meas. J. Int. Meas. Confed.* **2016**, *91*, 276–287. [CrossRef]
3. Blistan, P.; Jacko, S.; Kovanič, Ľ.; Kondela, J.; Pukanská, K.; Bartoš, K. TLS and SfM Approach for Bulk Density Determination of Excavated Heterogeneous Raw Materials. *Minerals* **2020**, *10*, 174. [CrossRef]
4. Pavelka, K.; Šedina, J.; Matoušková, E.; Hlaváčová, I.; Korth, W. Examples of different techniques for glaciers motion monitoring using InSAR and RPAS. *Eur. J. Remote Sens.* **2019**, *5*, 219–232, ISSN 2279-7254. [CrossRef]
5. Kovanič, Ľ.; Blistan, P.; Urban, R.; Štroner, M.; Blišťanová, M.; Bartoš, K.; Pukanská, K. Analysis of the Suitability of High-Resolution DEM Obtained Using ALS and UAS (SfM) for the Identification of Changes and Monitoring the Development of Selected Geohazards in the Alpine Environment—A Case Study in High Tatras, Slovakia. *Remote Sens.* **2020**, *12*, 3901. [CrossRef]
6. Komárek, J.; Klouček, T.; Prošek, J. The potential of Unmanned Aerial Systems: A tool towards precision classification of hard-to-distinguish vegetation types? *Int. J. Appl. Earth Obs. Geoinf.* **2018**, *71*, 9–19. [CrossRef]
7. Klouček, T.; Komárek, J.; Surový, P.; Hrach, K.; Janata, P.; Vašíček, B. The Use of UAV Mounted Sensors for Precise Detection of Bark Beetle Infestation. *Remote Sens.* **2019**, *11*, 1561. [CrossRef]
8. Šedina, J.; Housarová, E.; Raeva, P. Using of rpas in precision agriculture. In *17th International Multidisciplinary Scientific Geoconference, Conference Proceedings Volume 17, Informatics, Geoinformatics and Remote Sensing Issue 22, Geodesy and Mine Surveying. Sofia: International Multidisciplinary Scientific GeoConference SGEM*; STEF92 Technology Ltd.: Sofia, Bulgaria, 2017; Volume 17, pp. 331–338, ISSN 1314-2704; ISBN 978-619-7408-02-7. [CrossRef]
9. Buffi, G.; Manciola, P.; Grassi, S.; Barberini, M.; Gambi, A. Survey of the Ridracoli Dam: UAV–based photogrammetry and traditional topographic techniques in the inspection of vertical structures. *Geomat. Nat. Hazards Risk* **2017**, *8*, 1562–1579. [CrossRef]
10. Kumhálová, J.; Moudrý, V. Topographical characteristics for precision agriculture in conditions of the Czech Republic. *Appl. Geogr.* **2014**, *50*, 90–98. [CrossRef]
11. Puniach, E.; Bieda, A.; C´wiąkała, P.; Kwartnik-Pruc, A.; Parzych, P. Use of Unmanned Aerial Vehicles (UAVs) for Updating Farmland Cadastral Data in Areas Subject to Landslides. *ISPRS Int. J. Geo-Inf.* **2018**, *7*, 331. [CrossRef]
12. Moudrý, V.; Beková, A.; Lagner, O. Evaluation of a high resolution UAV imagery model for rooftop solar irradiation estimates. *Remote Sens. Lett.* **2019**, *10*, 1077–1085. [CrossRef]
13. Kalvoda, P.; Nosek, J.; Kuruc, M.; Volařík, T.; Kalvodova, P. *Accuracy Evaluation and Comparison of Mobile Laser Scanning and Mobile Photogrammetry Data*; IOP Conference Series: Earth and Environmental Science; IOP Publishing Ltd.: Bristol, UK, 2020; pp. 1–10, ISSN 1755-1307. [CrossRef]
14. Tavani, S.; Pignalosa, A.; Corradetti, A.; Mercuri, M.; Smeraglia, L.; Riccardi, U.; Seers, T.; Pavlis, T.; Billi, A. Photogrammetric 3D Model via Smartphone GNSS Sensor: Workflow, Error Estimate, and Best Practices. *Remote Sens.* **2020**, *12*, 3613. [CrossRef]
15. Tavani, S.; Corradetti, A.; Granado, P.; Snidero, M.; Seers, T.D.; Mazzoli, S. Smartphone: An alternative to ground control points for orienting virtual outcrop models and assessing their quality. *Geosphere* **2019**, *15*, 2043–2052. [CrossRef]
16. Jaud, M.; Bertin, S.; Beauverger, M.; Augereau, E.; Delacourt, C. RTK GNSS-Assisted Terrestrial SfM Photogrammetry without GCP: Application to Coastal Morphodynamics Monitoring. *Remote Sens.* **2020**, *12*, 1889. [CrossRef]
17. Sanz-Ablanedo, E.; Chandler, J.H.; Rodríguez-Pérez, J.R.; Ordóñez, C. Accuracy of Unmanned Aerial Vehicle (UAV) and SfM Photogrammetry Survey as a Function of the Number and Location of Ground Control Points Used. *Remote Sens.* **2018**, *10*, 1606. [CrossRef]
18. Agüera-Vega, F.; Carvajal-Ramírez, F.; Martínez-Carricondo, P. Assessment of photogrammetric mapping accuracy based on variation ground control points number using unmanned aerial vehicle. *Measurement* **2017**, *98*, 221–227. [CrossRef]
19. Vacca, G.; Dessì, A.; Sacco, A. The Use of Nadir and Oblique UAV Images for Building Knowledge. *ISPRS Int. J. Geo-Inf.* **2017**, *6*, 393. [CrossRef]
20. Nesbit, P.R.; Hugenholtz, C.H. Enhancing UAV–SfM 3D Model Accuracy in High-Relief Landscapes by Incorporating Oblique Images. *Remote Sens.* **2019**, *11*, 239. [CrossRef]
21. Gerke, M.; Nex, F.; Remondino, F.; Jacobsen, K.; Kremer, J.; Karel, W.; Hu, H.; Ostrowski, W. Orientation of oblique airborne image sets–experiences from the isprs/eurosdr benchmark on multi-platform photogrammetry. *Int. Arch. Photogramm. Remote Sens. Spatial Inf. Sci.* **2016**, *41*, 185–191. [CrossRef]
22. Gerke, M.; Przybilla, H.J. Accuracy Analysis of Photogrammetric UAV Image Blocks: Influence of Onboard RTK-GNSS and Cross Flight Patterns. *Photogramm. Fernerkund. Geoinf.* **2016**, *2016*, 17–30. [CrossRef]
23. Forlani, G.; Diotri, F.; Morra di Cella, U.; Roncella, R. UAV Block GEOREFERENCING and control by on-board GNSS data. *Int. Arch. Photogramm. Remote Sens. Spatial Inf. Sci* **2020**, *XLIII-B2-2020*, 9–16. [CrossRef]
24. Jon, J.; Koska, B.; Pospíšil, J. Autonomous Airship Equipped with Multi-Sensor Mapping Platform. *ISPRS-Int. Arch. Photogramm. Remote Sens. Spat. Inf. Sci.* **2013**, *40*, 119–124, ISSN 2194-9034. [CrossRef]
25. Jaud, M.; Passot, S.; Le Bivic, R.; Delacourt, C.; Grandjean, P.; Le Dantec, N. Assessing the Accuracy of High Resolution Digital Surface Models Computed by PhotoScan®and MicMac®in Sub-Optimal Survey Conditions. *Remote Sens.* **2016**, *8*, 465. [CrossRef]
26. Tahar, K.N.; Kamarudin, S.S. Uav onboard gps in positioning determination. *Int. Arch. Photogramm. Remote Sens. Spatial Inf. Sci.* **2016**, *XLI-B1*, 1037–1042. [CrossRef]

27. Urban, R.; Štroner, M.; Kuric, I. The use of onboard UAV GNSS navigation data for area and volume calculation. *Acta Montan. Slovaca* **2020**, *25*, 361–374. [CrossRef]
28. Padró, J.C.; Munoz, F.J.; Planas, J.; Pons, X. Comparison of four UAV georeferencing methods for environmental monitoring purposes focusing on the combined use with airborne and satellite remote sensing platforms. *Int. J. Appl. Earth Obs. Geoinf.* **2019**, *75*, 130–140. [CrossRef]
29. Hung, I.-K.; Unger, D.; Kulhavy, D.; Zhang, Y. Positional Precision Analysis of Orthomosaics Derived from Drone Captured Aerial Imagery. *Drones* **2019**, *3*, 46. [CrossRef]
30. Křemen, T. Measurement and Documentation of St. Spirit Church in Liběchov. In *Advances and Trends in Geodesy, Cartography and Geoinformatics II*; Taylor & Francis Group: London, UK, 2020; pp. 44–49. ISBN 978-0-367-34651-5. [CrossRef]
31. Koska, B.; Křemen, T. The Combination of Laser Scanning and Structure from Motion Technology for Creation of Accurate Exterior and Interior Orthophotos of St. Nicholas Baroque Church. *ISPRS Int. Arch. Photogramm. Remote Sens. Spat. Inf. Sci.* **2013**, *XL-5/W1*, 133–138, ISSN 2194-9034. [CrossRef]
32. Moudrý, V.; Lecours, V.; Gdulová, K.; Gábor, L.; Moudrá, L.; Kropáček, J.; Wild, J. On the use of global DEMs in ecological modelling and the accuracy of new bare-earth DEMs. *Ecol. Model.* **2018**, *383*, 3–9, ISSN 0304-3800. [CrossRef]
33. Tomaštík, J.; Mokroš, M.; Surový, P.; Grznárová, A.; Merganič, J. UAV RTK/PPK Method—An Optimal Solution for Mapping Inaccessible Forested Areas? *Remote Sens.* **2019**, *11*, 721. [CrossRef]
34. Štroner, M.; Urban, R.; Reindl, T.; Seidl, J.; Brouček, J. Evaluation of the Georeferencing Accuracy of a Photogrammetric Model Using a Quadrocopter with Onboard GNSS RTK. *Sensors* **2020**, *20*, 2318. [CrossRef] [PubMed]
35. Peppa, M.V.; Hall, J.; Goodyear, J.; Mills, J.P. Photogrammetric Assessment and Comparison of DJI Phantom 4 Pro and Phantom 4 RTK Small Unmanned Aircraft Systems. *Int. Arch. Photogramm. Remote Sens. Spat. Inf. Sci* **2019**, *XLII-2/W13*, 503–509. [CrossRef]
36. Taddia, Y.; Stecchi, F.; Pellegrinelli, A. Coastal Mapping using DJI Phantom 4 RTK in Post-Processing Kinematic Mode. *Drones* **2020**, *4*, 9. [CrossRef]
37. Santise, M.; Fornari, M.; Forlani, G.; Roncella, R. Evaluation of DEM generation accuracy from UAS imagery. *Int. Arch. Photogramm. Remote Sens. Spatial Inf. Sci.* **2014**, *XL-5*, 529–536. [CrossRef]
38. Forlani, G.; Dall'Asta, E.; Diotri, F.; Cella, U.M.; Roncella, R.; Santise, M. Quality Assessment of DSMs Produced from UAV Flights Georeferenced with On-Board RTK Positioning. *Remote Sens.* **2018**, *10*, 311. [CrossRef]
39. Le Van Canh, X.; Cao Xuan Cuong, X.; Nguyen Quoc Long, X.; Le Thi Thu Ha, X.; Tran Trung Anh, X.; Xuan-Nam Bui, X. Experimental Investigation on the Performance of DJI Phantom 4 RTK in the PPK Mode for 3D Mapping Open-Pit Mines. *Inz. Miner. J. Pol. Miner. Eng. Soc.* **2020**, *46*, 65–74, ISSN 1640-4920. [CrossRef]
40. Teppati Losè, L.; Chiabrando, F.; Giulio Tonolo, F. Boosting the Timeliness of UAV Large Scale Mapping. Direct Georeferencing Approaches: Operational Strategies and Best Practices. *ISPRS Int. J. Geo-Inf.* **2020**, *9*, 578. [CrossRef]
41. Cramer, M.; Przybilla, H.-J.; Zurhorst, A. UAV cameras: Overview and geometric calibration benchmark. *Int. Arch. Photogramm. Remote Sens. Spatial Inf. Sci.* **2017**, *XLII-2/W6*, 85–92. [CrossRef]
42. Harwin, S.; Lucieer, A.; Osborn, J. The Impact of the Calibration Method on the Accuracy of Point Clouds Derived Using Unmanned Aerial Vehicle Multi-View Stereopsis. *Remote Sens.* **2015**, *7*, 11933–11953. [CrossRef]
43. James, M.R.; Robson, S. Mitigating systematic error in topographic models derived from UAV and ground-based image networks. *Earth Surf. Proc. Landf.* **2014**, *39*, 1413–1420. [CrossRef]
44. Zhang, H.; Aldana-Jague, E.; Clapuyt, F.; Wilken, F.; Vanacker, V.; Van Oost, K. Evaluating the potential of post-processing kinematic (PPK) georeferencing for UAV-based structure- from-motion (SfM) photogrammetry and surface change detection. *Earth Surf. Dynam.* **2019**, *7*, 807–827. [CrossRef]
45. Przybilla, H.-J.; Bäumker, M.; Luhmann, T.; Hastedt, H.; Eilers, M. Interaction between direct georeferencing, control point configuration and camera self-calibration for rtk-based uav photogrammetry. *Int. Arch. Photogramm. Remote Sens. Spatial Inf. Sci.* **2020**, *43*, 485–492. [CrossRef]

Article

Quality Assessment of Photogrammetric Methods—A Workflow for Reproducible UAS Orthomosaics

Marvin Ludwig [1,*], Christian M. Runge [2], Nicolas Friess [1], Tiziana L. Koch [3], Sebastian Richter [1], Simon Seyfried [1], Luise Wraase [1], Agustin Lobo [4], M.-Teresa Sebastià [2,5], Christoph Reudenbach [1] and Thomas Nauss [1]

1 Department of Geography, Philipps-University Marburg, Deutschhausstr. 10, 35037 Marburg, Germany; nicolas.friess@geo.uni-marburg.de (N.F.); richte3d@staff.uni-marburg.de (S.R.); seyfries@students.uni-marburg.de (S.S.); luise.wraase@geo.uni-marburg.de (L.W.); reudenbach@uni-marburg.de (C.R.); nauss@staff.uni-marburg.de (T.N.)

2 GAMES Group, Department of Horticulture, Fruit, Growing Botany and Gardening, University of Lleida, 25198 Lleida, Spain; cristian.mestre@hbj.udl.cat (C.M.R.); teresa.sebastia@ctfc.cat (M.-T.S.)

3 Swiss Federal Institute for Forest, Snow and Landscape Research WSL, Zürcherstr. 111, 8903 Birmensdorf, Switzerland; tiziana.koch@wsl.ch

4 Geoscience Barcelona (GEO3BCN—CSIC), 08028 Barcelona, Spain; Agustin.Lobo@geo3bcn.csic.es

5 Laboratory ECOFUN, Forest Science and Technology Centre of Catalonia (CTFC), 25280 Solsona, Catalonia, Spain

* Correspondence: marvin.ludwig@geo.uni-marburg.de

Received: 21 October 2020; Accepted: 19 November 2020; Published: 22 November 2020

Abstract: Unmanned aerial systems (UAS) are cost-effective, flexible and offer a wide range of applications. If equipped with optical sensors, orthophotos with very high spatial resolution can be retrieved using photogrammetric processing. The use of these images in multi-temporal analysis and the combination with spatial data imposes high demands on their spatial accuracy. This georeferencing accuracy of UAS orthomosaics is generally expressed as the checkpoint error. However, the checkpoint error alone gives no information about the reproducibility of the photogrammetrical compilation of orthomosaics. This study optimizes the geolocation of UAS orthomosaics time series and evaluates their reproducibility. A correlation analysis of repeatedly computed orthomosaics with identical parameters revealed a reproducibility of 99% in a grassland and 75% in a forest area. Between time steps, the corresponding positional errors of digitized objects lie between 0.07 m in the grassland and 0.3 m in the forest canopy. The novel methods were integrated into a processing workflow to enhance the traceability and increase the quality of UAS remote sensing.

Keywords: unmanned aerial systems; unmanned aerial vehicle; time series; accuracy; reproducibility; orthomosaic; validation; photogrammetry

---

1. Introduction

Unmanned aerial systems (UAS) are widely used in environmental research. Applications encompass the retrieval of crop yield [1] or drought stress [2] in agricultural areas or the mapping of plant species [3–5], biomass [6,7] or forest structure [8–10] in nature conservation tasks. Today, UAS allow an extensive spatial coverage with high resolution that provides detailed observations on the individual plant level, e.g., for the detection of pest infections in trees [11] or rotten stumps [12]. The flexibility of UAS is also beneficial for multi-temporal observations since flights can be scheduled on short notice based on specific events like bud burst or local weather conditions. Therefore, UAS are regarded as a key component for bridging the scales between space-borne remote sensing systems and in-situ measurements in environmental monitoring systems [13].

Applications of UAS can be structured into two main components: the acquisition of individual images—including the flight planning—with an unmanned aerial vehicle (UAV) for a particular region of interest and the processing of these images with photogrammetric methods to obtain georeferenced orthophoto mosaics [8,14–16] or digital surface models [8,17,18]. Studies on the development of workflows for UAS are sparse, often exclude the flight planning and are mostly tied to a very specific application [4,19]. A generalized, flexible and commonly accepted workflow is still missing [13].

Standardized protocols and quality assessments are needed for a better understanding and appropriate use of UAS imagery. Since the final product quality depends on the initial image capturing, flight planning is one important aspect in a common workflow scheme. For example, the flight height in conjunction with the used sensor (RGB, multispectral or hyperspectral) affect the ground sampling distance (GSD, i.e., pixel size or spatial resolution) of the images and in conjunction with the flight path affect the overlap of the individual images which is a key factor for the successful image processing [20]. A fully reproducible study therefore must include the flight path and parameters as well as the camera configuration metadata. For ready-to-fly consumer UAV, flight planning is usually done in software which is tied to the specific hardware (e.g., the DJIFlightPlanner for DJI drones). These commercial solutions often do not provide the full control of autonomous flights and access to metadata which makes it difficult to integrate flight planning in a generalized workflow. In this respect, open hardware/software solutions like Pixhawk based systems and the MAVlink protocol (mavlink.io) are advantageous. To complete the metadata, environmental conditions during the flight like sun angle and cloudiness that also have impacts on the image quality [21] should be recorded.

Equally or even more important for valid results and their use in subsequent data analysis or synthesis is a quality assessment of the resulting image products in terms of their spatial accuracy and reproducibility. The basic image processing workflow starts with the alignment of the individual images, which results in a projected 3D point cloud. Usually, this point cloud is georeferenced through the individual image coordinates or via the use of measured ground control points (GCPs) [22,23]. The point cloud is the basis for the generation of a surface model and the orthorectification and mosaicing of the individual images based on this surface. Commercial software such as Metashape (Agisoft LLC, St. Petersburg, Russia; formerly known as Photoscan) makes these complex photogrammetric methods accessible to a broad range of users and has been utilized in numerous studies [3,13,24].

The quality of photogrammetrically compiled orthomosaics is commonly expressed as their georeference accuracy [25–28] or statistical error metrics derived from the image alignment (e.g., the amount of points in the point cloud or the reprojection error [8,29]). However, these measures alone provide no comprehensive information about the quality of the orthomosaic since the subsequent steps of orthorectification and mosaicing are not taken into account. Image artifacts and distortions can occur during these processing steps that are not reflected in the georeference accuracy [30]. Especially in forest ecosystems, the complex and diverse structures and similar image patterns in the canopy can lead to erroneous imagery [20]. In addition, wind exposure leads to changes in the structure of the tree canopy and consequently causes problems in the alignment of individual images [31]. The quality assessment becomes even more important when time series are analyzed since actual changes of the observed environmental variables have to be separated from deviations which stem from the image processing itself. In addition, here low georeferencing errors are even more important since the errors of the individual time steps can accumulate. Studies utilizing time series relied on georeferencing errors below 10 cm of the individual time steps [26,32].

To evaluate the reproducibility and validity of orthomosaic time series, it is therefore necessary to: (1) optimize the positional accuracy of the individual orthophoto mosaics, (2) to evaluate the reproducibility of the photogrammetric processing of these orthophoto mosaics and (3) to evaluate the positional accuracy between features in the individual time steps of the series.

This study proposes (i) an optimization for the orthorectification of UAS images and (ii) an additional quality criterion for UAS orthomosaics that focuses on the reproducibility of the photogrammetric processing. The orthorectification is improved by an automated optimization of the checkpoint error based on an iteration of point cloud filters. The reproducibility of orthomosaics is quantified by the repeated processing of the same scene and a pixel-wise correlation analysis between the resulting orthophoto mosaics. We illustrate the importance of both methods using different orthorectification surfaces and two time series in a grassland and a forest area, respectively. To foster an error and reproducibility optimized orthomosaic processing, we incorporate the new methods into an impoved UAS workflow.

2. Materials and Methods

2.1. Multi-Temporal Flights

Two multi-temporal UAV-based image dataset were acquired as a test sample for this study (Table 1). The first dataset is a series of six consecutive flights over a small temperate forest patch (Wolfskaute, Hesse, Germany). The surveyed area covers 7 ha of a forested hill with an adjacent meadow and ranges from 283 m to 320 m a.s.l. (above sea level). with a canopy height of up to 37 m (Figure 1a). The six flights were performed on 2020-07-07 between 11:00 and 14:00 CEST using a 3DR Solo Quadrocopter (3D Robotics, Inc., Berkeley CA, USA) and a GoPro Hero 7 camera (GoPro Inc., San Mateo, CA, USA; Appendix B, Table A1). The flight plan was made with Qgroundcontrol and refined with a LiDAR derived digital surface model (DSM, provided by the Hessian Agency for Nature Conservation, Environment and Geology (HLNUG)) with the R-package uavRmp to achieve a uniform altitude of 50 m above the forest canopy (see Appendix A). For georeferencing and checkpoint error calculation, 13 ground control points (GCPs) were surveyed with the Real Time Kinematic (RTK) GNSS (Global Navigation Satellite System) device Geomax Zenith 35 (GeoMax AG, Widnau, Switzerland). The RTK GNSS measurements had an error of between 0.9 and 1.6 cm in the horizontal direction and between 1.9 and 3.7 cm in the vertical direction. Eight GCPs were used as controlpoints and five served as independent checkpoints to evaluate the georeferencing error (Figure 1a).

The second dataset is an inter-annual time series of a grassland area (La Bertolina, Eastern Pyrenees, Spain). Terrain altitudes range from 1237 m to 1328 m a.s.l. The flights were performed in spring or early summer of 2013, 2015 and 2017 using an octocopter with a Pixhawk controller. Cameras and flight plans in a fixed altitude differ between the dates (Table 1), detailed camera settings in Appendix B, Table A1). The flights took place in the morning with sub-optimal illumination angles below 35 degrees [21] and partly scattered light conditions due to the presence of clouds. Five to eight GCPs were measured in each year with a conventional GPS device without RTK, from which three were used as checkpoints (Figure 1b).

Both datasets were used to empirically determine the georeferencing accuracy and reproducibility of the photogrammetrically retrieved orthomsoaics. For a better understanding of the newly introduced approaches, the following chapters first outline the general UAS image processing workflow.

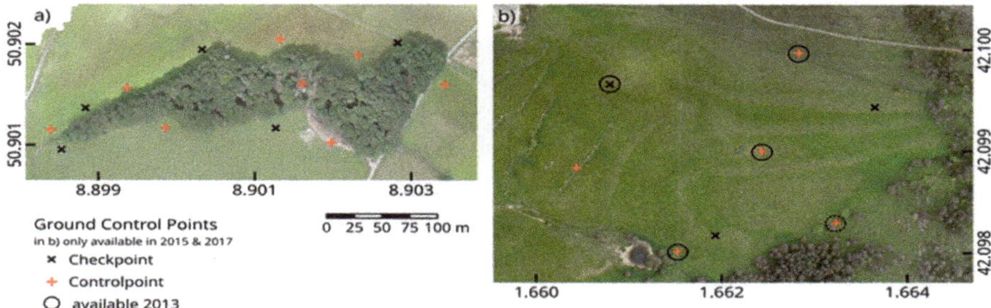

Figure 1. Overview of the two study areas and the location of ground control points. (**a**) Forested area in Wolfskaute, Hesse, Germany. (**b**) Grassland area in La Bertolina, Eastern Pyrenees, Spain. Both maps are projected in UTM but with geographic coordinates for a better overview.

Table 1. Overview of the flight missions used to acquire the two test datasets in forest and grassland environments. The cameras were triggered by time interval. In the forest flights, altitude refers to a uniform height above the canopy. In the grassland flights, altitude referes to a fixed relative height above the take-off point. Overlap referes to both forward (F) and side (S) overlap.

Mission	Date	Sun Angle (°)	Conditions	Camera	Area (ha)	GSD (cm/px)	Alt. (m)	Overlap, F/S (%)	Images
Forest 01	2020/07/07 11:20 a.m.	38.1	cloud free	GoPro Hero 7	7	2.58	50	> 90/75	630
Forest 02	2020/07/07 11:42 a.m.	44.33	cloud free	GoPro Hero 7	7	2.58	50	> 90/75	630
Forest 03	2020/07/07 12:11 a.m.	50.33	partially cloudy	GoPro Hero 7	7	2.58	50	> 90/75	630
Forest 04	2020/07/07 12:40 a.m.	56.13	partially cloudy	GoPro Hero 7	7	2.58	50	> 90/75	630
Forest 05	2020/07/07 13:10 a.m.	61.76	cloudy	GoPro Hero 7	7	2.58	50	> 90/75	630
Forest 06	2020/07/07 13:43 a.m.	67.28	cloudy	GoPro Hero 7	7	2.58	50	> 90/75	630
Grassland 2013	2013/06/01 11:17 a.m.	41.34	partially cloudy	Sony NEX-SN	7.68	3.32	111	70/75	27
Grassland 2015	2015/05/22 09:26 a.m.	19.99	cloud free	Sony NEX-7	14.2	3.68	169	75/75	57
Grassland 2017	2017/05/18 08:53 a.m.	13.58	partially cloudy	Sony ILCE-7RM2	32.4	3.97	132	75/75	57

2.2. Image Georeferencing

Very high resolution orthomosaics such as those resulting from UAV flights require precise positioning to avoid the introduction of complex errors in the image processing [33]. The standard GNSS receivers that are built in cameras do not provide sufficient accuracy. There are two alternative strategies for georeferencing the UAS products: direct georeferencing of the images with a RTK on the UAV or the use of GCPs. Direct georeferencing requires the accurate time synchronization between the RTK device and the camera, which has been reported as a major source of error [33–35].

The use of GCP implies that the study area is accessible in order to install visible ground markers before the flight and precisely measure their position. Ideally, the GCP should be equally distributed over the study area to avoid distortions during processing [33]. During the orthomosaic processing, the ground markers need to be interactively identified in the images. Despite these drawbacks, georeferencing through GCP with general-purpose GNSS receiving systems—that are nowadays standard equipment for surveying—is still far more widespread and potentially more cost-effective than the direct georeferencing with RTK [34].

In any case, GCPs are also required for the independent validation of the referencing accuracy during the processing [24] and therefore essential for a proper accuracy assessment. The geolocation accuracy is usually given as the checkpoint root mean squared error (checkpoint error, Equation (1)) that quantifies the distance between the position of the measured GCP (XYZ_{gcp}) and the estimated positions of these coordinates in the photogrammetric processing (XYZ_{est}). It can be calculated for each direction individually.

$$Checkpointerror = \sqrt{mean\left(\left(XYZ_{gcp} - XYZ_{est}\right)^2\right)} \tag{1}$$

2.3. Photogrammetric Processing

The Metashape software (Agisoft LLC, St. Petersburg, Russia; formerly known as Photoscan) is widely used for UAS image processing. The standard photogrammetric workflow includes the image alignment, the generation of a digital surface model (called Digital Elevation Model in Metashape) and the orthorectification and mosaicing (Figure 2). The image alignment starts with the automatic identification of distinct features in the individual images. This process is enhanced and requires less computation time if the individual images already contain GNSS information. Those features which appear in more than one image, the so called tie points, are matched and projected in a 3D space, forming the sparse cloud that is georeferenced using the surveyed GCPs. The georeferenced sparse cloud is subsequently used to compute a digital surface model, either through a dense pointcloud or a mesh interpolation of the sparse cloud (Figure 2a,b). The surface model is finally used for rectifying the georeferenced images.

For each processing step, a multitude of parameters and options are available that affect the results in terms of georeferencing accuracy and orthomosaic quality. While Metashape offers default values for these parameters, the methods described below aim to optimize and alter the standard workflow to obtain high quality and reproducible orthomosaics.

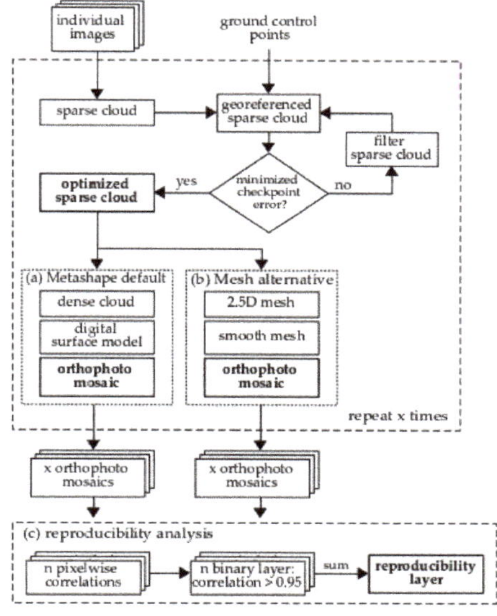

Figure 2. Overview of the workflow and reproducibility analysis. The georeferenced sparse point cloud is iteratively filtered until the checkpoint error reaches its minimum. A surface model and orthomosaic can either be retrieved through (**a**) the dense point cloud or (**b**) the creation of a mesh. For the reproducibility analysis (**c**) the whole photogrammetric process is repeated x times, leading to x orthomosaics from the same image source. Pixel-wise correlation analysis between pairs of orthomosaics leads to n correlation layers and n binary layers based on a correlation coefficient threshold of >0.95. The final reproducibility layer is the sum of all binary layers.

2.4. *Optimizing the Georeferencing*

Each point in the sparse cloud has four accuracy attributes: the reconstruction accuracy (RA), the reprojection error (RE), the projection accuracy (PA) and the image count [35]. In particular, the RE is suggested as the quality measure of tie points [20,36]. It is the deviation of the positions of identified features in the original image from positions of the same features in the calculated 3D space. The removal of points with a high error and the subsequent optimization of the camera positions can improve the georeferencing. However, by removing too many points in the individual sparse clouds, images no longer align and the checkpoint error increases. An iterative approach is used to find the optimal RE threshold for the dataset by filtering the sparse cloud using different RE threshold values in order to minimize the checkpoint error (Figure 2). This method was applied to each flight of the two multi-temporal datasets.

The initial pointclouds are the direct results from the feature identification and matching algorithms from Metashape. In order to account for the inherited randomness of these processes and the slight differences of the checkpoint error due to the manual alignment of the GCP, the optimization was repeated five times and the standard deviation of the checkpoint error was calculated.

2.5. Orthomosaic Reproducibility

Since the orthomosaics are the result of complex photogrammetric methods, its reproducibility has to be assessed. In this context, reproducibility is a measure of how identical individual orthomosaics are, if they are computed from the same image source and with identical photogrammetric processing parameters. This way, the reproducibility of the photogrammetric process itself is evaluated without the influence of changes in the surveyed environment. For this purpose, a set amount of orthomosaics is computed with identical settings (Figure 2). To quantify the reproducibility, the pixel-wise correlation coefficient of the RGB values is calculated between each pair of the computations (*raster* R package, corLocal function). Pixels with a correlation coefficient of 0.95 or higher are considered identical between two orthomosaics. This leads to a binary layer for all pair-wise correlations marking reproducible and non-reproducible pixels. By summing up the binary layers, regions of high and low reproducibility can be identified (Figure 2c). High values then denote a high level of reproducibility of a pixel.

The more orthomosaics are computed, the more correlation layers can be calculated (Equation (2)) and the more likely a layer is to receive a non-correlating pair of pixels. Therefore, a preliminary test with an arbitrarily high number of 25 orthomosaics (x = 25) was done, which leads to n = 300 correlation layers.

$$n_{correlations} = \frac{x!}{2! * (x-2)!} \tag{2}$$

In practice, computing 25 orthomosaics is, in most cases, unreasonable regarding the computation time and processing resources. Therefore, the reproducibility analysis was also done with only 5 identical orthomosaic computations (i.e., 10 pairwise correlations). The comparison of both reproducibility layers revealed that summing up 10 correlation layers ($x = 5$) is sufficient to identify most pixels which are also denoted as non-reproducible when using 300 binary layers. The analysis of the time series and the full forest set were therefore done with only 5 identical orthomosaic computations.

In addition, the edges of the orthomosaic are heavily distorted and have a lower positional accuracy due to less image overlap [37]. Therefore, the orthomosaic should be cropped to the central area with a sufficient overlap. In the R package uavRmp provided with this study, this crop mask is automatically generated from spatial polygons defined by the seamlines (i.e., the outline of the individual image parts) of the mosaic. The outermost polygons are identified using a concave hull of the seamlines and are discarded from the orthomosaic.

All computations were done in R (Version 4.0.2; [38]). All presented methods are provided as the R-package uavRmp (https://gisma.github.io/uavRmp/) and the Metashape Python Scripts (https://github.com/envima/MetashapeTools).

2.6. Assessing the Orthorectification Surface

The standard workflow in Metashape suggests a DSM created from a dense pointcloud as the orthorectification surface (Figure 2a) [16]. In vegetation free areas, this DSM is mostly equivalent to a digital elevation model (DEM) [39] or digital terrain model (DTM) and therefore suitable for the creation of orthomosaics. In areas with vegetation, the DEM requires the classification of ground points in the dense pointcloud which is currently not viable in Metashape for structurally rich environments like forests or grasslands in the phases of maturation and flowering. Alternatively, a 2.5D mesh can be created from the sparse cloud on which the images are projected [40]. By smoothing the mesh to eliminate sharp edges, the surface can be regarded as an approximation of a DEM. This approach requires far less computational ressources since the creation of a densecloud is skipped. It is therefore more suitable for low-budget UAS setups. To demonstrate and validate its usage, the mesh surface was compared to the DSM for one of the forest scenes with respect to the reproducibility of the derived orthomosaics using the pixel-wise correlation method described above.

2.7. Time Series Accuracy

To assess the overall reproducibility of time series, reproducibility masks have been computed for each time step and overlaid to identify pixels that are reproducible over the multi-temporal data and suitable for time series analyses. To differentiate between positional errors from the photogrammetric processing and actual environmental changes between the time steps, identifiable objects and trees were digitized in each individual orthomosaic (7 geometries in the forest, 4 in the grassland). The positional shift of the bounding boxes for each digitized object was calculated between each time step. This provides a more critical assessment of time series than the individual checkpoint errors alone, since relative position differences and environmental changes between the time steps are taken into account.

3. Results

3.1. Optimized Georeferencing Accuracy

To evaluate the optimized georeferencing approach, the sparse clouds were iteratively filtered with decreasing RE thresholds. The sparse clouds of the six forest flights originally included between 560,000 and 680,000 tie points (Figure 3b) after the image alignment with a maximum initial checkpoint error of 3 m. The checkpoint errors were minimized to values between 0.021 and 0.046 m if a RE threshold of 0.4 m was used (Figure 3a). The corresponding pointclouds consisted of about 150,000 tie points. Further reducing the RE threshold to 0.1 m increased the checkpoint error (Figure 3b) due to an insufficient number of tie points for image alignments.

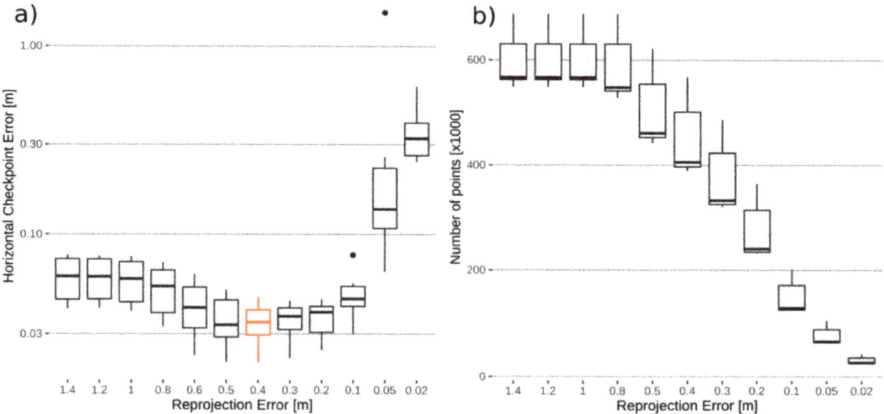

Figure 3. (**a**) Checkpoint error in horizontal direction of the six forest flights with different reprojection error thresholds of the pointcloud filters. A reprojection error threshold of 0.4 m (red) led to the optimal checkpoint error of 0.067 m in the horizontal direction. The initial checkpoint error values of the sparse clouds without a camera optimization were 3 m on average. For better visibility, the y-Axis uses a log10 scale. (**b**) Number of points in the sparse clouds of the six forest flights with different reprojection error thresholds.

In order to test the robustness of the method, the determined optimal RE threshold of 0.4 m was used to filter the sparse clouds of five identical computations of the six forest flights. The average controlpoint error in the horizontal direction was consistently below 0.02 m in all six flights and deviated less than 0.001 m in each of the five computations. The horizontal checkpoint error was between 0.02 m and 0.06 m over all six flights and deviated less than 0.01 m within the five computations. The error in the vertical direction (Z in Table 2) was up to five times higher; however, the reproducibility in each flight is still stable with a maximum deviation of 0.03 m over the five computations.

In the grassland area, the iterative point cloud filtering only marginally improved the checkpoint error since almost all tie points had already very low RE of less than 0.4 m. The final checkpoint errors for the years 2013, 2015 and 2017 were 0.29 m, 0.18 m and 0.07 m, respectively. These errors are up to 10 times higher than in the forest time series, which is mostly due to the use of a conventional GNSS measurements for the GCP in the grassland compared to the RTK GNSS measurement in the forest. Nevertheless, the five computations of the grassland time series led to very consistent checkpoint errors with standard deviations close to 0 (Table 2).

Table 2. Controlpoint and checkpoint error of the five computations of the six forest flights. The images of from each flight were computed five times with identical settings.

	Controlpoint Error (m)				Checkpoint Error (m)			
Flight	**XYmean**	**XYsd**	**Zmean**	**Zsd**	**XYmean**	**XYsd**	**Zmean**	**Zsd**
Forest 01	0.0149	0.0003	0.0207	0.0003	0.0220	0.0002	0.0591	0.0010
Forest 02	0.0082	0.0008	0.0217	0.0006	0.0377	0.0008	0.1989	0.0019
Forest 03	0.0140	0.0001	0.0264	0.0012	0.0565	0.0003	0.1765	0.0030
Forest 04	0.0112	0.0004	0.0122	0.0006	0.0529	0.0034	0.0861	0.0090
Forest 05	0.0176	0.0005	0.0215	0.0005	0.0362	0.001	0.1344	0.0022
Forest 06	0.0120	0.0002	0.0173	0.0003	0.0595	0.0053	0.1845	0.0223
Grassland 2013	0.0001	<0.0001	0.0001	<0.0001	0.2917	0.0012	2.0691	0.0041
Grassland 2015	0.0009	<0.0001	0.0001	<0.0001	0.1837	0.0006	1.3638	0.0095
Grassland 2017	0.0040	<0.0001	0.0001	<0.0001	0.0700	<0.0001	0.0007	<0.0001

3.2. Orthomosaic Reproducibility

To evaluate the reproducibility of orthomosaics, the images of the 4th forest flight were computed 25 times with identical photogrammetric parameters. The 300 pixel-wise correlation analysis between the 25 orthomosaics were performed within a testing area of 600 by 650 pixels showing the forest canopy (Figure 4a). Pixels with correlation coefficients higher than 0.95 were considered reproducible between the orthomosaics. Highly reproducible pixels are characterized by consistently high correlation coefficients and therefore high values in the summed up layer shown in Figure 4b. Non-reproducible regions appear mainly in forest clearings or at dead trees. The actual canopy appears stable across multiple computations.

Using only five identical orthomosaics (i.e., 10 pairwise correlation analyses, Figure 4c) revealed the same patterns as the 300 correlation layers. Hence, only five repetitions are considered enough for subsequent reproducibility analyses.

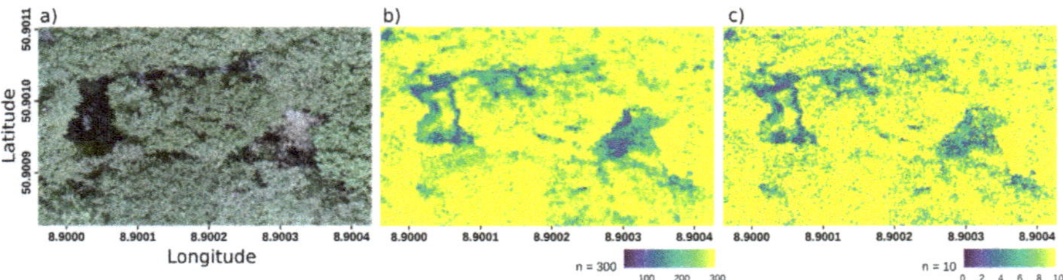

Figure 4. Pixel-wise correlations of the RGB values of a 600 by 650 pixels area of the canopy (**a**) of identical orthomosaic processings. (**b**) The sum of a binary classification over 25 identical computations, hence 300 pairwise correlations. For the binary classification, a pixel-wise correlation coefficient of 0.95 or greater was used. High values (yellow) denote high reproducibility of the RGB values in this pixel over the 25 images. Low values (blue) indicate non-reproducible orthomosaics since the correlation coefficient between the 25 computations is consistently below 0.95. (**c**) The results of only 5 identical computations, hence the sum of 10 correlation layers.

3.3. Comparison of Mesh and DSM Surface-Based Orthomosaics

To evaluate to which degree the reproducibility of orthomosaics depends on the use of the underlying mesh and DSM surface in the forest environment, both surfaces have been used in otherwise identical computation workflows. Using 5 identical orthomosaic computations, 82% and 85% of the pixels were considered reproducible using the DSM and mesh, respectively. All non-reproducible pixels were found in the forest clearing areas of the images. The surrounding meadow did not differ between the orthomosaics (Figure 5). When only the forested area is considered, the number of reproducible pixels decreased to 69% in the DSM and 74% in the mesh-based processing.

While being nearly identical in their amounts of reproducible pixels, there is still a large contrast between the orthomosaic reproducibility of the two surfaces. If a pixel in the DSM-based orthomosaics is non-reproducible between two images, there is a high probability that this pixel is non-reproducible in all images (Figure 5). The mesh-based orthomosaics show significantly more pixels that are non-reproducible between only one or two different orthomosaics, but were stable between the other computations.

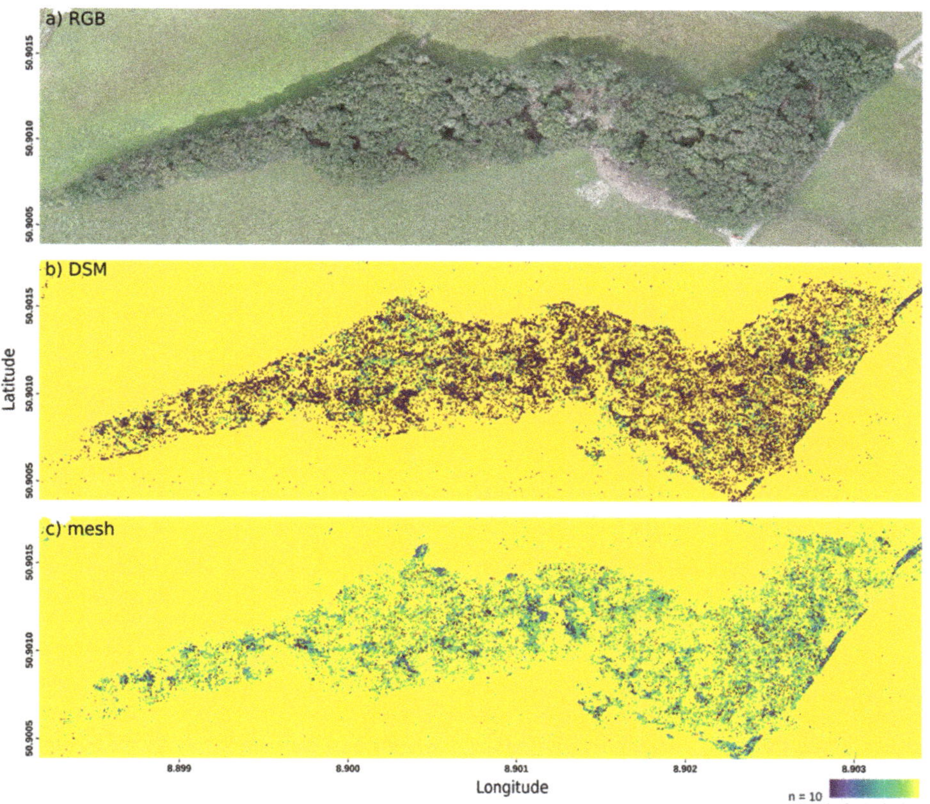

Figure 5. Comparison of DSM and mesh-based orthomosaics in terms of their reproducibility. High values (yellow) denote high reproducibility of the RGB values in this pixel over the 5 orthomosaics. (**a**) RGB orthomosaic of the forested area with the mesh as its orthorectification basis. (**b**) and (**c**) show the sum of the 10 correlation layers of the 5 identical computations using the DSM (**b**) or the mesh (**c**).

3.4. Forest Time Series Reproducibility

The same canopy part of the orthomosaics as in Figure 4 was used for the assessment of the reproducibility along a time series. Figure 6 reveals that non-reproducible areas are mostly consistent between the flights. They concentrate around clearings and around the branches of dead crowns visible in Figure 4a. The actual forest canopy is reproducible. Flight conditions also seem to have an impact on the overall reproducibility. Forest 03 to 06 which were performed in cloudy conditions show less deviations between computations than Forest 01 and 02 where cloud-free conditions and low solar elevations are present.

Summing up all the correlation layers of the six flights (Figure 7) leads to a quality mask for the whole time series. This confirms that the canopy region is reproducible and stable even across the time series.

3.5. Grassland Time Series Reproducibility

In the grassland time series, the reproducibility of each orthomosaic was also tested with five identical computations. Only 1% of the pixels in the grassland area deviated between the computations of each year. In 2013 and 2015, the non-reproducible areas occur mainly in the area of a micrometeorological station in the middle of the meadow (Figure 8). In 2017 some singular pixels also deviated in the meadow areas.

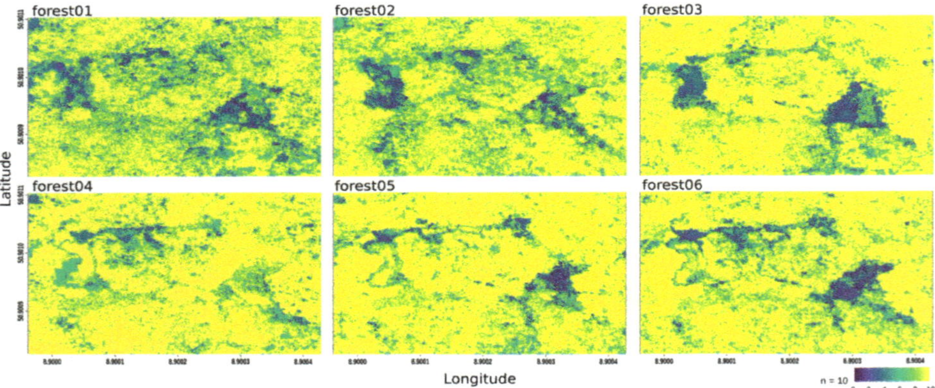

Figure 6. Comparison of the orthomosaic reproducibility of the six flights. High values (yellow) denote high reproducibility of the RGB values in this pixel over the 5 orthomosaics. Low values (blue) indicate non-reproducible pixels since the 10 pairwise correlation coefficients between the 5 computations is consistently below 0.95.

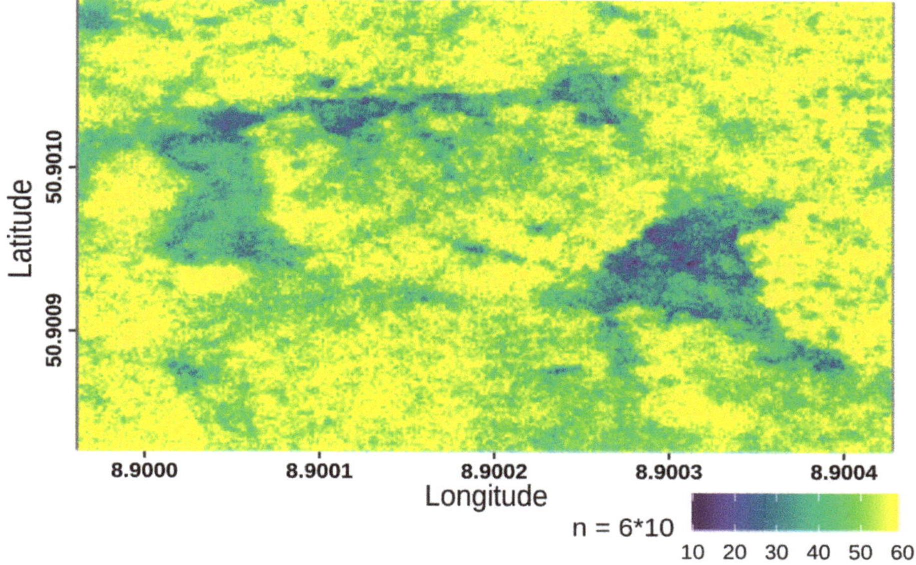

Figure 7. Combination of the orthomosaic reproducibility of the six forest flights.

Figure 8. Inter-annual time series of the grassland area. In red are pixels which had a correlation coefficient lower than 0.95 in one of the 10 pairwise correlation layers of 5 identical orthomosaic computations.

3.6. Time Series Positional Accuracy

To further assess the validity of UAS time series, the positional shift between 7 digitized tree crowns in the forest and four visible objects in the grassland were calculated. Tree crowns moved by 0.3 m on average with a maximum shift of 0.75 m of one tree between forest flight 02 and 03. During the image acquisition of these two flights, the lighting conditions changed due to the presence of clouds and changing wind speeds. In Figure 9, the positional shift of 0.3 m to the left of the marked tree is visible as well as slight differences in the geometry of the crown due to different lighting conditions and wind.

The solar panel visible in Figure 6 is one of four objects which were digitized to measure the positional accuracy of the grassland time series. Between the individual time steps, on average, the polygons differed by 0.03 m in their position. The largest deviation occurred between the orthomosaics of 2013 and 2017 with a maximum shift of 0.07 m between one object. Hence, environmental changes in the grassland have less impact on the time series.

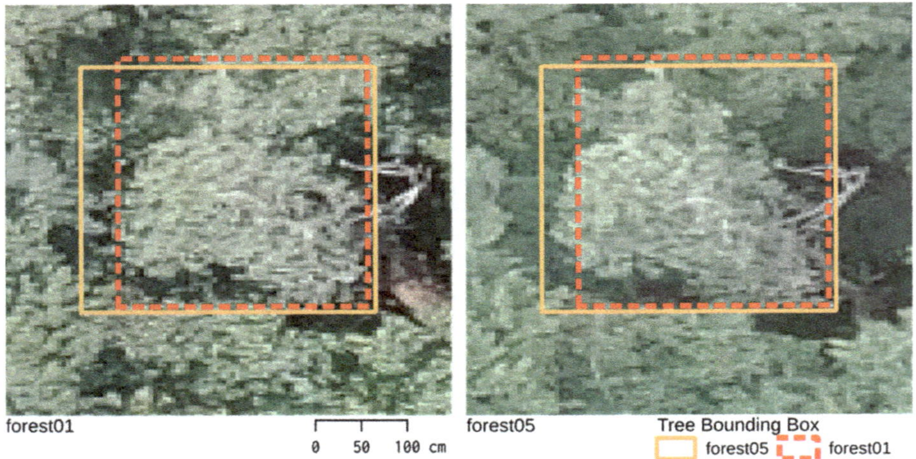

Figure 9. Example of the positional change of a tree between the forest flights 01 and 05.

4. Discussion

The increasing use of UAS imagery in science and science-related services demands a operational processing and reliable validation techniques in the commonly used photogrammetric workflows. This study introduces two optimizations to the conventional photogrammetric workflow: (1) a new optimization for the georeferencing workflow and (2) a novel technique aiming to evaluate the repeatability of photogrammetrically retrieved orthomosaics. The application of these methods demonstrated the possibility to acquire accurately referenced UAS orthomosaic time series with low-cost UAVs and RGB cameras for both forest and grassland environments. The reproducibility of orthomosaics was highly dependent on the vegetation structure of the survey area.

4.1. Optimized Georeferencing Accuracy

The determination of optimal tie point filters leads to positional precisions of less than 6 cm in forested areas. Regarding the GSD of 2.58 cm/px the resulting orthomosaics have a positional error of up to three pixels. This error is stable over multiple computations and different sets of images from the six flights over the forest. This suggests that the iterative filtering approach leads to robust RE thresholds and only needs to be computed one time.

The difference of 0.04 m in the checkpoint error between the six flights could come from different GNSS satellite constellations or cloud conditions [41] over the three hours the flights took place, but, most likely, these small differences come from slight inaccuracies during the manual alignment of the GCP. This suggests that the operational workflow consistently leads to viable orthomosaics with resolutions of less than 10 cm, which is more than sufficient for detailed spatio-temporal structural analysis of forests [9,10]. In Belmonte et al. [8], a checkpoint error of 1.4 m and a GSD of 15 cm led to validated object-based analysis even in moderately dense canopies. The accuracy in the experimental forest areas even keep up with the checkpoint error in the grassland time series (between 0.04 m and 0.08 m). This also compares very well with other studies in structurally sparse landscapes where checkpoint errors tend to be very low [33,42].

The grassland time series further demonstrates that the proposed methods of optimization and validation work outside of the experimental setup. Differences in flight planning, low quality GNSS measurements at the GCP and the usage of different cameras still led to consistent and accurate orthomosaics. Hence, the provided workflow can be used as a fully operational method in grassland and agricultural contexts.

4.2. Pixel-Wise Reproducibility of Orthomosaics

The pixel-wise correlation of identically computed RGB orthomosaics leads to a quantitative measurement of reproducibility. This is a necessary addition to assess the photogrammetic processing of images, especially considering the "black box" nature of non-open-source software like Metashape. With the pixel-wise approach, deviations between computations are assigned to certain spatial regions of the orthomosaic. The mesh and DSM as orthorectification surfaces in the forest time series showed similar amounts of reproducible pixels (DSM: 69%, mesh: 74%). However, the mesh is considered superior since it leads to a better reproducibility in canopy areas. The calculation of the mesh is also less time consuming than the computation of the DSM. Both digital surfaces failed to reproduce fine structures like single tree branches or forest gaps. This can be problematic, since these structures are most likely the ones researchers aim to observe with UAS imagery [11,16,43].

The results also suggest that non-reproducibility can be tracked down to uncertainties in the initial step of the photogrammetric process, the feature identification and feature matching of the individual images [31]. These uncertainties increase with the presence of fine structures in the images since they are prone to move even under light wind conditions. It is therefore more likely that their position changes in consecutive images. In particular, Döpper et al. (2020) [44] recently demonstrated that acquiring UAV data for forest, grasslands and crop environments in low-light conditions such as low Solar Elevation Angle or high cloud cover causes problems in matching characteristics in the image alignment process.

Although this study declares these areas as not reproducible, the structures are still apparent in the orthomosaics. Image analysis methods (e.g., an object-based classification) of these areas might still lead to viable results and consistent geometries. This should be investigated in subsequent studies.

4.3. Time Series

The combination of multiple reproducibility layers enables the validation of UAS derived orthomosaic timeseries. The high reproducibility of multi-temporal grassland orthomosaics confirms the valid analysis of vegetation dynamics in grassland and agricultural studies. Forested area time series are also possible, however non-reproducible regions have to be considered.

The checkpoint error of each orthomosaic in the time series alone gives no insight into the positional relation between the individual time steps. Image acquisition with the UAV, analysis tools (processing software), field experiment designs and environmental conditions have a strong impact on the geometric accuracy of photogrammetric products [6,45,46]. Hence, it is essential to quantify the geometric accuracy on aerial imagery when combining UAS data from different flights, dates and sources [47]. We suggest the addition of geometry-based deviations from digitized objects. In case of the grassland time series, a maximum positional shift of 0.07 m between the time steps is tolerable for most use cases such as the modelling of the temporal dynamics of biophysics and biochemical variables of the meadow canopy or even analyze the variability in size and distribution of vegetation patterns (Lobo et al. in prep). This error also lies in the range of the individual checkpoint errors (0.04 to 0.08 m). In the forest time series, similar accuracies were achieved in the surrounding meadow areas. However, the canopy showed positional deviations of up to 0.5 m in digitized trees. A proportion of this error comes from the actual movement of the canopy due to wind and changes in the lighting conditions [44]. The non-reproducibility of some parts of the canopy, especially at forest clearings may also contribute to this error. We suggest object-based analysis instead of pixel-based approaches when high-resolution forest time series are regarded.

4.4. Improved UAS Workflow

The suggested methods of checkpoint error optimization and reproducibility validation complement the general UAS workflow. In order to make these methods more accessible to users, we provide a Python module—MetashapeTools (https://github.com/envima/MetashapeTools)—

which utilizes the Metashape API for an improved photogrammetric workflow. The orthomosaic processing in form of a script-based workflow ensures the documented parameterization of all the modules in Agisoft Metashape. The workflow is therefore shareable and can be easily integrated into a version control system, making UAS research more transparent. Apart from the manual alignment of the GCP, the photogrammetric process is fully automated. The default parameters in the MetashapeTools are the results of the experimental forest flights and a starting point for a multitude of flight areas. The script-based framework provides flexibility to alter different parts of the workflow and, e.g., integrate alternative processing steps for time series like in Cook et al. [48].

In the future, the general workflow should utilize only open-source software. Currently, Agisoft Metashape is the de facto standard and the most promising software in affordable UAS image processing [3,13]. The development of open-source photogrammetry projects like OpenDroneMap are promising and will be integrated once they are fully operational. The transition from proprietary software towards open and transparent workflows is an ongoing trend worth supporting in spatial analyses [49]. For now, publications utilizing Metashape or other "black box" software should at least include the checkpoint error and the full parameterization of the processing modules. Ideally, the parameterization can be provided as a script, e.g., as a supplementary material or published in a repository. Although the computation of the reproducibility layer can be intensive, its inclusion in studies provide the necessary transparency about the quality and interpretation of the orthomosaics. The documented and evaluated orthomosaics are a big contribution to environmental mapping and monitoring system [50].

5. Conclusions

The rising popularity of UAS imagery in all fields of spatial research led to a variety of processing approaches. The supposedly ease of use and low cost of ready-to-fly UAS opened up some pitfalls in the image acquisition and processing which this study addressed. The evaluation of the orthomosaic accuracy aimed at the reproducibility of the final product. The presented optimization of the georeferencing accuracy based on the checkpoint error and the quantification of the orthomosaic reproducibility enhance the UAS workflow with the necessary quality assessment. This complements the standardized acquisition of high quality UAS time series.

In forest environments, there are still some shortcomings of UAS orthomosaic reproducibility that quantitative analyses need to consider. In grassland environments, these issues are marginal, which supports the validity of UAS in agricultural applications. The novel approaches of this study and their incorporation into a workflow are promising for validated and transparent UAS reserach.

Supplementary Materials: https://github.com/envima/UASreproducibility.

Author Contributions: M.L. analyzed data, contributed to the study design, developed the methodology, and was the main writer of the manuscript; C.M.R. analyzed data (grassland), and contributed to the manuscript; N.F. contributed to the study design (forest), and reviewed the manuscript; T.L.K. contributed to the study design, and reviewed the manuscript; S.R. provided methods; S.S. performed the flights (forest), and provided methods; L.W. provided methods; A.L. contributed to the study design (grassland), analyzed data, reviewed the manuscript, and supervised the project; M.-T.S. contributed to the grassland study design, reviewed the manuscript, and supervised the project; C.R. provided methods, contributed to the study design, contributed to the manuscript, and supervised the project; T.N. contributed to the study design, reviewed the manuscript, and supervised the project; All authors have read and agreed to the published version of the manuscript.

Funding: This research was funded by the Hessian State Ministry for Higher Education, Research and the Arts, Germany, as part of the LOEWE priority project Nature 4.0—Sensing Biodiversity. The grassland study was funded by the Spanish Science Foundation FECYT-MINECO through the BIOGEI (GL2013- 49142-C2-1-R) and IMAGINE (CGL2017-85490-R) projects, and by the University of Lleida; and supported by a FI Fellowship to C.M.R. (2019 FI_B 01167) by the Catalan Government.

Acknowledgments: Thanks to Wilfried Wagner for letting us use his meadow as a landing site.

Conflicts of Interest: The authors declare no conflict of interest.

Appendix A

Flight mission planning is the basis for all UAS derived orthomosaics and therefore crucial for high quality and reproducible image processing. The planning requires the consideration of hardware limitations like UAS speed or the image sampling rate of the camera as well as the aspired ground sampling distance. Further, individual images need to overlap sufficiently in order to process the orthomosaics.

The provided R-package uavRmp strives for the automated and reproducible creation of flight tracks. The package helps users by suggesting image sampling rates and UAS speed with the given camera parameters and the required overlap and GSD. Rectangular study areas can directly be planned in R. Furthermore, uavRmp provides a high resolution surface following mode if a digital elevation model is provided. This makes it possible to follow detailed structures like forest canopies and areas with steep terrain. The camera is also oriented in a fixed direction for the whole mission. The flight is automatically split into multiple MAVlink protocols according to a provided battery lifetime including a safety buffer for proper operations.

Appendix B

Table A1. Details about the cameras and settings.

Camera Model	Sony NEX-SN	Sony NEX-7	Sony ILCE-7RM2	GoPro Hero 7
Image Width	4912 pix	6000 pix	7952 pix	4000 pix
Image Height	3264 pix	4000 pix	5304 pix	3000 pix
Sensor Width	23.5 mm	23.5 mm	35.9 mm	6.17 mm
Sensor Height	15.6 mm	15.6 mm	24 mm	4.63 mm
Focal l Length	16 mm	18 mm	15 mm	17 mm
Resolution	16.7 megapixels	24.3 megapixels	43.6 megapixels	12 megapixels
ISO	100–125	400	1000–1600	400
Shutter	1/640	1/1000	1/1000	Auto

References

1. Nebiker, S.; Lack, N.; Abächerli, M.; Läderach, S. Light-weight Multispectral UAV Sensors and their capabilities for predicting grain yield and detecting plant diseases. *ISPRS-Int. Arch. Photogramm. Remote Sens. Spat. Inf. Sci.* **2016**, *41*, 963–970. [CrossRef]
2. Gago, J.; Douthe, C.; Coopman, R.; Gallego, P.; Ribas-Carbo, M.; Flexas, J.; Escalona, J.; Medrano, H. UAVs challenge to assess water stress for sustainable agriculture. *Agric. Water Manag.* **2015**, *153*, 9–19. [CrossRef]
3. Dash, J.P.; Watt, M.S.; Paul, T.S.H.; Morgenroth, J.; Hartley, R. Taking a closer look at invasive alien plant research: A review of the current state, opportunities, and future directions for UAVs. *Methods Ecol. Evol.* **2019**, *10*, 2020–2033. [CrossRef]
4. Laliberte, A.S.; Goforth, M.A.; Steele, C.M.; Rango, A. Multispectral Remote Sensing from Unmanned Aircraft: Image Processing Workflows and Applications for Rangeland Environments. *Remote Sens.* **2011**, *3*, 2529–2551. [CrossRef]
5. Lu, B.; He, Y. Species classification using Unmanned Aerial Vehicle (UAV)-acquired high spatial resolution imagery in a heterogeneous grassland. *ISPRS J. Photogramm. Remote Sens.* **2017**, *128*, 73–85. [CrossRef]
6. Poley, L.G.; McDermid, G.J. A Systematic Review of the Factors Influencing the Estimation of Vegetation Aboveground Biomass Using Unmanned Aerial Systems. *Remote Sens.* **2020**, *12*, 1052. [CrossRef]
7. Possoch, M.; Bieker, S.; Hoffmeister, D.; Bolten, A.; Schellberg, J.; Bareth, G. Multi-temporal Crop Surface Models Combined with the Rgb Vegetation Index from Uav-based Images for Forage Monitoring in Grassland. *ISPRS-Int. Arch. Photogramm. Remote Sens. Spat. Inf. Sci* **2016**, *41*, 991–998. [CrossRef]
8. Belmonte, A.; Sankey, T.; Biederman, J.A.; Bradford, J.; Goetz, S.J.; Kolb, T.; Woolley, T. UAV-derived estimates of forest structure to inform ponderosa pine forest restoration. *Remote Sens. Ecol. Conserv.* **2019**. [CrossRef]

9. González-Jaramillo, V.; Fries, A.; Bendix, J. AGB Estimation in a Tropical Mountain Forest (TMF) by Means of RGB and Multispectral Images Using an Unmanned Aerial Vehicle (UAV). *Remote Sens.* **2019**, *11*, 1413. [CrossRef]
10. Bourgoin, C.; Betbeder, J.; Couteron, P.; Blanc, L.; Dessard, H.; Oszwald, J.; Roux, R.L.; Cornu, G.; Reymondin, L.; Mazzei, L.; et al. UAV-based canopy textures assess changes in forest structure from long-term degradation. *Ecol. Indic.* **2020**, *115*, 106386. [CrossRef]
11. Lehmann, J.; Nieberding, F.; Prinz, T.; Knoth, C. Analysis of Unmanned Aerial System-Based CIR Images in Forestry—A New Perspective to Monitor Pest Infestation Levels. *Forests* **2015**, *6*, 594–612. [CrossRef]
12. Puliti, S.; Talbot, B.; Astrup, R. Tree-Stump Detection, Segmentation, Classification, and Measurement Using Unmanned Aerial Vehicle (UAV) Imagery. *Forests* **2018**, *9*, 102. [CrossRef]
13. Manfreda, S.; McCabe, M.; Miller, P.; Lucas, R.; Madrigal, V.P.; Mallinis, G.; Dor, E.B.; Helman, D.; Estes, L.; Ciraolo, G.; et al. On the Use of Unmanned Aerial Systems for Environmental Monitoring. *Remote Sens.* **2018**, *10*, 641. [CrossRef]
14. Kattenborn, T.; Lopatin, J.; Förster, M.; Braun, A.C.; Fassnacht, F.E. UAV data as alternative to field sampling to map woody invasive species based on combined Sentinel-1 and Sentinel-2 data. *Remote Sens. Environ.* **2019**, *227*, 61–73. [CrossRef]
15. Park, J.Y.; Muller-Landau, H.C.; Lichstein, J.W.; Rifai, S.W.; Dandois, J.P.; Bohlman, S.A. Quantifying Leaf Phenology of Individual Trees and Species in a Tropical Forest Using Unmanned Aerial Vehicle (UAV) Images. *Remote Sens.* **2019**, *11*, 1534. [CrossRef]
16. Bagaram, M.B.; Giuliarelli, D.; Chirici, G.; Giannetti, F.; Barbati, A. UAV Remote Sensing for Biodiversity Monitoring: Are Forest Canopy Gaps Good Covariates? *Remote Sens.* **2018**. [CrossRef]
17. Fawcett, D.; Azlan, B.; Hill, T.C.; Kho, L.K.; Bennie, J.; Anderson, K. Unmanned aerial vehicle (UAV) derived structure-from-motion photogrammetry point clouds for oil palm (Elaeis guineensis) canopy segmentation and height estimation. *Int. J. Remote Sens.* **2019**, *40*, 7538–7560. [CrossRef]
18. Mohan, M.; Silva, C.; Klauberg, C.; Jat, P.; Catts, G.; Cardil, A.; Hudak, A.; Dia, M. Individual Tree Detection from Unmanned Aerial Vehicle (UAV) Derived Canopy Height Model in an Open Canopy Mixed Conifer Forest. *Forests* **2017**, *8*, 340. [CrossRef]
19. Wijesingha, J.; Astor, T.; Schulze-Brüninghoff, D.; Wengert, M.; Wachendorf, M. Predicting Forage Quality of Grasslands Using UAV-Borne Imaging Spectroscopy. *Remote Sens.* **2020**, *12*, 126. [CrossRef]
20. Seifert, E.; Seifert, S.; Vogt, H.; Drew, D.; van Aardt, J.; Kunneke, A.; Seifert, T. Influence of Drone Altitude, Image Overlap, and Optical Sensor Resolution on Multi-View Reconstruction of Forest Images. *Remote Sens.* **2019**, *11*, 1252. [CrossRef]
21. Pepe, M.; Fregonese, L.; Scaioni, M. Planning airborne photogrammetry and remote-sensing missions with modern platforms and sensors. *Eur. J. Remote Sens.* **2018**, *51*, 412–435. [CrossRef]
22. Sanz-Ablanedo, E.; Chandler, J.; Rodríguez-Pérez, J.; Ordóñez, C. Accuracy of Unmanned Aerial Vehicle (UAV) and SfM Photogrammetry Survey as a Function of the Number and Location of Ground Control Points Used. *Remote Sens.* **2018**, *10*, 1606. [CrossRef]
23. Martinez-Carricondo, P.; Aguera-Vega, F.; Carvajal-Ramirez, F.; Mesas-Carrascosa, F.J.; Garcia-Ferrer, A.; Perez-Porras, F.J. Assessment of UAV-photogrammetric mapping accuracy based in variation of ground control points. *Int. J. Appl. Earth Obs. ans Geoinf.* **2018**, *72*, 1–10. [CrossRef]
24. Latte, N.; Gaucher, P.; Bolyn, C.; Lejeune, P.; Michez, A. Upscaling UAS Paradigm to UltraLight Aircrafts: A Low-Cost Multi-Sensors System for Large Scale Aerial Photogrammetry. *Remote Sens.* **2020**, *12*, 1265. [CrossRef]
25. Capolupo, A.; Kooistra, L.; Berendonk, C.; Boccia, L.; Suomalainen, J. Estimating Plant Traits of Grasslands from UAV-Acquired Hyperspectral Images: A Comparison of Statistical Approaches. *ISPRS Int. J. Geo-Inf.* **2015**, *4*, 2792–2820. [CrossRef]
26. van Iersel, W.; Straatsma, M.; Addink, E.; Middelkoop, H. Monitoring height and greenness of non-woody floodplain vegetation with UAV time series. *ISPRS J. Photogramm. Remote Sens.* **2018**, *141*, 112–123. [CrossRef]
27. Näsi, R.; Viljanen, N.; Kaivosoja, J.; Alhonoja, K.; Hakala, T.; Markelin, L.; Honkavaara, E. Estimating Biomass and Nitrogen Amount of Barley and Grass Using UAV and Aircraft Based Spectral and Photogrammetric 3D Features. *Remote Sens.* **2018**, *10*, 1082. [CrossRef]

28. Michez, A.; Lejeune, P.; Bauwens, S.; Herinaina, A.; Blaise, Y.; Muñoz, E.C.; Lebeau, F.; Bindelle, J. Mapping and Monitoring of Biomass and Grazing in Pasture with an Unmanned Aerial System. *Remote Sens.* **2019**, *11*, 473. [CrossRef]
29. Barba, S.; Barbarella, M.; Benedetto, A.D.; Fiani, M.; Gujski, L.; Limongiello, M. Accuracy Assessment of 3D Photogrammetric Models from an Unmanned Aerial Vehicle. *Drones* **2019**, *3*, 79. [CrossRef]
30. Gross, J.W.; Heumann, B.W. A Statistical Examination of Image Stitching Software Packages for Use with Unmanned Aerial Systems. *Photogramm. Eng. Remote Sens.* **2016**, *82*, 419–425. [CrossRef]
31. Iglhaut, J.; Cabo, C.; Puliti, S.; Piermattei, L.; O'Connor, J.; Rosette, J. Structure from Motion Photogrammetry in Forestry: A Review. *Curr. For. Rep.* **2019**, *5*, 155–168. [CrossRef]
32. Guerra-Hernández, J.; González-Ferreiro, E.; Monleón, V.; Faias, S.; Tomé, M.; Díaz-Varela, R. Use of Multi-Temporal UAV-Derived Imagery for Estimating Individual Tree Growth in Pinus pinea Stands. *Forests* **2017**, *8*, 300. [CrossRef]
33. Padró, J.C.; Muñoz, F.J.; Planas, J.; Pons, X. Comparison of four UAV georeferencing methods for environmental monitoring purposes focusing on the combined use with airborne and satellite remote sensing platforms. *Int. J. Appl. Earth Obs. Geoinf.* **2019**, *75*, 130–140. [CrossRef]
34. Ekaso, D.; Nex, F.; Kerle, N. Accuracy assessment of real-time kinematics (RTK) measurements on unmanned aerial vehicles (UAV) for direct geo-referencing. *Geo-Spat. Inf. Sci* **2020**, 1–17. [CrossRef]
35. *Agisoft Metashapel*; Version 1.6; Agisoft, L.L.C.: St. Petersburg, Russia, 2020.
36. Fretes, H.; Gomez-Redondo, M.; Paiva, E.; Rodas, J.; Gregor, R. A Review of Existing Evaluation Methods for Point Clouds Quality. In Proceedings of the 2019 Workshop on Research, Education and Development of Unmanned Aerial Systems (RED UAS), Cranfield, UK, 25–27 November 2019; pp. 247–252. [CrossRef]
37. Hung, I.K.; Unger, D.; Kulhavy, D.; Zhang, Y. Positional Precision Analysis of Orthomosaics Derived from Drone Captured Aerial Imagery. *Drones* **2019**, *3*, 46. [CrossRef]
38. R Core Team. *R: A Language and Environment for Statistical Computing*; R Foundation for Statistical Computing: Vienna, Austria, 2020.
39. Doughty, C.; Cavanaugh, K. Mapping Coastal Wetland Biomass from High Resolution Unmanned Aerial Vehicle (UAV) Imagery. *Remote Sens.* **2019**, *11*, 540. [CrossRef]
40. Laslier, M.; Hubert-Moy, L.; Corpetti, T.; Dufour, S. Monitoring the colonization of alluvial deposits using multitemporal UAV RGB -imagery. *Appl. Veg. Sci.* **2019**, *22*, 561–572. [CrossRef]
41. Dandois, J.; Olano, M.; Ellis, E. Optimal Altitude, Overlap, and Weather Conditions for Computer Vision UAV Estimates of Forest Structure. *Remote Sens.* **2015**, *7*, 13895–13920. [CrossRef]
42. Turner, D.; Lucieer, A.; de Jong, S. Time Series Analysis of Landslide Dynamics Using an Unmanned Aerial Vehicle (UAV). *Remote Sens.* **2015**, *7*, 1736–1757. [CrossRef]
43. Dash, J.P.; Watt, M.S.; Pearse, G.D.; Heaphy, M.; Dungey, H.S. Assessing very high resolution UAV imagery for monitoring forest health during a simulated disease outbreak. *ISPRS J. Photogramm. Remote Sens.* **2017**, *131*, 1–14. [CrossRef]
44. Döpper, V.; Gränzig, T.; Kleinschmit, B.; Förster, M. Challenges in UAS-Based TIR Imagery Processing: Image Alignment and Uncertainty Quantification. *Remote Sens.* **2020**, *12*, 1552. [CrossRef]
45. Xiang, H.; Tian, L. Method for automatic georeferencing aerial remote sensing (RS) images from an unmanned aerial vehicle (UAV) platform. *Biosyst. Eng.* **2011**, *108*, 104–113. [CrossRef]
46. Tmuši'c, G.; Manfreda, S.; Aasen, H.; James, M.R.; Gonçalves, G.; Ben-Dor, E.; Brook, A.; Polinova, M.; Arranz, J.J.; Mészáros, J.; et al. Current Practices in UAS-based Environmental Monitoring. *Remote Sens.* **2020**, *12*, 1001. [CrossRef]
47. Oliveira, R.A.; Näsi, R.; Niemeläinen, O.; Nyholm, L.; Alhonoja, K.; Kaivosoja, J.; Jauhiainen, L.; Viljanen, N.; Nezami, S.; Markelin, L.; et al. Machine learning estimators for the quantity and quality of grass swards used for silage production using drone-based imaging spectrometry and photogrammetry. *Remote Sens. Environ.* **2020**, *246*, 111830. [CrossRef]
48. Cook, K.L.; Dietze, M. Short Communication: A simple workflow for robust low-cost UAV-derived change detection without ground control points. *Earth Surf. Dyn.* **2019**, *7*, 1009–1017. [CrossRef]
49. Brunsdon, C.; Comber, A. Opening practice: Supporting reproducibility and critical spatial data science. *J. Geogr. Syst.* **2020**. [CrossRef]

50. Haase, P.; Tonkin, J.D.; Stoll, S.; Burkhard, B.; Frenzel, M.; Geijzendorffer, I.R.; Häuser, C.; Klotz, S.; Kühn, I.; McDowell, W.H.; et al. The next generation of site-based long-term ecological monitoring: Linking essential biodiversity variables and ecosystem integrity. *Sci. Total Environ.* **2018**, *613–614*, 1376–1384. [CrossRef]

Article

3D Reconstruction of Power Lines Using UAV Images to Monitor Corridor Clearance

Elżbieta Pastucha [1,2], Edyta Puniach [1,*], Agnieszka Ścisłowicz [1], Paweł Ćwiąkała [1], Witold Niewiem [1] and Paweł Wiącek [1,3]

1 Faculty of Mining Surveying and Environmental Engineering, AGH University of Science and Technology, al. Mickiewicza 30, 30-059 Cracow, Poland; elpa@mmmi.sdu.dk (E.P.); ascis@agh.edu.pl (A.Ś.); pawelcwi@agh.edu.pl (P.Ć.); wniewiem@agh.edu.pl (W.N.); pwiacek@agh.edu.pl (P.W.)

2 The Maersk Mc-Kinney Moller Institute, University of Southern Denmark, Campusvej 55, 5230 Odense, Denmark

3 FlyTech UAV sp. z o.o., ul. Balicka 18A, 30-149 Cracow, Poland

* Correspondence: epuniach@agh.edu.pl

Received: 1 October 2020; Accepted: 10 November 2020; Published: 11 November 2020

Abstract: Regular power line inspections are essential to ensure the reliability of electricity supply. The inspections of overground power submission lines include corridor clearance monitoring and fault identification. The power lines corridor is a three-dimensional space around power cables defined by a set distance. Any obstacles breaching this space should be detected, as they potentially threaten the safety of the infrastructure. Corridor clearance monitoring is usually performed either by a labor-intensive total station survey (TS), terrestrial laser scanning (TLS), or expensive airborne laser scanning (ALS) from a plane or a helicopter. This paper proposes a method that uses unmanned aerial vehicle (UAV) images to monitor corridor clearance. To maintain the adequate accuracy of the relative position of wires in regard to surrounding obstacles, the same data were used both to reconstruct a point cloud representation of a digital surface model (DSM) and a 3D power line. The proposed algorithm detects power lines in a series of images using decorrelation stretch for initial image processing, the modified Prewitt filter for edge enhancement, random sample consensus (RANSAC) with additional parameters for line fitting, and epipolar geometry for 3D reconstruction. DSM points intruding into the corridor are then detected by calculating the spatial distance between a reconstructed power line and the DSM point cloud representation. Problematic objects are localized by segmenting points into voxels and then subsequent clusterization. The processing results were compared to the results of two verification methods—TS and TLS. The comparison results show that the proposed method can be used to survey power lines with an accuracy consistent with that of classical measurements.

Keywords: unmanned aerial vehicles; power lines; image-based reconstruction; 3D reconstruction

---

1. Introduction

Power lines are a typical part of urban and rural landscapes. Due to the need for power, national and regional networks cover most of the world and continue to expand. They require regular monitoring and maintenance work. Monitoring power lines features two aspects: power line components and occlusions of the line corridor. Both are important and interconnected and are thus often addressed simultaneously.

The power line corridor is a 3D buffer around the wires and is defined by the set distance from the wires. It is thus necessary to find the precise position of the wires [1]. Regular inspections of vegetation inside and near the power line corridor are needed to identify trees or branches that need to be cut due to safety concerns, as the direct proximity of trees in a line corridor might trigger, for example, a bushfire. There are monitoring methods that are used to identify branches or canopies that endanger the inviolability of the power line corridor [2] or detect and classify trees to help evaluate the impact on the line [3]. Other techniques focus on volumetric analysis in order to evaluate the impact of the vegetation and its progression by calculating a differential map of a digital surface model (DSM) for two epochs [4].

A range of techniques have been implemented to solve the above problem. They vary from mundane and time-consuming methods of classical surveying to technologically advanced and highly expensive ones. Among the most popular inventory methods for inspecting power lines are airborne laser scanning [5–8] and mobile terrestrial scanning [9,10]. The dense point clouds generated by laser scanning can be used to form models of 3D power lines, in the context of surrounding vegetation, and a survey network. The typical workflow features classifying point clouds, creating the digital terrain model (DTM), and 3D line modeling [11]. Algorithms based on point position, the intensity of response, multiple echoes, and 2D projections enable automated data processing [6,12,13]. Although Light Detection and Ranging (LiDAR) is an effective and robust method, it has some drawbacks. Some, such as problems with suitable weather conditions, are partially shared with passive methods. Additional drawbacks include problems with identifying towers and simply the cost of the equipment, survey, and processing [14]. Some research has also dealt with a combination of unmanned aerial vehicle (UAV) technology and LiDAR for surveying power lines [5,15,16]. The availability of lighter LiDAR scanners and developments in the UAV platforms have contributed to improving efficiency. Although this technology has potential for further development, the cost of a device in combination with the high risk of failure decreases its economic efficiency.

Many applications use optical images and computer vision systems [17]. Satellite, airborne, and UAV images have been employed [18,19]. Satellite images, owing to their low resolution, are limited to providing generalized information on terrains and vegetation [20]. Aerial images rely significantly on manual stereo measurements [19]. UAV-based optical images can provide accurate and high-resolution data [21], and their use with a range of stereomatching algorithms is a promising solution. Attempts have been made to consider photogrammetry as a source of the point cloud and to analyze and filter data similarly to the procedure in LiDAR [22]. Automatic software for dense matching, where the geometry of power lines is reconstructed, could be a fast and convenient solution. In addition to its clear drawbacks relating to optical images, such as the sensitivity to changes in lighting conditions, and the atmospheric influence, the radiometric differences between lines and a background make it even less effective. The lines usually occupy a small part of a photo; otherwise, the time needed for the surveying, the size of the survey data, and the processing time increase. Furthermore, the complexity and the variability of the background is an obstacle for the 3D line reconstruction [23]. The aforementioned reasons and the use of the outliers' approach might preclude the operation of dense matching algorithms. For these reasons, this solution is not feasible.

Other solutions have been proposed for clearance monitoring using aerial images, but methods that use UAVs as the main source of data are becoming increasingly popular. The biggest advantages to using UAVs are their low altitude of flight and the flexibility and economy of the method in comparison with airborne photogrammetry [24,25]. Both multi-rotors and fixed wings are used for this task. The former constructions are especially useful for precise surveys at a low altitude, but the latter approaches are more efficient and have a greater fly range [18].

Considerable research has been dedicated to power line monitoring using UAVs. Most of it has focused solely on wire detection in images [26–34], but a few studies have taken a more holistic approach by considering not only the position of the power line, but also the line corridor and obstacles. The means of data processing and 3D power line reconstruction are different and depend on the aim of the calculation. Such an approach was proposed in two papers [26,35]. One focused on dense matching algorithms and the automation of obstacle detection, whereas power line reconstruction was performed manually, and aided by epipolar images [23]. The other study implemented the results of past work together with fully automated line detection algorithms based on images [35]. This method is based on changes in the gradient combined with the high gray response of the power line and assumes multiple thresholds. This solution provides good results, although it has yet to be proven to work with more versatile data. Similarly, research on epipolar imagery was presented in another piece of research [36], where a real-time system was developed for obstacle avoidance by UAVs as they monitored power lines. To calculate the relative 3D position of the power lines, several steps were implemented, including TopHat transform and the cross-based arbitrary shape support region method, to create a depth map. The study assumed that the background was not complex, and featured either the sky or some treetops. Another holistic approach used semantic segmentation based on fully convolutional neural networks to enhance depth maps and accurately reconstruct linear objects [37]. The research also used enhanced dense matching to detect obstacles in the power line corridor. The neural networks were also used successfully for line segmentation in another study [38]. The effectiveness of these algorithms is at least 80%. However, their major drawback is the demand for a vast learning dataset [3]. Additionally, there are hardly any cases of a comprehensive methodology for detecting and reconstructing power lines in the literature using neural networks [39]. Additional research on the 3D reconstruction of power lines was presented in two papers by the same authors [40,41]. In both, the lines are initially detected from epipolar images using a simple extraction template. Three-dimensional reconstruction is performed differently in each, however. One introduces a 3D grid based on the expected ground sampling distance (GSD) and the positions of the utility poles [40]. The grid is then reprojected on images to validate the detected power lines and establish their relative correspondence. In the second paper, all combinations of wires detected in both images in a stereopair are considered and then reprojected on a third image to validate the choice [41].

This paper aims to develop a comprehensive and robust method for occlusion monitoring in the power line corridor (Figure 1). The main goals are to minimize the time needed to perform the survey and the user input in subsequent processing. The same dataset was thus used to reconstruct power lines and acquire a point cloud representation of a DSM for a possible occlusion check. However, images suitable for the creation of a dense point cloud representation of the terrain usually record power lines as barely distinguishable objects that are only a few pixels wide. Therefore, additional processing is required to successfully extract and model power lines.

Figure 1. A visualization of the goal of the research, i.e., the results of the 3D reconstruction of power lines (source: FlyTech unmanned aerial vehicle (UAV) test flights)—power line over the UAV-derived point cloud representation of the digital surface model (DSM).

The remainder of this paper is structured as follows: Section 2 describes the proposed method, including the data acquisition process, initial data processing, 2D image processing, 3D reconstruction of power lines, and obstacle detection. The datasets acquired to create and test the proposed method are also presented in Section 2. Section 3 describes the results of UAV image processing along with the assessment of their accuracy. A discussion of errors and their possible sources is in Section 4. Section 5 offers the conclusions of this paper. In Appendix A, the additional results of a threshold sensitivity analysis are included.

2. Materials and Methods

The proposed method features the following steps (Figure 2):

- Data acquisition—the terrain was imaged according to certain principles of photogrammetry to ensure high accuracy and automated processing.
- Bundle adjustment and data ordering—the image data were processed using photogrammetric software to estimate the exterior orientation elements (EOE) and the interior orientation elements (IOE) for DSM reconstruction. The data were then ordered into consecutive stereopairs.
- Power line detection in images and reconstruction of 3D geometry—the process uses several techniques, both on separate images (2D) as well as stereopairs (3D). The approximate position of each power line was calculated, either from manual input or 3D points projected on the image. The images were then processed using the modified Prewitt operator, automatic thresholding, and binarization. The random sample consensus (RANSAC) algorithm with additional parameters was then used to calculate the adjusted position of the detected power line. Three-dimensional reconstruction was then performed using the detected power lines and principles of epipolar geometry.
- Detection of obstacles within the power line corridor—a simple procedure where the distance between reconstructed power lines and the point cloud representation of DSM is computed; occluding points are then bound into voxels [37] and clustered into objects [42].

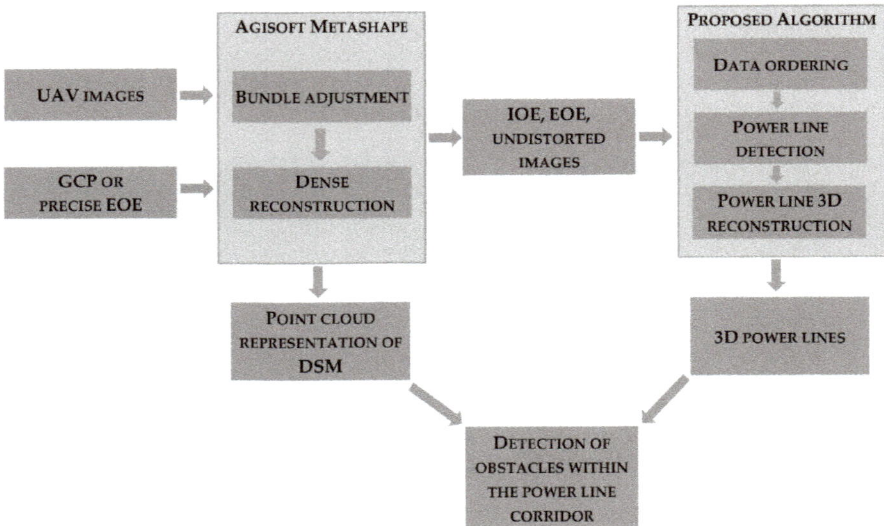

Figure 2. A general schema of the proposed method.

All the steps are described in detail below.

2.1. Data Acquisition

Appropriate data acquisition can enable the detection of power lines and is as important as the methods for the subsequent processing of the data. Some requirements need to be satisfied to ensure the reliable operation of the algorithm. While maintaining the highest efficiency, the measurement requirements of the data should be universal enough to allow for the use of different cameras and UAVs.

First, it is important to define image quality requirements. In this case, the most significant parameter is resolution as defined by the GSD. The maximal GSD to ensure that the wires are detected must be smaller than their diameters. However, to increase the efficiency of detection, the GSD should be half the diameter. Another important aspect is the camera's exposure settings, which must ensure that the wires can be distinguished from the background (Table 1). Owing to the small size of the wires, the ISO setting should be as small as possible. To avoid blurred images, the shutter speed should be adjusted to UAV flight speed, and should not be higher than the ratio of the GSD to the speed of the UAV. If the camera settings allow for the disabling of the low-pass filter, this should be done. Moreover, for the high accuracy of the end product, the use of a global shutter camera would be advised.

Table 1. Recommended camera settings.

ISO	100–400
Minimal shutter speed	$\frac{GSD}{flight\ speed}$
Aperture	with highest geometrical resolution (for most cameras F/5.6)
Focus	manual
Low-pass filter (if changeable)	disabled

Secondly, data acquisition must ensure an appropriate data structure for the algorithm. Because a major function of the algorithm is to transfer detection between stereopairs, the flight plan has to ensure the visibility of the wires for all subsequent images between poles. To meet this requirement, the flight path must be linear, parallel to the power line, and consist of at least two rows, placed on opposite sides of the surveyed corridor (Figure 3). Another advantage of a linear flight plan is the possibility of measuring power lines over distances ranging from a few (multi-rotor) to dozens of kilometers (fixed wing) in one flight.

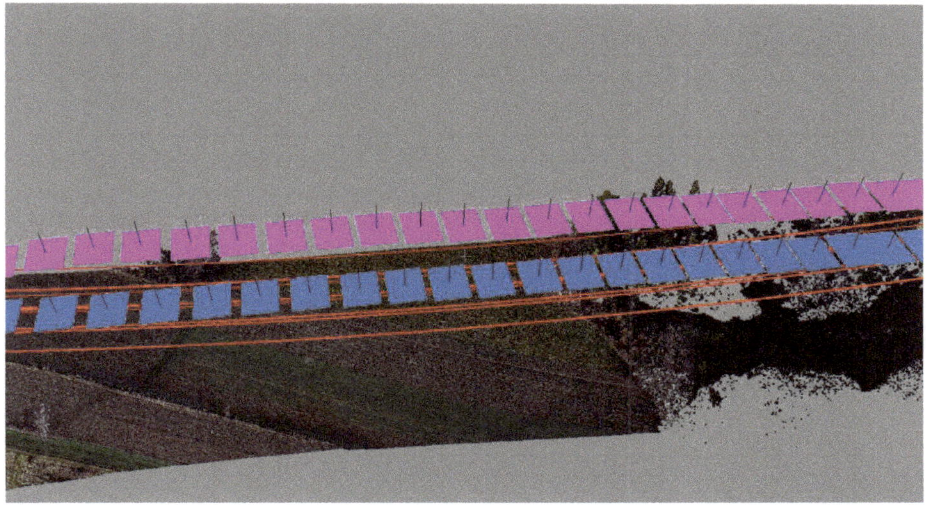

Figure 3. A model flight plan against point cloud representation of the terrain. Power lines in red, opposite stripes (left and right) in pink and blue.

Third, image overlap and the field of view should be carefully planned. The side overlap should allow for the wires to be visible in both neighboring strips. The front overlap should not be lower than 50% (at the height of the wires, overlap on the ground is higher) to ensure the correct transfer of detection between stereopairs, but this should also not exceed the minimum duration of the acquisition of the camera. The front overlap should thus be adjusted according to flight speed and altitude.

The last thing to take into consideration while planning a survey is the global accuracy of the resultant product. Photogrammetric blocks, in the case of corridor mapping, have highly unstable geometry. To maintain high accuracy and prevent errors that could occur in self-calibration due to potentially high correlation between interior and exterior camera orientation, additional measures have to be introduced. A network of ground control points (GCPs) can be introduced to stabilize the block [43]. However, it might not be feasible in remote terrains and it also increases the time and cost of the survey. Equipping the UAV with a global navigation satellite system (GNSS) post-processing kinematic (PPK) receiver to achieve centimeter accuracy of EOE can also help to mitigate the problem [44]. However, the PPK receiver increases the overall cost of the platform. To ensure correct results, one of those measures must be introduced.

2.2. Bundle Adjustment and Data Ordering

The Agisoft Metashape Professional (version 1.6.0 build 9128) software was used to estimate both the IOE and EOE for the obtained UAV images. All images collected during the mission were processed together. The processing, which included automatic bundle adjustment with self-calibration, was mostly automated. A typical set of parameters were estimated during self-calibration [45], utilizing the Brown distortion model [46]:

- principal point position: x_0, y_0,
- focal length: f,
- parameters of radial distortion: r_1, r_2, r_3,
- parameters of tangential distortion: p_1, p_2.

The global accuracy was ensured either by the use of GCPs or precise EOE (GNSS PPK processing). Ultimately, the resulting data contained undistorted images (achieved using the calculated parameters of calibration), along with their corresponding EOE and calibrated focal length.

The data ordering process, as shown in Figure 4, involved firstly automatically assigning the images to flight lines (strip) and then subsequently assigning stereopairs, consisting of the two closest images from their opposite strips (Figure 3), which were subjected to further processing. A set of ordered stereopairs is crucial for the seamless operation of the algorithm. It was necessary to process the images in a defined order (in the direction of processing). Therefore, stereopairs were selected and sorted in ascending order in the direction of processing. This direction was defined by the ordered coordinates of the poles acquired through Agisoft Metashape (or obtained from the operator of the power line).

For the seamless operation of the algorithm, a power line fragment targeted in a UAV flight survey mission was divided into survey sections that were subjected to further processing. The section was defined by a starting utility pole, transfer utility poles, and an ending utility pole. One survey section might have consisted of several power line spans (Figure 5). The order of the poles in the survey section determined the direction of subsequent image processing, and thus had a crucial effect on the entire process of power line detection.

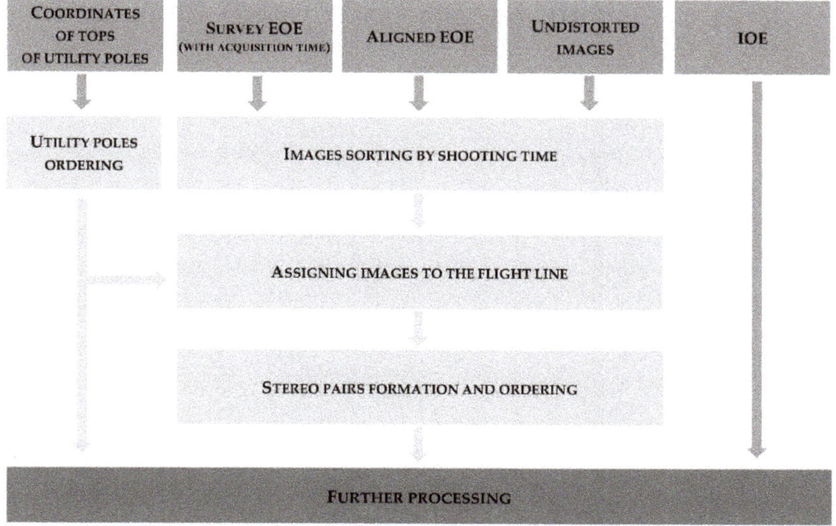

Figure 4. The chart of the first step of algorithm processing. This step of the algorithm is aimed at data ordering for further processing. The survey data (survey exterior orientation elements (EOE)) and the data acquired by means of Agisoft Metashape Professional (aligned EOE, interior orientation elements (IOE), undistorted images, and coordinates of tops of utility poles) are used.

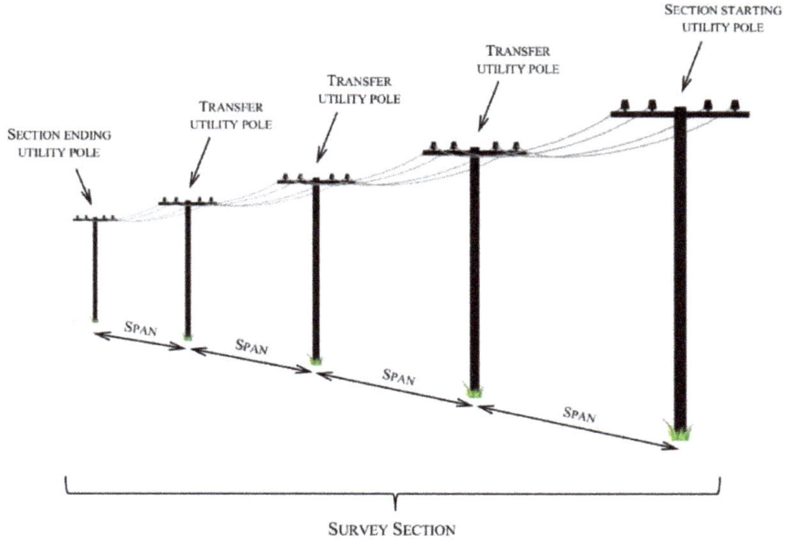

Figure 5. Sketch of basic elements of a survey section.

Defining the survey sections for image processing was an important stage in data ordering. Depending on the type of power line, they were defined differently. For high-voltage lines, one span was defined as one survey section. For medium-voltage lines, the survey section included rectilinear sections of the power line. To divide the survey section, each image containing a pole was analyzed. Three-dimensional lines connecting a pole, captured within the image, and its neighbors were reprojected on the image. When the in-line direction change was greater than 1°, the pole, captured within the image, was assumed to mark the end of the given survey section and the start of a new one. Data thus prepared and ordered were then subjected to further processing.

2.3. Power Line Detection in Images

Detection and reconstruction were performed separately, using dedicated algorithms. The process consists of multiple subsequent steps and was written in the Python programming language.

For effective automation, it was necessary to detect the power lines continuously through a sequence of images. The simplest and most accurate way to transfer detection between neighboring photographs is through 3D space, where the relevant geometry was reconstructed and then projected onto the next image. Such reconstruction was possible by using two across-track neighboring images. Thus, the processing unit in the detection algorithm was a single stereopair. Detection was then performed separately on both images while keeping track of the respective wires. Both images were then used to reconstruct the 3D position of the given power line.

The process can be divided into several steps (Figure 6) that vary depending on the processed image: detection (Case 1—images capturing the survey sections from the starting utility pole) or continued detection (Case 2—all remaining images).

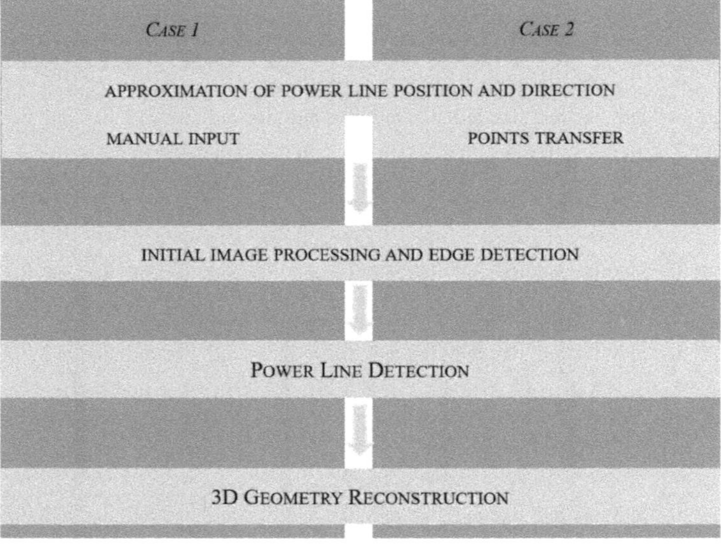

Figure 6. Outline of power line detection process.

Due to the wire's relative position in the image and its sag, the wire was almost never recorded as a straight line in an image. The discrepancies between the fitted line and the empirically captured position of each wire varied from two to dozens of pixels. Moreover, due to varying backgrounds, neither the color of the wire nor the contrast in the image remained constant. To accommodate change in both position and contrast, a local, constantly adapting approach was chosen. For each wire, the image was divided into small, ordered sections within which the power line could be approximated by a straight line. To define the position and order of the processing windows, past information about the approximate positions and directions of the power lines was needed. This was acquired either by a manual initialization procedure in the image at the beginning of the survey section or by using projected points from the preceding stereopair.

2.3.1. Approximation of Position and Direction of Power Line

To approximate the position and direction of each power line, two approaches were adopted according to the given case.

Case 1—starting detection required a manual input in the form of two points per power line in the image. This was sufficient to calculate the general direction and position of the given wire, and important information was obtained regarding the correspondence of the wires between image stereopairs. Owing to the different modes of construction of the utility poles, the wires could be captured in different orders between images.

To transfer detection between images (Case 2), the 3D coordinates of the power lines were used and were then projected onto images in the subsequent stereopair. The line was fitted to points that were within the boundaries of the image. The line parameters defined sought after the position and direction of the power line.

2.3.2. Initial Image Processing and Edge Detection

To detect power lines in images, a method to enhance their visibility was needed. For this purpose, a decorrelation stretch of histograms of the images was used [47]. The enhancement algorithm was based on principal component analysis (PCA). The covariance matrix was calculated from the three RGB bands, and its eigenvalues were found to form coefficients of the transformation of the principal component. After the normalization of the transform bands, new image bands were created. The result was compared with the weighted arithmetic mean of all bands (Figure 7) [48]. The third band of the processed image was chosen as it delivered the highest visibility of the power line.

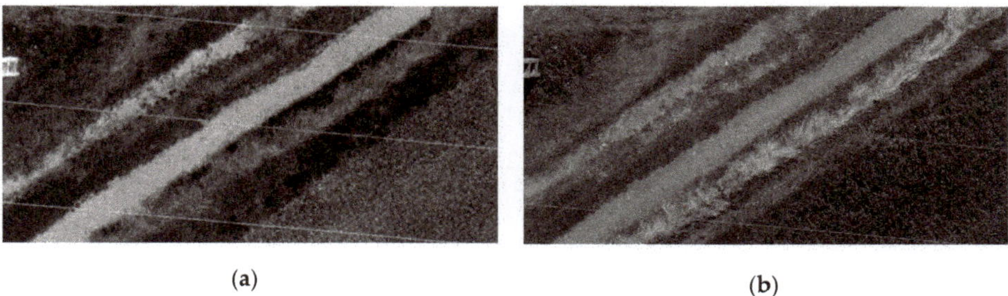

(a) (b)

Figure 7. The same part of an example image showing power lines: (**a**) the band corresponding to the original blue band, obtained by the decorrelation stretching of the raw image; (**b**) weighted arithmetic mean of all bands ($0.2989 \times R + 0.5870 \times G + 0.1140 \times B$).

The edge detection operator needed to be set to enhance long, linear edges, and presumably only ones that aligned with the approximated direction of the power line. The modified Prewitt operator was used for this. The operator was expanded from its base 3×3 form to 31×31. Then, the rotation to the direction of the power line was computed using the previously acquired approximate parameters of the power line. The final operator scheme is shown in Figure 8.

0	0	0	-1	0	1	0	0	0	0	0
0	0	0	-1	0	1	0	0	0	0	0
0	0	0	-1	0	1	0	0	0	0	0
0	0	0	-1	0	1	1	0	0	0	0
0	0	0	0	-1	0	1	0	0	0	0
0	0	0	0	-1	0	1	0	0	0	0
0	0	0	0	-1	0	1	0	0	0	0
0	0	0	0	-1	-1	0	1	0	0	0
0	0	0	0	0	-1	0	1	0	0	0
0	0	0	0	0	-1	0	1	0	0	0
0	0	0	0	0	-1	0	1	0	0	0

Figure 8. An excerpt from the modified Prewitt operator.

The convolution of the grayscale image was calculated using the modified Prewitt operator. The resulting image was not normalized and had both negative and positive values.

All the highest values occurred along the edges on the right side, and all the lowest values occurred along the edges on the left side. Two images were created: one to identify edges on the right side and the other to identify those on the left side. Both images were then normalized to eight-bit unsigned integer space and subsequently binarized (Figure 9).

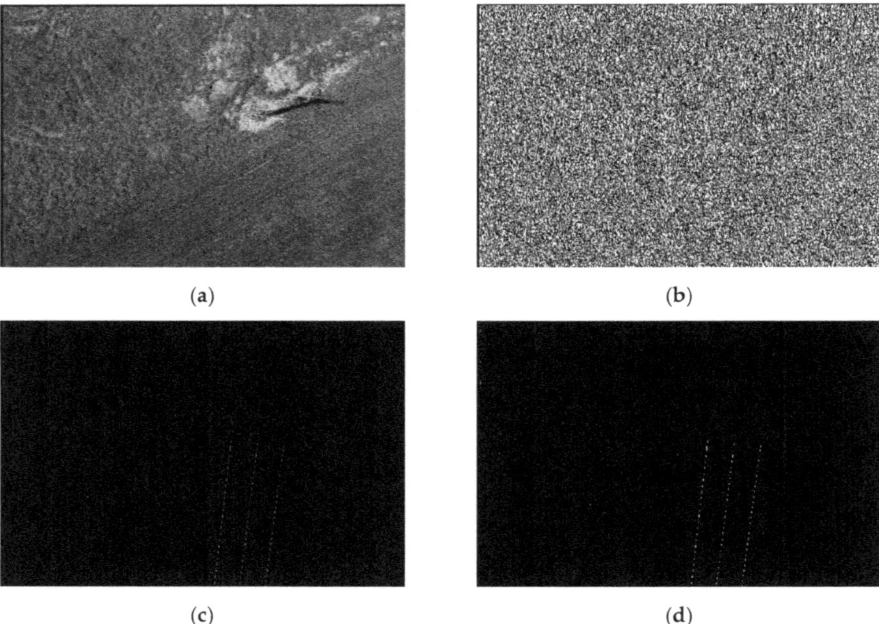

(**a**) (**b**) (**c**) (**d**)

Figure 9. Four stages of image processing: (**a**) original image; (**b**) pre-processed image; (**c**) normalized image; (**d**) binarized image.

2.3.3. Power Line Detection

Separately, for each wire in the image, the process of detection was conducted over small image segments. The position and order of the segments were calculated according to the approximated position and direction of the wire, and they were evenly spaced, with at least a 10% overlap, along the line between the captured utility pole (if present) and the boundary of the image (Figure 10). The order of the segments was set along the direction of the power line.

Figure 10. Original image with the calculated image segments in red.

For each segment, the detection was performed as follows. In images of both the left and the right edges, a line was fitted within the data using the RANSAC algorithm [49]. Then, detected symmetric lines were used to determine the final position of the power line. A list of parameters of detection was provided, together with the results, to assess the correctness of the process:

- c_r—right edge coherence, a quotient of inliers in RANSAC to all positive pixels in the image segment;
- c_l—left edge coherence, a quotient of inliers in RANSAC to all positive pixels in the image segment;
- e_dist_{max}—the maximum distance between lines of the right and left edges within the image segment;
- p—parallelism coefficient, the quotient of the minimal and maximum distances between lines of the right and left edges within the image segment.

Depending on the values of the above parameters, the detection was judged to be successful or incorrect/implausible. If the detection was accepted, the parameters of the line were calculated for the line between the right and left images, and involved the following:

- image coordinates of both ends of the detected line segment,
- a, b parameters of line equation $y = ax + b$.

The rejected detection was replaced by an extension of parameters detected in the previous image segment. The process was repeated until the end of the image was reached. All parameters of the line for all segments were then saved for further processing.

2.3.4. Three-Dimensional Geometry Reconstruction

The last stage of the power line detection process was 3D geometry reconstruction. The global coordinates of the points along the power line were determined using the spatial intersection of homologous rays. The problem was to identify homologous points on wires between the left and the right images in a given stereopair. Previous knowledge of corresponding wires and epipolar geometry was invoked to perform this task. Instead of identifying corresponding points on wires using feature descriptors, a purely geometric approach was chosen.

Each wire captured within the left and right images in a stereopair was represented by line segments established in the detection step. Two hundred evenly spaced points were chosen along the wire on the left image. Their coordinates were derived directly from the parameters of segments of the line. Then, for the nodal points, the respective epipolar lines on the right image were computed (Figure 11a) using a fundamental matrix (Equation (1)):

$$x' \cdot F = k'', \tag{1}$$

where:

x' = [x y 1]—focal coordinates of a point on the left image,
F—fundamental matrix, calculated from the positions and rotations of the left and right images,
k'' = [A B C]—parameters of line equation in a general form.

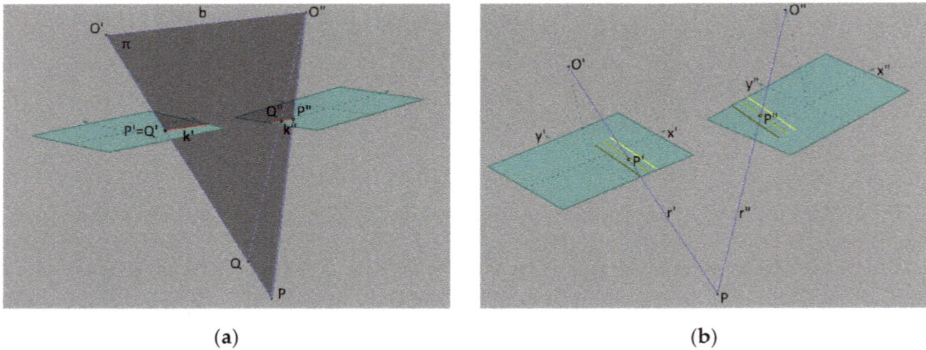

(a) (b)

Figure 11. Three-dimensional reconstruction procedure: (**a**) calculated epipolar line in red; (**b**) the obtained key points and their resection. Symbol descriptions: O′, O″—left and right image projection centers; b—baseline, k′, k″—epipolar lines, respectively, on the left and the right image; π—epipolar plane; Q, P—points in 3D space; P′, P″, Q′, Q″—P, Q points projections on the left and the right image; r′, r″—homologous rays; x′, y′, x″, y″—image coordinates axis on the left and the right image.

The intersections of the epipolar lines and line segments representing the wire on the right image were then calculated to determine key points in it. Finally, the corresponding key points in both images were used to compute the spatial intersections and determine the terrain 3D coordinates (Figure 11b). Together, they provided a discrete representation of the wire.

2.3.5. Catenary Curve Fitting

The wire in discrete representation was not sufficient for assessing power line diagnostics. The sag of the wire, which is acquired from a fitted catenary curve, was also needed. The expression for it describes the geometry of a wire hanging under its weight when supported only at its ends:

$$y = k \cdot cosh\left(\frac{x}{k}\right), \tag{2}$$

where:

$k = \frac{F_x}{q}$—the catenary constant, where

F_x—horizontal force on the cable [kG/mm^2],
q—the weight of the cable per unit arclength [kG/(m·mm^2)].

The catenary curve was fitted to previously obtained points representing the wires using the classical approach. It assumes that the points are represented in the local coordinate system, where the x-axis runs along the wire, while the coordinates of the y-axis correspond to the height of points on the wire. This coordinate system was defined independently for each wire, and the catenary equation was determined in the following form:

$$y - w = k \cdot cosh\left(\frac{x-u}{k}\right), \tag{3}$$

where:

w, u—a parallel offset of the terrain coordinate system from that of the catenary curve.

The solution to Equation (3) was obtained by using the least squares method. Three randomly selected points from a set of representative points on the wire were used to calculate the approximate values of the unknowns (i.e., w_0, u_0, and k_0). Together with the x coordinate of each point, they were used to calculate the deviations in the y coordinates and, subsequently, the adjusted parameters of the catenary curve that best fitted a series of data points. Using this, the maximum value of the sag of the wire was calculated as follows:

$$f_s = y_A + \frac{b}{a} \cdot (x_S - x_A) - k \cdot cosh\frac{x_S}{k}, \tag{4}$$

where:

$x_S = k \cdot arcsinh\frac{b}{a}$,
$a = x_B - x_A$, $b = y_B - y_A$, and
x_A, y_A, x_B, y_B—coordinates of the beginning and end of a catenary curve in the local coordinate system.

Owing to the large number of points representing each wire and the random nature of their selection to calculate approximate values of parameters of the catenary curve, unsatisfactory results of curve fitting were possible (Figure 12a). To avoid such errors and obtain the best possible result, two additional assumptions were introduced: only solutions where the value of the sag of the wire (Equation (4)) was positive were considered, and the choice of the best solution was based on the RANSAC algorithm (Figure 12b).

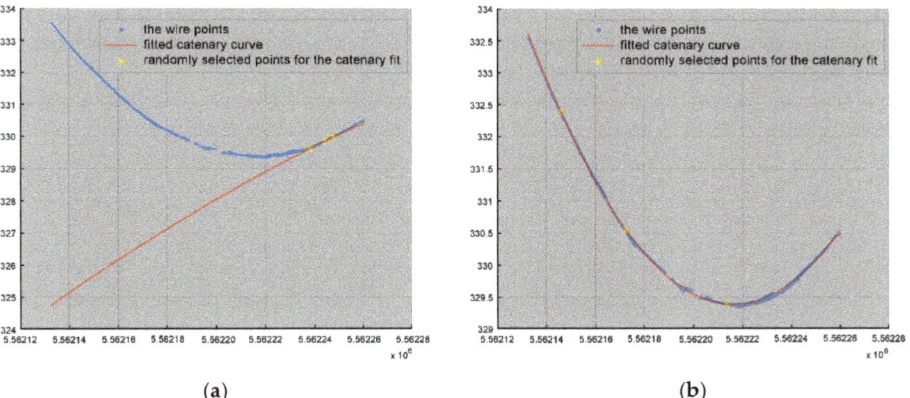

Figure 12. The catenary curve fitting based on the randomly selected points: (**a**) without control, (**b**) with the implementation of the RANSAC algorithm.

The final curve parameters and discrete representation of the wire were saved. The latter consisted of 1000 equally spaced points along each curve per wire.

2.4. Detection of Obstacles within the Power Line Corridor

One objective of the 3D reconstruction of power lines and the creation of the DSM is to monitor the separation of the wires from elements of land cover. We had to check whether the UAV-derived data allowed for the correct identification of obstacles too close to the power lines.

The purpose of the analysis here was to detect DSM fragments (points in the point cloud) that were within the power line corridor. Cloud Compare software and its cloud-to-cloud distance function were used for this task. For each point in the point cloud constituting the DSM, the spatial distance to the nearest point representing the wire was determined.

Next, all the points recorded within the corridor distance were segmented into set size voxels (3D pixels) and subsequently clustered into objects using neighborhood connections. Objects consisting of only one voxel were removed from the dataset. For each object, its volume, center, and bounding box were recorded for further analysis by the power line operator (Figure 13).

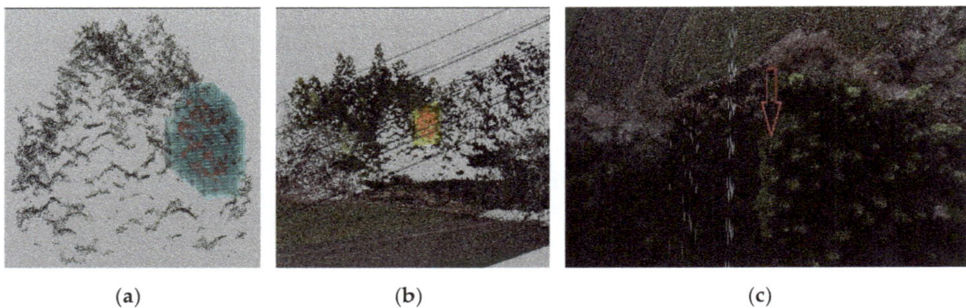

(a) (b) (c)

Figure 13. Detected occlusion—positioned at 49°55′09.3″N, 20°43′09.2″E with a volume of 39.25 m^3. A close look at a point cloud with voxels containing occlusion (**a**), bounding box containing the object with a point cloud representation of DSM (**b**), and source image with an arrow pointing to the detected occlusion (**c**).

2.5. Verification and Quality Assessment

To assess the algorithm, three approaches were adopted. First, the quality of the power line detection and subsequent reconstruction was evaluated. All data used in this assessment were firstly processed and subsequently manually checked for correct and incorrect detection. All errors were marked, and a presumed source of error was noted.

Secondly, the global accuracy of the reconstruction of the power lines was compared to established methods—total station (TS) and terrestrial laser scanning (TLS). The comparison was conducted in three ways. For each wire, both horizontal and vertical position discrepancies were calculated. Additionally, sag values were compared between proposed and reference methods. Though TS and TLS accuracy capabilities reach millimeter levels, this is not the case for a highly dynamic object, where small environmental influences change its geometry. Wire sag changes substantially with a change in temperature, while wind gusts cause oscillating vibrations. Considering the time needed to carry out both TS and TLS surveys, one has to expect significant changes within the surveyed power line. This notably diminishes the accuracy of both methods. Thus, the data were not treated as ground truths, but as reference, comparison data. Thirdly, corridor occlusions detected using the proposed method and TLS data were compared.

2.6. Experimental Data

To conduct the assessment, data were collected for several fragments of medium- and high-voltage power lines. The power lines were located in Małopolskie Voivodeship (Poland), in areas with varied relief.

The data were divided into four datasets denoted as Dataset I, Dataset II, Dataset III, and Dataset IV. Dataset I was used to develop and optimize the algorithm. The effectiveness and the feasibility of the algorithm were analyzed on an independent dataset (Dataset II). The accuracy of the measurements of the power lines and corridor obstacle detection were assessed on Datasets III and IV. The datasets were independent, and therefore assessments of the reliability and accuracy of the proposed method were reliable.

2.6.1. Datasets I and II

Dataset I was used for threshold sensitivity analysis and algorithmic optimization. The survey area contained 13 middle-voltage power line spans (14 utility poles (Figure 14a)), over 1.2 km in a rather flat area. The photogrammetric data were collected in March 2017 using a GRYF octocopter (FlyTech UAV, Krakow, Poland), fitted with a precision positioning system based on a single-frequency GNSS receiver Emlid Reach M+ (Table 2). Dataset I contained 225 images captured with an a6000 camera (Sony, Tokyo, Japan) (Table 3) and Nakton 40 mm (Voigtlander, Braunschweig, Germany). The flight was fully autonomous and was conducted linearly along the power line in two strips. The flight altitude was 60 m above ground level and 50 m above the top of the poles, which yielded 6 and 5 mm of GSD, respectively (Table 4).

(a) (b)

Figure 14. Example images of utility poles supporting power lines: (**a**) for Dataset I; (**b**) for Dataset II.

Table 2. UAV on-board global navigation satellite system (GNSS) receiver parameters.

Model	Emlid Reach M+
Frequency bands	Single-band
Receiver type	72-channel u-blox M8 engine GPS L1C/A, GLONASS L1OF, BeiDou B1I
Max navigation rate	5 Hz
PPK Horizontal position accuracy	7 mm + 1 ppm
PPK Vertical position accuracy	14 mm + 1 ppm

Table 3. Camera parameters.

Model	Sony a6000	Sony RX1RM2
Image Sensor	APS-C (15.6 × 23.5 mm)	FF (35.9 × 24 mm)
Resolution	24 MP (4000 × 6000)	42 MP (7952 × 5304)
Pixel size	15.28 μm^2 (3.9 × 3.9 μm)	20.43 μm^2 (4.5 × 4.5 μm)
Shutter	Mechanical curtain (with rolling shutter effect)	Mechanical central (without rolling shutter effect)
Interchangeable lens	YES	NO
Focusing system	mechanical	electronic
Aperture setting	F/5.6	F/4.0
Shutter setting	1/1000 s	1/1600 s
ISO setting	Auto 100–400	Auto 100–400

Table 4. Basic parameters of UAV missions.

Dataset	Number of Images	Flight Altitude	GSD	Side/Front Overlap
Dataset I	225	60 m	6 mm	70%/50%
Dataset II	238	70 m	9 mm	75%/75%
Dataset III	282	60 m	8 mm	75%/75%
Dataset IV	203	124 m	15 mm	75%/70%

Dataset II was used to validate the feasibility, reliability, and efficiency of the proposed method. It consisted of 238 images captured with an a6000 camera (Sony) (Table 3) and 30 mm Sigma lens (Sigma Corporation, Kawasaki, Japan) in November 2017. The flight was performed using a GRYF octocopter (FlyTech UAV). It was fully autonomous and was conducted in three strips parallel to the power line. The flight altitude was 70 m above ground and 40 m above the top of the poles (Table 4). Both the side and front overlap between the images were 75%. The flight area covered four spans of high-voltage power lines (five utility poles (Figure 14b)) with a total length of 1.4 km. Only two outer strips were used in the processing.

2.6.2. Datasets III and IV

Datasets III and IV were used to verify the accuracy of the proposed method. They included UAV imagery as well as the results of the survey of power lines through TLS and classic TS measurements. The 3D geometry of the wires is constantly changing as a consequence of changing weather conditions (especially changes in temperature). It was thus assumed that the data for the power lines needed to be collected using different methods simultaneously. Unfortunately, the time needed to perform each of the surveys varied greatly from a couple of hours (UAV) to a couple of days (TLS).

The data marked as Datasets III and IV concern power lines located in hilly and difficult to access areas. Dataset III contained data on a segment of a medium-voltage power line consisting of 10 spans, with a total length of 1.3 km. The segment was located in an area with a maximum height difference of 60 m. The power line was fitted with three transmission wires set at the same height. A 400 kV high-voltage power line was another object of research, and data related to it were collected in Dataset IV. The tested section of the power line with a total length of 1.55 km consisted of four spans. The height difference between the beginning and end of the analyzed power line section was 100 m, with a height difference of 60 m for one of the spans. It was fitted with 12 transmission wires and two ground wires, all positioned at varying heights.

The photogrammetric data in Datasets III and IV included high-resolution digital images of the power lines taken with a DSC-RX1RM2 (35 mm) non-metric camera (Sony) (Table 3) from the GRYF octocopter (FlyTech UAV) on 6 December 2018 (Table 4). The images were captured in two strips parallel to the power line that was the subject of the measurement. A network of GCPs and checkpoints (CPs) was established for each dataset. The coordinates of markers were determined by the GNSS real time network (RTN) method with a Leica GS16 receiver (Leica Geosystems AG, Sankt Gallen, Switzerland) that had a horizontal accuracy of 3 cm and a vertical accuracy of 5 cm.

The segments of power lines represented by Datasets III and IV were measured using other methods such as TLS and TS measurements for reference. Fieldwork was performed on 1–3 and 6 December 2018. The weather conditions during the measurements were variable. The temperature was between −5 °C and +5 °C, consequently changing the power line sag. In addition, on 2 and 3 December, the wind blew at a speed of up to 15 m/s (gusts), which caused the oscillating movement of the wires.

For TS measurements, a Nova MS50 (Leica Geosystems AG) was used. Each span (all its wires) was measured from a single instrument station approximately located in the middle of the span. A single wire point at its beginning and end, and several points along its entire length, were measured. The coordinates of the stations and reference points were determined by the GNSS RTN method. Using these data, the spatial coordinates of points representing the individual wires were determined.

In addition to the TS measurements, the power lines were measured using TLS. The Leica ScanStation C10 laser scanner (Leica Geosystems AG), together with a set of HDS (High Definition Survey) 6″ targets (Leica Geosystems AG), were used for this purpose. The TLS measurements were carried out using the three-tripod method with a traverse workflow. The resolutions of the TLS were 10 and 7.5 mm at a distance of 10 m. This was connected with the maximum reduction in measurement time while maintaining satisfactory scanning results and was preceded by tests to determine the optimum scanning resolution depending on the type of power line and the distance between stations. The data were collected at 39 stations for Dataset III and 29 stations for Dataset IV. Finally, the point clouds from all scanner stations were registered and georeferenced (Figure 15) using Leica Cyclone v.9.4.2 (Leica Geosystems AG) software. The point cloud registration was performed using HDS 6″ targets and their coordinates, as determined by the GNSS RTN method. The final root mean-squared errors (RMSEs) of the registration of the point clouds was 1.2 cm for Dataset III and 1.0 cm for Dataset IV.

Figure 15. High-resolution terrestrial laser scanning (TLS) point cloud of a high-voltage power line (Dataset IV). The color of the points is determined by the intensity (the strength of the reflected laser beam).

The reference data did not fully reflect the state of the power lines when measured using the UAV because different durations were needed to record the data using different means. UAV missions to survey sections for Datasets III and IV were performed over one day and lasted four hours (including preparatory work). The TLS and TS surveys were time consuming and fieldwork using these methods lasted four days.

3. Results

This section presents the results of the validation of the proposed method for using UAV images to detect power lines, provide a 3D reconstruction, and localize obstacles in the power lines corridor.

3.1. Bundle Adjustment

The photogrammetric data were processed using Agisoft Metashape Professional. Aerotriangulation was performed at a high level of accuracy, whereby the software worked with images at their original sizes. Optimization was performed in the next stage and included a realignment of the image block and the determination of the parameters for camera calibration. In the case of Dataset I and Dataset II, the GCPs were not used in bundle adjustment; instead, only the precise coordinates of the projection centers of the images, measured using GNSS PPK, were used. The GNSS PPK calculations were done using RTKLIB software, utilizing measurements from a GNSS base station set up for the duration of the survey. However, in the bundle adjustment of images from Datasets III and IV, the GCPs were used. The a priori accuracy of the GCPs was taken into account in this process. Table 5 presents data on the accuracy of the aerotriangulation of Dataset III and Dataset IV. The last stage involved generating dense point clouds at a high level of detail, which means that the software determined the spatial coordinates for each group, consisting of four pixels in the image (2 × 2 pixels).

Table 5. Root mean-squared errors (RMSEs) of coordinates of the GCPs and checkpoints.

Data Set	Ground Control Points					Check Points				
	m_x (mm)	m_y (mm)	m_z (mm)	*Pix Error*	*Number of GCP*	m_x (mm)	m_y (mm)	m_z (mm)	*Pix Error*	*Number of GCP*
Dataset III	4.8	4.2	7.9	0.787	12	13.2	20.4	42.1	0.673	9
Dataset IV	8.5	5.1	20.0	0.454	19	17.4	30.9	38.4	0.494	11

Information on the location of each pole within sections of the power lines was manually obtained through Agisoft Metashape Professional. Finally, the data necessary for further processing were exported to reconstruct the 3D geometry of the power lines based on the UAV images (i.e., final EOE, IOE, undistorted images, and coordinates of the poles and the dense point cloud).

3.2. Results of Processing Datasets I and II

The proposed method for 3D reconstruction of power lines in UAV images used multiple thresholds. To establish appropriate values, a threshold sensitivity analysis was conducted. The process is described in Appendix A. The established set of thresholds was then used in the processing of all experimental data (filter size—30, c_r—0.4, c_l—0.4, e_dist_{max}—10, p—10).

Owing to a lack of reference data for Datasets I and II, as well as the use of a camera with a rolling shutter, a check was performed manually to verify only the efficiency of the proposed detection algorithm. Errors were recorded whenever a detected line segment did not overlay an empirically determined line within an image segment. As a summary of the validation, the success rate was calculated. Each wire in an image was considered a case. A complete and correct case detection was regarded as a success, and any other outcome was considered a failure. The success rate here was the ratio of the number of cases of success to the total number of cases in the dataset.

Dataset I contained 225 images, covering 13 power line spans (14 utility poles (Figure 14a)). Three wires were spaced equally at the same height above ground. The survey was conducted in a rural area, where the background consisted mostly of farm fields, meadows, backyards, and occasional roads. Using Dataset I, the proposed algorithm derived three survey sections (straight sections) between the utility poles 1–6, 6–7, and 7–14. The processing was smooth and detection was uninterrupted (Figure 16). The detection was rejected in 13 images (detection parameters were below set thresholds). Upon closer inspection, six images with less than perfect detection were obtained. A total of seven errors were recorded in cases where the power line was incorrectly detected. Six of these occurred due to an unfavorable background, and the other was in the vicinity of a utility pole. The calculated success rate was 98.96% (Table 6).

Figure 16. Detection results of the wires: (**a**) original image; (**b**) processed image with highlighted wires detected.

Table 6. The results of processing Datasets I and II.

Dataset	I	II
Utility poles	14	5
Survey sections	3	4
No. of images	225	238
Stopped detection	none	none
No. of images	-	-
No. of errors	0	0
Power lines overlay	none	yes
No. of images	-	6
No. of errors	0	6
Detection transfer	yes	yes
No. of images	13	10
No. of errors	7	18
Success rate	98.96%	92.16%

Dataset II contained 238 images covering four spans of high-voltage power lines (Figure 14b), featuring six transmission wires and two ground wires. As the wires were at different heights, they were recorded in different configurations in the right and left side images, with some wires lying only a few pixels from one another (Figure 17). The background varied greatly, from forest to industrial. Owing to the high sag of the wires, all spans were processed separately.

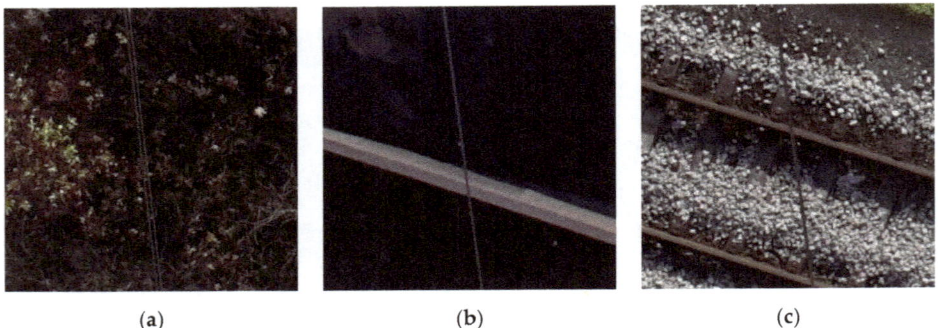

Figure 17. Example problematic images: (**a**) closely recorded wires; (**b**,**c**), two wires recorded as one line.

In this dataset, far more errors were observed. They mostly occurred due to unfavorably placed wires in images (placed close to or overlying wires). Moreover, in a few cases, mistakes were caused by a challenging background. A success rate of 92.16% was calculated (Table 6). A lower success rate was expected because Dataset II consisted of far more challenging images than Dataset I. The wires were barely visible, while the background was very unfavorable, containing mostly industrial waste or dried vegetation.

3.3. Results of Processing Datasets III and IV

The data collected and pre-processed from Datasets III and IV enabled the assessment of the accuracy of the 3D reconstruction of the power lines using UAV imagery. Attention was paid here to the comparison of the resultant catenary curves obtained from the proposed method and reference methods.

The UAV images were processed using the method proposed in Section 2, using thresholds established in Appendix A, and the 3D geometry of the wires was saved for comparison. A detailed manual check of the detection was not performed. Processing for Dataset III was uninterrupted, while processing for Dataset IV was manually restarted once to detect the ground wires. The RMSEs of fitting the catenary curve to the photogrammetric data varied from ±1.0 cm to ±23.8 cm for Dataset III, and from ±2.1 cm to ±19.1 cm for Dataset IV, which shows sufficient accuracy for the purpose of corridor clearance monitoring [7,8]. The largest errors were noted in the detection of the ground wires in Dataset IV and were related to their diameters and the difference in sag from the transmission wires. Most of them were fixed by a second iteration of the detection for problematic images.

The catenary was also fitted into the data collected using TLS and TS. A single wire was usually represented by several thousand TLS-derived points and a few dozen points obtained from TS measurement. The RMSEs of the curve fitting into TLS-derived points varied between ±0.4 cm and ±3.9 cm for Dataset III. For Dataset IV, it varied from ±0.7 cm to ±9.3 cm. In the case of TS data, the RMSEs of the catenary fitting were between ±0.2 cm and ±9.2 cm for medium-voltage power lines (Dataset III), and between ±2 cm and ±15.1 cm for high-voltage power lines (Dataset IV). Multiple discrepancies were observed in the data, most likely due to small momentary vibrations in the wires. During two days of the survey, there was a strong wind, which led to high-amplitude vibrations in the wires. This had a direct impact on the quality of the results obtained.

The accuracy of the proposed method was verified in three ways: comparisons in both the vertical and the horizontal plane, and a comparison of the parameters of the catenary curve.

A comparison of the calculated heights of the corresponding points was carried out between the proposed and the reference methods. The catenary curves were represented by 1000 points for the UAV-based method, with 1000 points calculated with respect to the XY positions for the TS, and source data for the TLS. The corresponding comparative points were found by locating the nearest points in between points from TLS and UAV datasets.

As a measure of accuracy, the RMSEs of the differences in height for individual wires were calculated. The results of the comparison are presented in Table 7.

Table 7. The accuracy of wire reconstruction using the proposed method.

Dataset	Error Type	RMSE [cm]		
		TS–TLS	**TS–UAV**	**TLS–UAV**
Dataset III	Height	±2.3	±6.9	±6.7
	Horizontal	±0.9	±1.0	±1.0
	Maximum sag	±9.7	±14.5	±16.5
Dataset IV	Height	±6.1	±11.2	±10.8
	Horizontal	±9.6	±3.8	±12.3
	Maximum sag	±16.4	±26.3	±29.6

TS—total station measurements, TLS—terrestrial laser scanning, UAV—proposed method based on UAV imagery; TS–TLS—difference between TLS and TS; TS–UAV—difference between UAV and TS; TLS–UAV—difference between UAV and TLS.

The results of the recorded wire geometry using TS and TLS measurements were highly consistent. The mean difference in height for Dataset III was −2.0 cm, and for Dataset IV it was −3.0 cm, with RMSEs of ±2.3 cm and ±6.1 cm, respectively. After removing outliers, the mean differences in height between the reference data and the UAV-derived data varied from −13.0 to −11.0 cm for Dataset III, and from −17.2 to −14.3 cm for Dataset IV. This means that, within the vertical profile, wires reconstructed using UAV imagery were placed higher than those determined by means of the reference methods. The discrepancies can be attributed to differences in the densities of points at wires and the continuity of TLS point clouds representing the wires. This was especially prominent in the longest span—a 500 m long span in Dataset IV. The span length is one of the most important factors in changes to the sag of the wire due to temperature differences. Since the TLS survey took a significant amount of time, discrepancies were to be expected. In light of this, as well as the expected accuracy of the 3D reconstruction for the provided data, we can conclude that the proposed method achieved the expected accuracies.

To assess the proposed method along the horizontal plane, discrete descriptions of the UAV-derived catenary curves were used. For the TS and TLS, line representations of the catenary curves in the XY plane were utilized. The distances between UAV-derived points and horizontal lines fitted to TS and TLS data were calculated for each point regardless of its position toward the reference lines (absolute value).

For Dataset III, the results of the analysis were promising. The mean horizontal distance between the UAV-derived data and the reference data was, on average, 2.8 cm, with a maximum value not exceeding 7.8 cm. A comparison of the TLS and TS data gave similar results. In the case of Dataset IV, the horizontal accuracy of the geometry of the wires determined using UAV imagery was worse. The mean value of the parameter analyzed was 12.0 cm for TS–UAV differences and 26.1 cm for TLS–UAV differences. RMSE values were ±3.8 cm for TS–UAV differences and ±12.3 cm for TLS–UAV differences (Table 7). The maximum value obtained for the ground wires reached up to 1.2 m and was significantly higher compared to transmission wires. The discrepancies can be attributed to many factors: the complexity of the power line captured within Dataset IV, significantly longer spans (which translated into larger geometrical changes), and the height of wires above the ground. The root of the problem probably lay within the bundle adjustment procedure, where most of the tie points were located on the ground. The longer the distance to the ground, the larger the errors are to be expected in the 3D reconstruction.

Another parameter used for comparison was the value of the maximum sag f_s of the wire calculated for each method, according to Equation (4). This parameter is required for power line inspections in many countries. The differences in the maximum sag of the wires Δf_s were calculated among the measurement methods. The RMSEs of Δf_s are listed in Table 7. The occurrence of outliers was connected with incomplete TS and TLS data, collected during wire measurements. The results are presented here for medium- (Dataset III) and high-voltage (Dataset IV) power lines.

For Dataset III, the results obtained using UAV data differed on average by −3.6 cm and −6.1 cm from the results of TLS and TS data processing, respectively. The RMSEs of these differences were ±16.5 cm and ±14.5 cm, respectively.

There was no significant decrease in the accuracy of the reconstruction of the vertical geometry of the wire for spans of the medium-voltage power line (Dataset III) located in forested areas, which occurred when the maximum values of the sag of the wires were determined. In the case of Dataset IV, a cross-comparison of all measurement methods used gave similar results, both in terms of the values of mean difference and RMSEs. Therefore, the accuracy of determining the maximum sag of the wire using UAV images did not significantly differ from the accuracy of calculating this parameter by means of TS and TLS measurements.

3.4. Validation of Detection of Obstacles within the Power Line Corridor

Reference data were used to validate obstacle detection in data processed by the created solution. The TLS data were chosen as a reference owing to the great detail of geometry reconstruction, both for the wires and in the vicinity of the power line.

Two analyses were conducted, visual and quantitative. The visual analysis included comparing data resulting from distance analysis in Cloud Compare software. A check was performed to see if similar places were included in the resultant occlusion sets. The exemplary results are shown in Figure 18, where points in the point cloud are colored according to the distance to the power lines.

The results obtained using the two methods (TLS, UAV) were consistent, especially in places where abnormalities related to the maintenance of an appropriate separation between the wires and elements of land cover were significant.

Then, for quantitative analysis, occlusion points were submitted to a voxelization procedure and, lastly, clustered into separate objects (Figure 19). For each object summary, the approximated volume was calculated as a sum of the volume of all voxels that formed it. Then, all the data were summed up and compared to UAV and TLS data results (Table 8).

Table 8. Summary of detected occlusions within the power lines corridor. The corridor for Dataset III was 5 m distance from the power lines, while for dataset IV it was 15 m.

	Dataset III		Dataset IV	
	TLS	**UAV**	**TLS**	**UAV**
No. of objects	264	98	214	709
No. of voxels (0.5 m^3)	62,794	4217	228,573	93,451
Volume (m^3)	7849.25	527.125	28,571.625	11,681.375

Big differences can be observed between TLS and UAV data. This was to be expected. TLS and UAV products due to their acquisition procedures being widely different. Photogrammetric products do not penetrate greenery, while TLS does. Thus, TLS data have the advantage of a more continuous representation of lattice objects. This can cause an effect of one object in TLS data being represented by multiple objects in UAV data (Figure 19e,f). As a consequence, the recorded volume of obstacles determined using TLS data will be far higher than using UAV data.

Similarly, very small vegetation elements could have been picked up by TLS but would not come up within dense point cloud reconstruction (Figure 19a,b). The importance of such objects is negligible. Nonetheless, that also causes differences in the total volume of occlusions. The date of the data acquisition was also not without consequence. Bare trees are quite difficult to reconstruct in photogrammetric data, so—for better results—data should be acquired during the growing season.

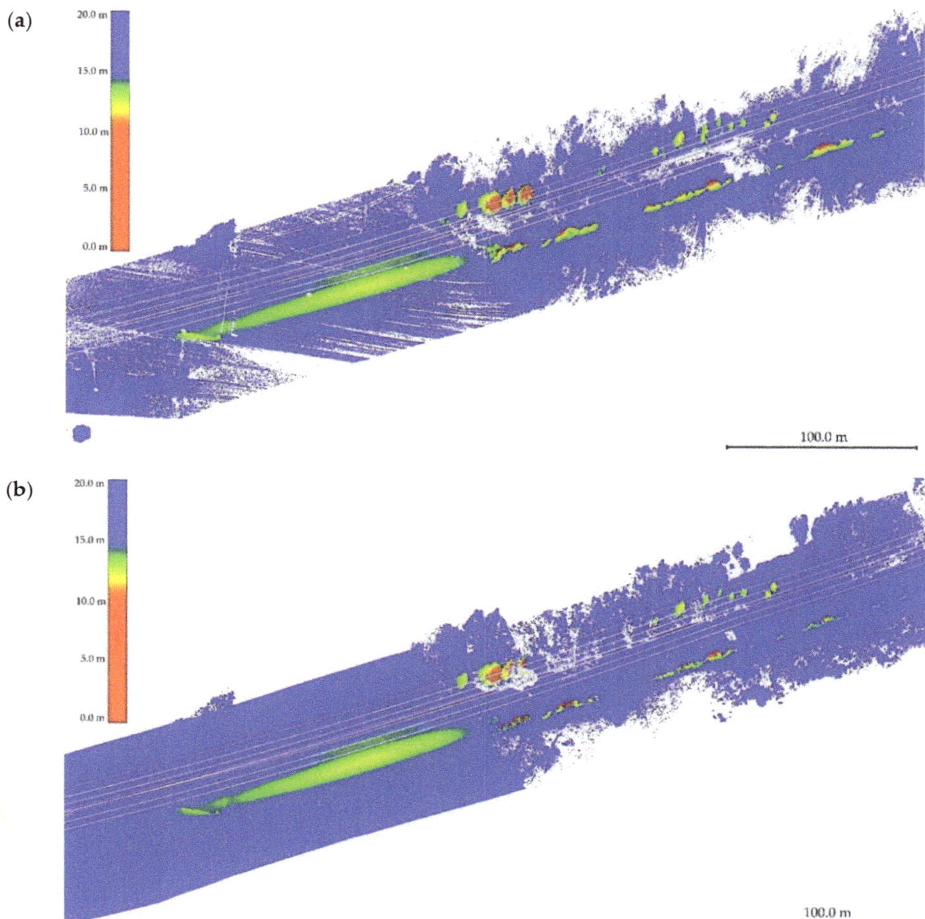

Figure 18. Results of the monitoring of occlusions of a high-voltage power line corridor (Dataset IV): (**a**) based on TLS data; and (**b**) based on UAV data. The red color indicates points less than 10 m from the wires and the blue color indicates those more than 15 m away.

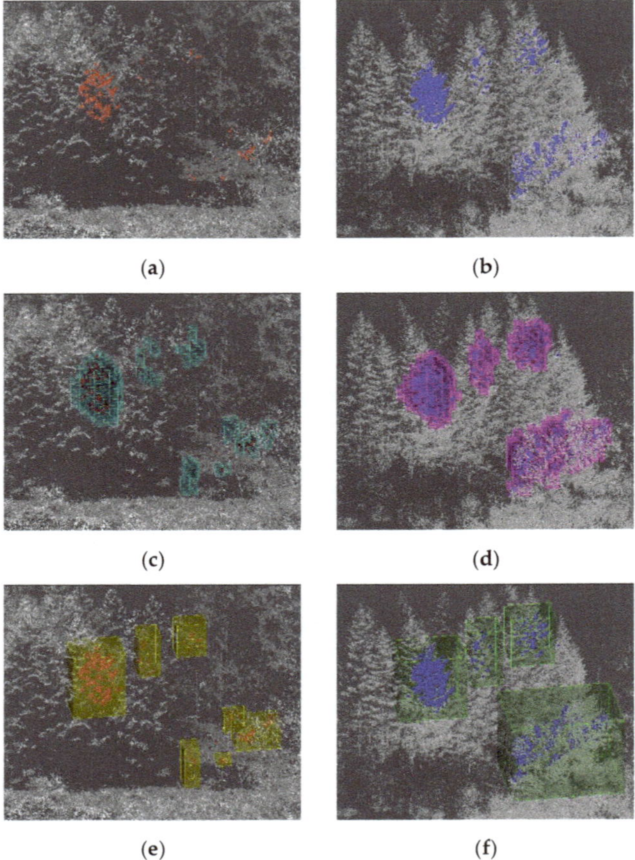

Figure 19. Exemplary results of obstacle detection for proposed methods and TLS reference data. Points recorded within the corridor (**a**) UAV, (**b**) TLS, calculated voxels (**c**) UAV, (**d**) TLS, clustered objects (**e**) UAV, (**f**) TLS.

4. Discussion

This paper presents a comprehensive method of processing UAV images to detect and reconstruct 3D power lines, then subsequently compare them with a point cloud representation of DSM to localize any objects threatening the safety of the power lines.

Power line inspections are a topical issue these days. Modeling wires in 3D space is essential for the assessment of power line safety. Thus, their reconstruction has received much attention. However, unlike other studies on the subject, this paper presents the entire workflow for corridor clearance monitoring. Each step of the proposed method, starting with data acquisition requirements and ending with obstacle detection, is described comprehensively. The wire detection in UAV images was performed using a decorrelation stretch for initial image processing, the modified Prewitt filter for edge enhancement, and RANSAC with additional parameters for line fitting. This classic approach causes the line extraction in the UAV images to take place in a controlled way and, if necessary, the user can modify the processing parameters (thresholds) and even manually restart the process when detection errors occur. The combination of these elements creates a solution that is robust to low-quality input data (images). Despite a variety of backgrounds and the dubious visibility of wires, the created solution managed to consistently detect power lines in a series of images. Its highest achieved success rate exceeded 98% and remained above 90% for more challenging data. Good performance in the highly changeable environment can be attributed to complete disregard for the wire color, as well as the implementation of a local, ordered approach, which made the method adaptable to both contrast change as well as angle change in the wire positions.

The algorithm does not require an extensive learning set, which makes it different from many deep learning methods [37,38], which are currently gaining popularity. Similar to some other solutions [23,35,40,41], wire reconstruction in 3D space is performed using epipolar geometry. However, the proposed RANSAC-based approach to fit the catenary curve to previously obtained points representing the wires minimizes the noise.

Despite the fact that the algorithm is not completely autonomous, it is relatively flexible and robust. The minor user intervention allows the system to be applied in a variety of cases, either to a low or high voltage. However, there is room for improvement in the proposed method. Such errors as pixel discrepancies due to detection transfer are removed while fitting the chain curve to the 3D data. Others, such as a complete loss of detection, or overlay with neighboring wires, can be solved by implementing a two-step approach: after initial processing, the algorithm automatically directs the user to problematic sections where the process can be restarted or corrected manually.

There are multiple thresholds set in the method, though it must be said that, after an initial threshold sensitivity analysis, none were changed for any of the tested sets, nor in further commercial exploitation.

Another drawback might be the mandatory initial user input. However, it must be stated that this is limited to two points per wire for the whole survey section, which takes no more than a couple of minutes. Many methods can be applied instead of initial user input. The approximate direction based on the positions of the utility poles and the Hough transform [50] was used to detect the wires in a test phase in this research. The initial results were encouraging for medium-voltage power lines. However, with the diversification of the background and the introduction of wires at multiple heights in high-voltage lines, this method failed. The problem of the relative positioning of wires between images in stereopairs is complex. It is the process of recognizing the same wire in images within a stereopair. As wires are made of the same material, and due to the large distance between the background and them, there is no information linking two images that capture the same wire. An alternative here would be to include a third verification strip of images or to create multiple 3D reconstructions containing all combinations of wires and then deciding on the pairing. The first process would extend the duration of the survey and necessitate different mission planning methodologies. The second would also require either manual input or prior knowledge of the number of wires and types of utility poles.

In many studies on power line inspections, datasets have been small or numerical quality analyses have been omitted [23,40,41]. The method presented in this paper has been profusely tested beyond its scope since it has already been commercially implemented. This study demonstrates that the accuracy of the proposed method for the 3D reconstruction of power lines is consistent with that achieved using classical measurements. Similar accuracies are reported by Oh and Lee [40], but their accuracy analyses were limited to only two wires whose geometry was determined using the reference method.

It should be noted that none of the verification methods used for accuracy assessment in this study were free from errors or significantly more accurate than any other. As the true value of the measured sag of the wires or their 3D geometry was unknown, it was only possible to assess the mutual compatibility of the methods used, and not their absolute accuracy.

The results of the accuracy assessment described in this paper were also influenced by the fact that surveys of power lines using three methods (UAV, TLS, TS) were not carried out simultaneously, or under the same weather conditions. This was due to the technical feasibility of surveys. For most spans, the TS and TLS measurements were performed in parallel (except for one span of a high-voltage power line) and lasted four days. On the contrary, UAV flights over power lines in Datasets III and IV were completed in one day. For this reason, different conditions of the power lines were recorded between the methods.

The mapping of vegetation around power lines is also an issue of great importance. In this paper, we assessed whether UAV-derived point cloud representations of DSM, generated using a classic approach, are sufficient to monitor power line corridors. In the case of the UAV-based photogrammetric products, the most common error was missing data in representing the DSM. The solution to this was to take UAV images in more than two strips. However, this did not significantly increase the correctness of tree reconstruction for the DSM, which, in turn, was the main reason for incorrect DSM extraction. One way to solve this problem is to capture UAV-based photogrammetric data during the vegetation period. However, as a rule, in the immediate vicinity of the power line (up to ~10 m from the line axis), the quality of the UAV-derived DSM was adequate to analyze the separation between the wires from and elements of the land cover.

5. Conclusions

The aim of this research was to create a novel method based on UAV imagery for occlusion monitoring in the corridor of power lines. The proposed method mainly consists of three parts: the reconstruction of the geometry of the wires in 3D space, the reconstruction of point cloud representation of the DSM, and subsequently detecting obstacles in the power lines corridor. Power line reconstruction, which was carried out using images captured for the DSM calculation, is its essential part. Well-known computer vision algorithms and epipolar geometry were adopted for this task. This makes the proposed method user-friendly and allows for image processing to be performed in a fully controlled way. There are other merits: no training data are required, the method is robust to low-quality input data, and the RANSAC-based approach to model the wires reduces the influence of the noise.

An integral part of the proposed method is a workflow for the detection of obstacles in the power line corridor. Obstacles are selected by calculating the distance between power lines and each point in the point cloud representation of the DSM and simplified into voxels and then objects. The analyzed data are georeferenced. Thus, the parts of the power line corridor where the maintenance work has to be performed are documented using both the precise information of their locations and images.

The feasibility of the proposed UAV-based method for the 3D reconstruction of power lines and corridor clearance monitoring was confirmed by reference surveys. They achieved results similar to those obtained using other available solutions. The method's relatively high accuracy, comparable with that obtained by means of the reference measurements, was also verified. The accuracy of its 3D reconstruction for medium-voltage power lines was 15 cm. In the case of high-voltage power lines, it did not exceed 30 cm. The proposed method allows for measurement data to be collected in a relatively short time, and is cheaper than other commonly used methods in the area.

In the case of corridor clearance monitoring, the results are also satisfying. Visual analysis proved that obstacles were detected in the same places for both UAV and TLS data. However, there were big differences in the volume of calculated obstacles in between methods. This was expected due to the properties of both data acquisition methods. However, this does not disprove the usability of the method. Crucial obstacles were identified, and the presence of obstacles is of the utmost importance. In future works, more focus should be placed on DSM.

The results of this study were implemented for commercial use by FlyTech UAV. The algorithm has already been used to measure several hundreds of kilometers of power lines.

Author Contributions: Conceptualization, E.P. (Elżbieta Pastucha), A.Ś., W.N. and E.P. (Edyta Puniach); methodology, E.P. (Elżbieta Pastucha), A.Ś. and E.P. (Edyta Puniach); software, E.P. (Elżbieta Pastucha) and A.Ś.; validation, E.P. (Edyta Puniach) and P.Ć.; formal analysis, E.P. (Elżbieta Pastucha), E.P. (Edyta Puniach), A.Ś., P.Ć., W.N. and P.W.; investigation, E.P. (Elżbieta Pastucha) and E.P. (Edyta Puniach); survey, E.P. (Edyta Puniach), P.Ć. and P.W.; writing—original draft preparation, E.P. (Elżbieta Pastucha), E.P. (Edyta Puniach), A.Ś., P.Ć., W.N. and P.W.; writing—review and editing, E.P. (Elżbieta Pastucha) and E.P. (Edyta Puniach); project administration, E.P. (Elżbieta Pastucha) and E.P. (Edyta Puniach); funding acquisition, P.Ć. All authors have read and agreed to the published version of the manuscript.

Funding: The research for and publication of this paper were completed as part of the project "Development of an unmanned aerial vehicle with diagnostic and surveying capabilities for use in the operation of power grids, especially high and medium-voltage grids", co-financed by the European Regional Development Fund as part of the Regional Operational Programme of Małopolskie Voivodeship for 2014–2020, Priority Axis 1–Knowledge-based economy, Measure 1.2, Research and Innovation in Enterprises, and Sub-measure 1.2.1, R&D projects of enterprises. The APC was funded by AGH University of Science and Technology under the Initiative for Excellence Research University project.

Conflicts of Interest: The authors declare no conflict of interest.

Appendix A

To establish appropriate threshold values, a set of seven images was chosen from Dataset I. The choice was made to maintain the maximum diversity of backgrounds within the images:

- ID 8—low vegetation (crops),
- ID 25—road and low vegetation,
- ID 32—low vegetation, buildings,
- ID 36—low vegetation, bare soil, car (Figure A1b),
- ID 192—bare soil, low vegetation
- ID 193—low vegetation (crops), bare soil, and
- ID 217—sparse crops, bare soil (Figure A1a).

Threshold sensitivity analysis was performed on the given images. As all thresholds were connected in a sense, it was not possible to perform separate tests, and not all values were tested numerically to simplify the process.

Multiple thresholds were used. They can be divided into two categories. The first, and less important, consisted of thresholds that were dependent on the size of the image or had been arbitrarily chosen. This group included segment window size, segment overlap and binarization threshold. It was decided to use five segments per image, though—depending on whether the sag of the wires was big or small—the number could be increased or decreased. The overlap value was set between 0% and 10% to avoid excessive calculations. The binarization threshold depended on the width of the wires in the images and, as a consequence, on the approximated values of pixels in segments that captured the wire. Within the data used in this study, each wire was around two pixels in width; thus, at a segment size of 1000, the binarization threshold was set to 0.003.

(a) (b)

Figure A1. Sample images used for threshold sensitivity analysis: (**a**) ID 217, (**b**) ID 36.

The second group of thresholds consisted of thresholds that needed to be set based on the threshold sensitivity analysis. They included:

- filter size,
- c_r—right edge coherence, a quotient of RANSAC inliers of the right edge image for all pixels in the image segment,
- c_l—left edge coherence, a quotient of RANSAC inliers of the left edge image for all pixels in the image segment,
- e_dist_{max}—a maximum distance between the right and the left edge lines within the image segment,
- p—parallelism coefficient, a quotient of the minimal and maximum distances between the right and left edge lines within the image segment.

The filter size was tested first. Filtration was performed on all test images, and each one was then normalized. Their results were compared to find the maximal difference between the wires and the background. The following values were tested: 5, 10, 20, 30, 40, 50, and 60 (Figure A2). In the listed thresholds, the wires became more distinguishable with a filter size of up to 30.

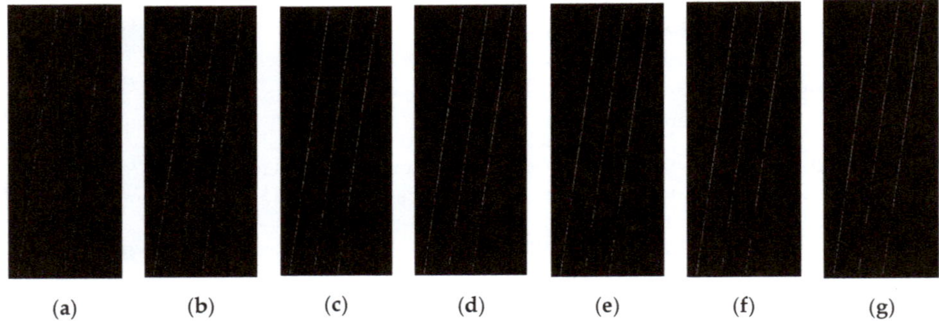

Figure A2. Results of filtration for filter sizes: (**a**) 5 pix., (**b**) 10 pix., (**c**) 20 pix., (**d**) 30 pix., (**e**) 40 pix., (**f**) 50 pix., (**g**) 60 pix.

To test the remaining thresholds, a different approach was used. The approximated positions of the wires were defined on the test images, and detection was performed to identify the relevant parameters that were saved in a file. A manual classification was then performed to sort the results into correct and incorrect detection groups. The values of the descriptive statistics were then calculated for each group (Figure A3).

A clear difference can be seen in results for both c_r and c_l. The distributions of those parameter values with resultant incorrect detection are quite narrow and focused below 0.25, with a couple of outliers. The spread for parameter values with resultant correct detection is wider, but only a few values fall below 0.5. The opposite can be observed for e_dist_{max} and p parameters. The parameter values with resultant correct detection have a narrow range, while the opposite ones are quite widespread. For e_dist_{max} only, outliers have a higher value than 10, while, for p only, outliers reach above seven. Parameter values for incorrect detection are far more widespread in the case of e_dist_{max} and p parameters. There is an overlap between values for the p parameter with resultant correct and incorrect detection. Thus, based on the p parameter value, only gross errors can be filtered out.

Following an analysis of the results, the same threshold was chosen for c_r and c_l. A more rigid approach was then chosen, and the threshold was set to 0.4. Parameters e_dist_{max} and p were more challenging because these thresholds relied on the size of the wire in the image. It was nearly impossible to keep the GSD on the power lines constant. In the case of Dataset I, the width of the power line varied from two to eight pixels. The threshold for e_dist_{max} and p was set to 10 pixels to provide enough space to pass all possible correct detections while filtering out gross errors.

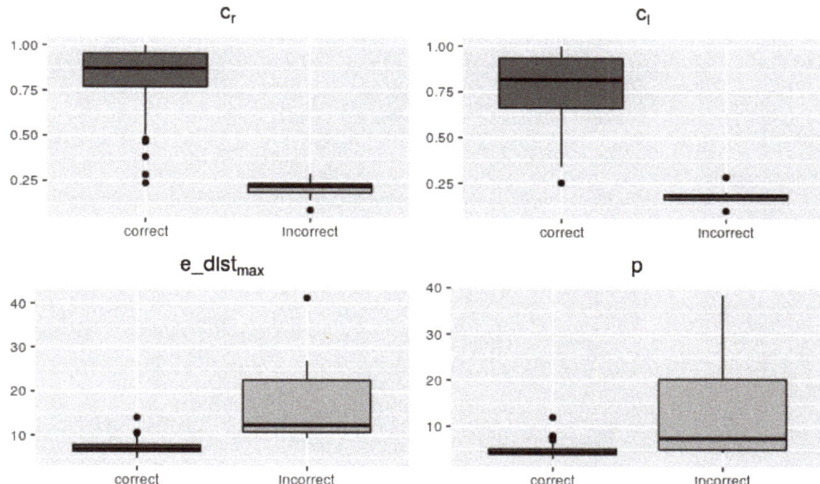

Figure A3. Boxplots of the parameter values against detection results.

References

1. Mirallès, F.; Pouliot, N.; Montambault, S. State-of-the-art review of computer vision for the management of power transmission lines. In Proceedings of the 3rd International Conference on Applied Robotics for the Power Industry, Foz do Iguassu, Brazil, 14–16 October 2014; pp. 1–6. [CrossRef]
2. Li, Z.; Walker, R.; Hayward, R.; Mejias, L. Advances in vegetation management for power line corridor monitoring using aerial remote sensing techniques. In Proceedings of the 1st IEEE International Conference on Applied Robotics for the Power Industry (CARPI), Montreal, QC, Canada, 5–7 October 2010; pp. 1–6. [CrossRef]
3. Li, Z.; Hayward, R.; Zhang, J.; Liu, Y. Individual Tree Crown Delineation Techniques for Vegetation Management in Power Line Corridor. In Proceedings of the Digital Image Computing: Techniques and Applications, Canberra, Australia, 1–3 December 2008; pp. 148–154. [CrossRef]
4. Stylianidis, E.; Akça, D.; Poli, D.; Hofer, M.; Gruen, A.; Sánchez Martín, V.; Smagas, K.; Walli, A.; Altan, O.; Jimeno, E. FORSAT: A 3D forest monitoring system for cover mapping and volumetric 3D change detection. *Int. J. Digit. Earth* **2019**, 1–32. [CrossRef]
5. Chen, C.; Yang, B.; Song, S.; Peng, X.; Huang, R. Automatic Clearance Anomaly Detection for Transmission Line Corridors Utilizing UAV-Borne LIDAR Data. *Remote Sens.* **2018**, *10*, 613. [CrossRef]
6. Guo, B.; Li, Q.; Huang, X.; Wang, C. An Improved Method for Power-Line Reconstruction from Point Cloud Data. *Remote Sens.* **2016**, *8*, 36. [CrossRef]
7. Jwa, Y.; Sohn, G.; Kim, H. Automatic 3D powerline reconstruction using airborne lidar data. *IAPRS* **2009**, *38*, 105–110.
8. Jwa, Y.; Sohn, G. A piecewise catenary curve model growing for 3D power line reconstruction. *Photogramm. Eng. Remote Sens.* **2012**, *78*, 1227–1240. [CrossRef]
9. Lehtomäki, M.; Kukko, A.; Matikainen, L.; Hyyppä, J.; Kaartinen, H.; Jaakkola, A. Power line mapping technique using all-terrain mobile laser scanning. *Autom. Constr.* **2019**, *105*, 102802. [CrossRef]
10. Wang, Y.; Chen, Q.; Liu, L.; Li, X.; Sangaiah, A.K.; Li, K. Systematic comparison of power line classification methods from ALS and MLS point cloud data. *Remote Sens.* **2018**, *10*, 1222. [CrossRef]
11. Sohn, G.; Jwa, Y.; Kim, H.B. Automatic powerline scene classification and reconstruction using airborne lidar data. *ISPRS Ann. Photogramm. Remote Sens. Spat. Inf. Sci.* **2012**, *I-3*, 167–172. [CrossRef]
12. Axelsson, P. Processing of laser scanner data—Algorithms and applications. *ISPRS J. Photogramm. Remote Sens.* **1999**, *54*, 138–147. [CrossRef]
13. Zhu, L.; Hyyppä, J. Fully-automated power line extraction from airborne laser scanning point clouds in forest areas. *Remote Sens.* **2014**, *6*, 11267–11282. [CrossRef]

14. Lu, M.L.; Kieloch, Z. Accuracy of Transmission Line Modeling Based on Aerial LiDAR Survey. *IEEE Trans. Power Deliv.* **2008**, *23*, 1655–1663. [CrossRef]
15. Ax, M.; Thamke, S.; Kuhnert, L.; Kuhnert, K.-D. UAV based laser measurement for vegetation control at high-voltage transmission lines. *Adv. Mater. Res.* **2012**, *614–615*, 1147–1152. [CrossRef]
16. Azevedo, F.; Dias, A.; Almeida, J.; Oliveira, A.; Ferreira, A.; Santos, T.; Martins, A.; Silva, E. LiDAR-Based Real-Time Detection and Modeling of Power Lines for Unmanned Aerial Vehicles. *Sensors* **2019**, *19*, 1812. [CrossRef]
17. Yan, G.; Li, C.; Zhou, G.; Zhang, W.; Li, X. Automatic extraction of power lines from aerial images. *IEEE Geosci. Remote Sens. Lett.* **2007**, *4*, 387–391. [CrossRef]
18. Matikainen, L.; Lehtomäki, M.; Ahokas, E.; Hyyppä, J.; Karjalainen, M.; Jaakkola, A.; Kukko, A.; Heinonen, T. Remote sensing methods for power line corridor surveys. *ISPRS J. Photogramm. Remote Sens.* **2016**, *119*, 10–31. [CrossRef]
19. Mills, S.J.; Castro, M.P.G.; Li, Z.; Cai, J.; Hayward, R.; Mejias, L.; Walker, R.A. Evaluation of Aerial Remote Sensing Techniques for Vegetation Management in Power-Line Corridors. *IEEE Trans. Geosci. Remote Sens.* **2010**, *48*, 3379–3390. [CrossRef]
20. Kobayashi, Y.; Karady, G.G.; Heydt, G.T.; Olsen, R.G. The Utilization of Satellite Images to Identify Trees Endangering Transmission Lines. *IEEE Trans. Power Deliv.* **2009**, *24*, 1703–1709. [CrossRef]
21. Jiang, S.; Jiang, W.; Huang, W.; Yang, L. UAV-Based Oblique Photogrammetry for Outdoor Data Acquisition and Offsite Visual Inspection of Transmission Line. *Remote Sens.* **2017**, *9*, 278. [CrossRef]
22. Jóźków, G.; Vander Jagt, B.; Toth, C. Experiments with UAS imagery for automatic modeling of power line 3D geometry. *ISPRS Int. Arch. Photogramm. Remote Sens. Spat. Inf. Sci.* **2015**, *XL-1/W4*, 403–409. [CrossRef]
23. Zhang, Y.; Yuan, X.; Fang, Y.; Chen, S. UAV Low Altitude Photogrammetry for Power Line Inspection. *ISPRS Int. J. Geo-Inf.* **2017**, *6*, 14. [CrossRef]
24. Li, Z.; Liu, Y.; Walker, R.; Hayward, R.; Zhang, J. Towards automatic power line detection for a UAV surveillance system using pulse coupled neural filter and an improved Hough transform. *Mach. Vis. Appl.* **2010**, *21*, 677–686. [CrossRef]
25. Toth, J.; Gilpin-Jackson, A. Smart view for a smart grid—Unmanned Aerial Vehicles for transmission lines. In Proceedings of the 1st International Conference on Applied Robotics for the Power Industry, Montreal, QC, Canada, 5–7 October 2010; pp. 1–6. [CrossRef]
26. Bhujade, R.; Vellaiappan, A.; Sharma, H.; Purushothaman, B. Detection of power-lines in complex natural surroundings. *Comput. Sci. Inf. Technol.* **2013**, *3*, 101–108. [CrossRef]
27. Cerón, A.; Mondragón, B.I.F.; Prieto, F. Power line detection using a circle based search with UAV images. In Proceedings of the International Conference on Unmanned Aircraft Systems (ICUAS), Orlando, FL, USA, 27–30 May 2014; pp. 632–639. [CrossRef]
28. Liu, X.; Hou, L.; Ju, X. A method for detecting power lines in UAV aerial images. In Proceedings of the 3rd IEEE International Conference on Computer and Communications (ICCC), Chengdu, China, 13–16 December 2017; pp. 2132–2136. [CrossRef]
29. Manlangit, C.; Green, R. *Automatic Power Line Detection for a UAV System*; University of Canterbury: Christchurch, New Zealand, 2012.
30. Nasseri, M.H.; Moradi, H.; Nasiri, S.M.; Hosseini, R. Power line detection and tracking using hough transform and particle filter. In Proceedings of the 6th RSI International Conference on Robotics and Mechatronics (IcRoM), Tehran, Iran, 23–25 October 2018; pp. 130–134. [CrossRef]
31. Sharma, H.; Bhujade, R.; Adithya, V.; Balamuralidhar, P. Vision-based detection of power distribution lines in complex remote surroundings. In Proceedings of the Twentieth National Conference on Communications (NCC), Kanpur, India, 28 February–2 March 2014; pp. 1–6. [CrossRef]
32. Yang, T.W.; Yin, H.; Ruan, Q.Q.; Han, J.D.; Qi, J.T.; Yong, Q.; Wang, Z.T.; Sun, Z.Q. Overhead power line detection from UAV video images. In Proceedings of the 19th International Conference on Mechatronics and Machine Vision in Practice (M2VIP), Auckland, New Zealand, 28–30 November 2012; pp. 74–79.
33. Yetgin, Ö.E.; Şentürk, Z.; Gerek, Ö.N. A comparison of line detection methods for power line avoidance in aircrafts. In Proceedings of the 9th International Conference on Electrical and Electronics Engineering (ELECO), Bursa, Turkey, 26–28 November 2015; pp. 241–245. [CrossRef]

34. Zhang, J.; Liu, L.; Wang, B.; Chen, X.; Wang, Q.; Zheng, T. High speed automatic power line detection and tracking for a UAV-based inspection. In Proceedings of the International Conference on Industrial Control and Electronics Engineering, Xi'an, China, 23–25 August 2012; pp. 266–269. [CrossRef]
35. Zhang, Y.; Yuan, X.; Li, W.; Chen, S. Automatic Power Line Inspection Using UAV Images. *Remote Sens.* **2017**, *9*, 824. [CrossRef]
36. Zhou, X.; Zheng, X.; Ou, K. Power line detect system based on stereo vision and FPGA. In Proceedings of the 2nd International Conference on Image, Vision and Computing (ICIVC), Chengdu, China, 2–4 June 2017; pp. 715–719. [CrossRef]
37. Maurer, M.; Hofer, M.; Fraundorfer, F.; Bischof, H. Automated inspection of power line corridors to measure vegetation undercut using UAV-based images. *ISPRS Ann. Photogramm. Remote Sens. Spat. Inf. Sci.* **2017**, *IV-2/W3*, 33–40. [CrossRef]
38. Saurav, S.; Gidde, P.; Singh, S.; Saini, R. Power Line Segmentation in Aerial Images Using Convolutional Neural Networks. In *Pattern Recognition and Machine Intelligence. PReMI 2019*; Lecture Notes in Computer Science; Deka, B., Maji, P., Mitra, S., Bhattacharyya, D., Bora, P., Pal, S., Eds.; Springer: Cham, Switzerland, 2019; Volume 11941. [CrossRef]
39. Jenssen, R.; Roverso, D. Automatic autonomous vision-based power line inspection: A review of current status and the potential role of deep learning. *Int. J. Electr. Power Energy Syst.* **2018**, *99*, 107–120. [CrossRef]
40. Oh, J.H.; Lee, C.N. 3D Power Line Extraction from Multiple Aerial Images. *Sensors* **2017**, *17*, 2244. [CrossRef]
41. Oh, J.H.; Lee, C.N. Automatic power line reconstruction from multiple drone images based on the epipolarity. *J. Korean Soc. Surv. Geod. Photogramm. Cartogr.* **2018**, *36*, 127–133. [CrossRef]
42. Hosoi, F.; Nakai, Y.; Omasa, K. 3-D voxel-based solid modeling of a broad-leaved tree for accurate volume estimation using portable scanning lidar. *ISPRS J. Photogramm. Remote Sens.* **2013**, *82*, 41–48. [CrossRef]
43. Griffiths, D.; Burningham, H. Comparison of pre- and self-calibrated camera calibration models for UAS-derived nadir imagery for a SfM application. *Prog. Phys. Geogr. Earth Environ.* **2018**, *43*, 215–235. [CrossRef]
44. Zhou, Y.; Rupnik, E.; Faure, P.-H.; Pierrot-Deseilligny, M. GNSS-Assisted Integrated Sensor Orientation with Sensor Pre-Calibration for Accurate Corridor Mapping. *Sensors* **2018**, *18*, 2783. [CrossRef]
45. Fraser, C.S. Digital camera self-calibration. *ISPRS J. Photogramm. Remote Sens.* **1997**, *52*, 149–159. [CrossRef]
46. Brown, D.C. Close-range camera calibration. *Photogramm. Eng.* **1971**, *37*, 855–866.
47. Gillespie, A.R.; Kahle, A.B.; Walker, R.E. Color enhancement of highly correlated images. I. Decorrelation and HSI contrast stretches. *Remote Sens. Environ.* **1986**, *20*, 209–235. [CrossRef]
48. Campbell, N.A. The decorrelation stretch transformation. *Int. J. Remote Sens.* **1996**, *17*, 1939–1949. [CrossRef]
49. Fischler, M.A.; Bolles, R.C. Random sample consensus: A paradigm for model fitting with applications to image analysis and automated cartography. *Commun. ACM* **1981**, *24*, 381–395. [CrossRef]
50. Duda, R.O.; Hart, P.E. Use of the Hough Transformation to Detect Lines and Curves in Pictures. *Commun. ACM* **1972**, *15*, 11–15. [CrossRef]

Publisher's Note: MDPI stays neutral with regard to jurisdictional claims in published maps and institutional affiliations.

Article

UAV-Based Terrain Modeling under Vegetation in the Chinese Loess Plateau: A Deep Learning and Terrain Correction Ensemble Framework

Jiaming Na [1,2,3,4,†], Kaikai Xue [4,5,†], Liyang Xiong [1,2,4,6,†], Guoan Tang [1,2,4,6,], Hu Ding [4], Josef Strobl [4,8] and Norbert Pfeifer [3]

1 Key Laboratory of Virtual Geographic Environment, Nanjing Normal University, Nanjing 210023, China; jiaming.na@njnu.edu.cn (J.N.); tangguoan@njnu.edu.cn (G.T.)
2 School of Geography, Nanjing Normal University, Nanjing 210023, China
3 Department of Geodesy and Geoinformation, TU Wien, 1040 Vienna, Austria; norbert.pfeifer@geo.tuwien.ac.at
4 Jiangsu Center for Collaborative Innovation in Geographical Information Resource Development and Application, Nanjing Normal University, Nanjing 210023, China; 171302105@stu.njnu.edu.cn (K.X.); josef.strobl@sbg.ac.at (J.S.)
5 Northwest Engineering Corporation Limited, Power Construction Corporation of China (POWERCHINA), Xi'an 710065, China
6 State Key Laboratory Cultivation Base of Geographical Environment Evolution, Nanjing Normal University, Nanjing 210023, China
7 School of Geography, South China Normal University, Guangzhou 510631, China; hu_ding@m.scnu.edu.cn
8 Department of Geoinformatics–Z_GIS, University of Salzburg, Salzburg 5020, Austria
* Correspondence: xiongliyang@njnu.edu.cn
† These authors have joint credit as first author.

Received: 8 September 2020; Accepted: 9 October 2020; Published: 12 October 2020

Abstract: Accurate topographic mapping is a critical task for various environmental applications because elevation affects hydrodynamics and vegetation distributions. UAV photogrammetry is popular in terrain modelling because of its lower cost compared to laser scanning. However, this method is restricted in vegetation area with a complex terrain, due to reduced ground visibility and lack of robust and automatic filtering algorithms. To solve this problem, this work proposed an ensemble method of deep learning and terrain correction. First, image matching point cloud was generated by UAV photogrammetry. Second, vegetation points were identified based on U-net deep learning network. After that, ground elevation was corrected by estimating vegetation height to generate the digital terrain model (DTM). Two scenarios, namely, discrete and continuous vegetation areas were considered. The vegetation points in the discrete area were directly removed and then interpolated, and terrain correction was applied for the points in the continuous areas. Case studies were conducted in three different landforms in the loess plateau of China, and accuracy assessment indicated that the overall accuracy of vegetation detection was 95.0%, and the MSE (Mean Square Error) of final DTM (Digital Terrain Model) was 0.024 m.

Keywords: UAV photogrammetry; terrain modeling; vegetation removal; deep learning

1. Introduction

Accurate topographic mapping is essential for various environmental applications because elevation affects hydrodynamics and vegetation distributions [1–3]. Small elevation changes can alter sediment stability, nutrient, organic matters, tides, salinity, and vegetation growth, and therefore might cause substantial vegetation transition in relatively flat wetlands [4–7]. Topography influences flow erosion and thus is a prerequisite for soil erosion studies, especially in the loess plateau of China [8,9]. The temporal dynamics of topography helps understand the erosion process and contributes to conservation planning.

Various remote sensing techniques, such as RADAR [10–13], light detection and ranging (LiDAR) [14–16], and stereo photogrammetry [17–20], were developed and applied to model terrains of various scales. However, accurate topographic mapping in gully areas in the loess plateau of China remains challenging due to complications of hydrodynamics, ever-changing terrains, and dense vegetation covers. The widely used LiDAR is the best method because it provides the highest accuracy of mean terrain error within 0.10 m to 0.20 m [21–23]. Meanwhile, terrestrial laser scanning is restricted for terrains with a strong relief [24]. The field measurement always fails in some certain areas because the complex terrain might influence the visibility from the sensor perspective. Airborne laser scanning is also limited under poor weather condition. Errors further increase in dense and tall vegetation conditions and might reach a challenging 'dead zone' when the marsh vegetation height is close to or beyond 2 m [4]. Moreover, laser scanning is expensive and hard to implement in developing countries [25]. Frequent deployment of LiDAR surveys is in such scenarios is cost-prohibited. Therefore, affordable methods for rapid and accurate measurements without relying on out-dated historical data are needed.

State-of-art unmanned aerial vehicle (UAV) provides a promising solution to general mapping applications. Remarkable progress was achieved in light-weight sensor and UAV system developments [26,27], improvement of data pre-processing [28], registration [29,30], and image matching [31–33]. The UAV-based terrain modeling has advantages of low costs, high spatial resolution, high portability, flexible mapping schedule, rapid response to disturbances, and convenient multi-temporal monitoring [34]. UAV has become a favourable surveying method in many areas with challenging mobility and accessibility. In particular, cameras are miniaturised and have low power consumption, making them ideal sensors for area-wise coverage from UAVs [35].

Despite various successful applications, challenges for UAV usage still remain, especially in areas with a dense vegetation condition. UAV terrain modeling is best suited to areas with sparse or no vegetation, such as sand dunes and beaches [36], coastal flat landscapes [37], and arid environments [38]. Establishing a satisfactory terrain model is hindered by difficulties in point-based ground-filtering. Some successful works for automatic ground-filtering were conducted in digital terrain model construction [39,40], the application of which remains 'pointless' due to difficulty in penetration and lack of points from ground [41]. Current developments in UAV communities provide no solution to these issues of terrain mapping in densely vegetated environments [42].

This study aimed to address the challenges in terrain mapping under vegetation cover by developing a UAV photogrammetry mapping solution that does not depend on historical data. The main objective was to propose an algorithmic framework correct terrain based on vegetation detection, by using deep learning (DL). First, image matching point cloud was generated by UAV photogrammetry. Second, vegetation points were identified based on U-net deep learning network. After that, ground elevation was corrected by estimating vegetation height to generate the digital terrain model (DTM). Two scenarios, namely, discrete and continuous vegetation areas were considered. The vegetation points in discrete area were directly removed and then interpolated, and terrain correction was applied for the points in continuous areas. Given that most photogrammetric UAV systems carry colour cameras, the possible application of the proposed method in photogrammetric UAV system for terrain mapping in vegetated environments was also explored.

2. Materials and Methods

The proposed approach involved the following four steps—(1) UAV photogrammetry; (2) DL-based vegetation detection, (3) terrain correction, and (4) DTM generation. Accuracy assessment was conducted through the comparison between check points generated by global navigation satellite system (GNSS) unit and produced DTM elevation.

2.1. Study Site

Three study areas, namely, Xining (SA1), Wangjiamao (SA2), and Wucheng (SA3) located in Qinghai, Shaanxi, and Shanxi, respectively, were selected in the Loess Plateau of China (Figure 1) and represent loess hill and gully, loess hill and loess valley area, respectively. Among them, Wangjiamao and Wucheng cover the complete catchments, and Xining covers a hillslope area. All three study areas were covered with vegetation since the implementation of the 'Grain for Green' project (changing the agriculture to conservation area) from late 1990s [43,44]. Vegetation status of three different study areas varied in their types and spatial distributions. The vegetation in Xining was manmade for ecological protection from the formal cultivation, with an average interval distance of 2 m in the terraced slopes. While in the Wuchenggou and Wangjiawao areas, vegetation are more natural but some cash crops like apples and jujubes (Chinese dates), were still planted, with more dense horizontal distance around 1 m in the slopes. The basic geographic information is listed in Table 1.

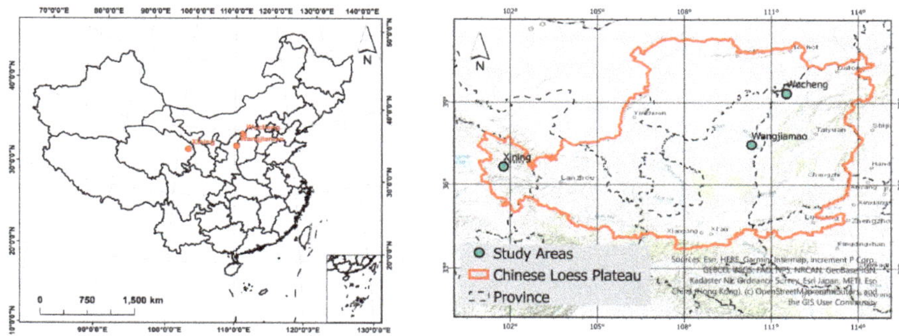

Figure 1. Study areas.

Table 1. Geography of study areas.

	Xining (SA1)	Wangjiamao (SA2)	Wucheng (SA3)
Location	36°39′N101°43′E	37°34′20″N~37°35′10″N 110°21′50″E~110°22′40″E	39°15′51″N~39°16′57″N 111°33′21″E~111°34′48″E
Area	0.07 km^2	2.21 km^2	3.17 km^2
Elevation	2266–2348 m	1011–1195 m	1238–1448 m
Landform	Loess hill and gully	Loess hill	Loess valley
Climate	Semi-arid *(BSh)*	Semi-arid *(BSh)*	Semi-arid *(BSh)*
Annual Temperature	6.5°C	9.7°C	8°C
Precipitation	327 mm/y	486 mm/y	~450 mm/y
Vegetation	Weed	Shrub	Arbor
Main vegetation type	*Rhamnus erythroxylon, Artemisia*	*Haloxylon ammodendron, Ziziphus jujuba*	*Hippophae, Malus domestica*
Vegetation height	0.5–2 m	0.5–6 m	0.5–6 m

2.2. Unmanned Aerial Vehicle (UAV) and Global Navigation Satellite System (GNSS) Field Data Collection

Image matching point clouds from UAV photogrammetry were used as the inputs for terrain modeling. Optical aerial photographs were captured using a DJI Inspire 1 microdrone [45] mounted with a digital camera system Zenmuse X5 [46] (15 mm focal length, RGB color, and 4096 × 2160 resolution), with a battery time of approximately 18 min, and could resist wind speeds of up to 10 m/s. Detailed flight information is shown in Table 2. Pix4D Capture flight planner software was used to plan a round-trip flight line along the study areas, and automatically collect images within certain designed flight distance. All flights were completed from 10 am to 2 pm, to ensure that the image quality would not be influenced by the shades. Ground control points (GCPs) in WGS-84 were obtained by the

Topcon HiperSR RTK GNSS unit [47] (10 mm horizontal positioning accuracy and 15 mm vertical positioning accuracy), with a tripod, to ensure horizontal and vertical accuracy. Bundle adjustment was implemented in Pix4D Mapper software [48]. The point clouds were finally generated and interpolated into the grid digital surface model (DSM).

Table 2. Unmanned aerial vehicle (UAV) flight information of three study areas.

	Xining	Wangjiamao	Wucheng
Flight date	2017.10.24	2019.08.20	2018.04.26
Flight height	50 m	150 m	200 m
Photo gained in total	80	420	680
Flight overlapping	80%	80%	80%
Side overlapping	70%	70%	70%
Ground sampling distance	2.31 cm	4.36 cm	8.06 cm
Ground Control Points in total	7	18	19
Mean RMS of GCPs	0.011 m	0.014 m	0.018 m
Point amount from dense matching	832341	7917617	9956200

Eight targets along the vegetation in Xining (SA1) were designated as check points (CPs) for the uncertainty assessment of the final terrain modeling results. These targets were 1-m-wide boards painted in black and white in a diagonal chessboard pattern.

2.3. Deep Learning (DL)-Based Vegetation Detection

Most DL networks connect simple layers for data distillation. Input information passes through a layer of filter that increases the purity in distillation to achieve the desired result [49]. Convolutional neural network (CNN) is one of the representative algorithm structures of the deep neural network structure and is a feed-forward neural network usually used in object recognition, target detection, semantic segmentation, and other issues [50,51]. A typical structure for a CNN network, U-Net [52], was implemented for vegetation detection, because of its effectiveness and simplicity. U-net adopts the principle of gradient descent, propagates data information forward, and reverses propagation to correct the parameter weights and deviations [53]. Certain layers were changed and adjusted to specific terrain modeling tasks, on the basis of the existing U-Net structure.

2.3.1. Training Data Generation

DL is usually used in datasets with a large amount of data, and convolutional neural networks are suitable for processing relevant image data. Therefore, the U-Net model can generate a large number of images as input data. Here, input data were randomly cropped to ensure proper representation and eliminate the influence of manual selection. Random coordinate points were expanded, based on the desired image size. The crop range was calibrated, and the crop operation fully utilized the cell size and projected coordinate information.

Data enhancement is the process of generating new data for training, based on image nature, without actually collecting new samples. Convolution operations have translational invariance, and similar transformations such as rotation and scaling of vegetation data do not change the information characteristics of the vegetation data. Here, similar data outside the sample area chart were provided to the model to ensure data diversity. Random similar transformation, scale transformation, Gaussian blur, and image enhancement were performed for the crop data, in which the rotation allocation transformation matrix and 2D Gaussian function were treated as follows—Equations (1) and (2).

$$M = \begin{pmatrix} cos\theta - sin\theta \\ sin\theta \; cos\theta \end{pmatrix} \tag{1}$$

$$G(x, y) = \frac{1}{2\pi\sigma^2} e^{-(x^2+y^2)/2\sigma^2} \tag{2}$$

where θ is the angle of rotation, and σ is the variance.

For the classification task, the training data was labeled as one-hot encoding logic category, namely, 1 for vegetation and 0 for non-vegetation. Manual work was first done for the labeling task at, based on the original point clouds. The RGB and additional elevation information of vegetation of the manmade labels of three study areas were then generated from the original image matching point cloud. All labels were divided into two groups for model training and validation. Forty percent of the dataset was randomly sampled as the training data. Since the DL requires a large amount of training samples, a tool was developed based on the ArcGIS Pro [54] software, using the python script for a multi-scale replicability of the training samples. Finally, 10,000 samples of 4 dimensions (R, G, B, Z) with 128 × 128 cells were automatically generated.

2.3.2. Feature Selection

The data for neural network represent a multidimensional feature array, also known as a tensor, a container for numerical data of images. All transformations learned by the neural network could be summed up as tensor operations for numerical data and formed matrix extension dimensions. Spectral information (R, G, and B values) and elevation provide theoretical feasibility for the division of vegetation. The training data generated by the original point clouds had an RGB value and underwent elevation, and the input data were normalized to reasonably eliminate the scale effect.

2.3.3. Design of the U-Net Network

An improved U-Net framework with a slightly altered structure was used for vegetation detection. The improved U-Net produced split maps of the same size as the input data and preserved the continuity of the resolution.

The predictive model describes the relationship between input x (features) and desired output (answer) y. The system 'learns' the relationship between data and output repeatedly through differential equations and random deviations and obtains the values of a series of unknown parameters, thus, forming a set of rules on its own. These rules are applied to a set of untrained data to allow the model to predict the corresponding set of answers. This process is the core architecture of the image segmentation task. With the use of the FCN (Fully Convolutional Networks, [55]) architecture, the simple representation of the relationship between the predictive output and the input is as follows—Equation (3).

$$y = f\left(\sum_{j=1}^{m}\left(w_j\left(\sum_{i=1}^{n} w_i x_i - \theta_n\right) - \theta_m\right)\right) \tag{3}$$

where x is the input; $\hat{y}$ is the forecast output; m is the number of hidden layers that determines the depth of the network to a certain extent and represents the complexity of the network; n is the number of neurons in each layer of network, and each neuron in the convolutional neural networks is represented as a filter (nine neurons in this study); w is expressed as a weight assigned to a neuron to connect input information for signaling; and f is an activation function for nonlinear mapping.

Three specific network structures architecture with different hyper-parameters were designed (Figure 2) for the vegetation detection tasks. In the down-sampling procedure, convolution was performed to extract features and activation values at different levels. Each convolution was based on the result of the previous layer of convolution, thus, bringing the model to a certain depth. Some convoluted feature values were de-dimensionalized from the input, through pooling, to reduce a large amount of computational consumption. The vegetation characteristics were summarized, and a wide range of features were extracted. The data were easily learned, and the model learning ability was enhanced. In the upper-sampling, the image size was expanded layer-by-layer to interpolate the feature maps at all levels. Details on the three model hyper-parameters are shown in Table 3.

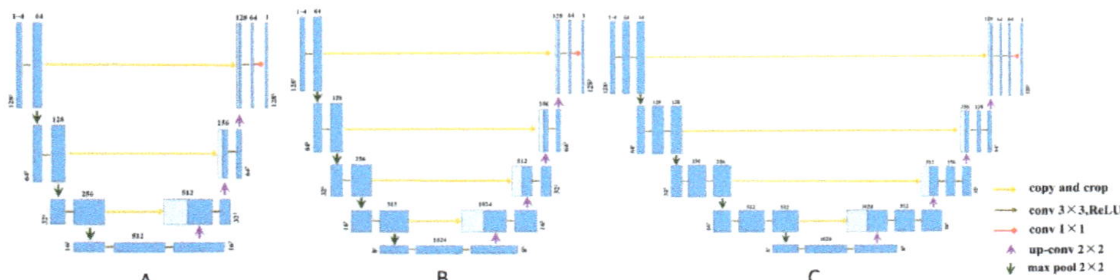

Figure 2. Three designed U-net model structures.

Table 3. Comparison of model hyper-parameters.

Network	A	B	C
Layers	5	6	10
Down-sampling	3× 3 × 64 (×128, ×256, ×512, ×512)	3 × 3 × 64 (×128, ×256, × 512, ×1024, ×1024)	Double B
Up-sampling	3 × 3 × 256 (×128,×64)	3×3×512(×256, ×128, ×64)	Double B
Pooling	(2 × 2) × 3	(2 × 2)× 4	(2 × 2) × 4
Jump connection	3 times	4 times	4 times

2.3.4. Vegetation Detection Accuracy Assessment

The detection accuracy was assessed through a comparison with the reference. The reference data were manually interpolated from the original point cloud. The confusion matrix was applied to calculate the accuracy in the rasterized results.

2.4. Terrain Correction

After vegetation detection, the terrain information could be modified using the vegetation results. In terrain modeling, the ability to reasonably eliminate the vegetation points, determined the accuracy of the DTM result. In urban areas, a cross-section is usually used to completely eliminate the vegetation point and then interpolate the complement point to obtain the DTM [56]. The ground is fitted in a 2D terrain plane, and the points higher than the plane are removed. However, this trend approach always fails, because the planes are difficult to estimate, due to the dramatic reliefs of the mountainous terrains (e.g., the Loess Plateau). The alternative practice for mountainous areas is usually to universally lower the vegetation points, based on the estimation of vegetation average height [37]. This method is effective for continuous vegetation in mountainous areas and maintaining the original terrain fluctuation, but is restricted for discrete vegetation in mountain areas, due to elevation fragmentation or convex terrain [57,58]. To solve this problem, this study divided the terrain correction into two scenarios, namely, discrete and continuous vegetation areas (Figure 3). The vegetation points in the discrete area were directly removed and then interpolated, and terrain correction was applied for the points in continuous areas.

Step 1: Identification of discrete and continuous vegetation areas.

The vegetation detection result was firstly rasterized then converted into polygon by the Raster to Polygon tool in ArcGIS Pro software [54]. A threshold of 30 m^2 by expertise was then used to identify discrete and continuous vegetation areas. Vegetation areas of less than 30 m^2 were classified as discrete, and those greater than 30 m^2 were labeled as continuous.

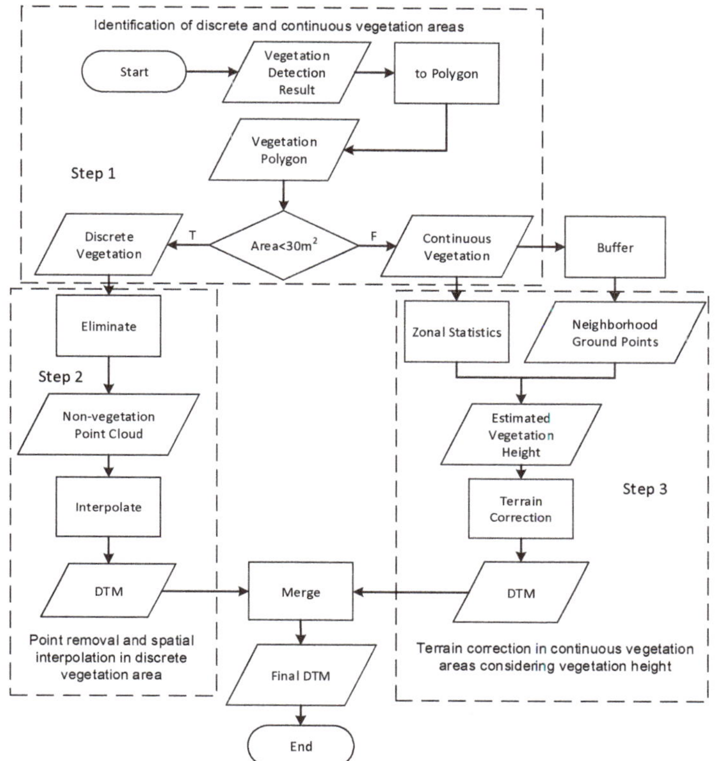

Figure 3. Workflow of terrain correction.

Step 2: Point removal and spatial interpolation in discrete vegetation area.

The original point cloud obtained for the UAV photogrammetry represents a surface model including the vegetation information. To achieve a terrain model, all these vegetation points should be excluded. The points in the discrete vegetation area could be directly eliminated. Since the 'holes' after the removal were relatively small, it would not affect the overall trend of the terrain. Therefore, the terrain could then be interpolated.

Step 3: Terrain correction in continuous vegetation areas when considering vegetation height.

The commonly used local polynomial interpolation ignores its own terrain fluctuations. Thus, the elevation information would be lost when the points in the continuous vegetation area are simply removed. A possible solution was to estimate the terrain elevation and then modify the elevation of the vegetation points in the point cloud. With regard to the varying heights for each individual continuous vegetation area, an adaptive process with less human interaction was proposed. Vegetation height was estimated by the elevation in the 0.5 m buffer zone of each polygon. This could be achieved by the Zonal Statistics tool by ArcGIS Pro [54], using the original point clouds. The difference between the vegetation elevation point and the ground elevation in the polygonal area from DSMs was treated as the unified elevation value of the area, and the final fine DTM was obtained by subtracting the estimated mean height of each polygon.

2.5. Terrain Modeling Result Validation

Evaluating the elevated generated DTM is the key to measuring accurate terrain modeling results. To achieve the validation, a comparison between the final generated DTM with CPs from field survey by GNSS unit was conducted. The Xining area was selected for the validation.

3. Results

3.1. Vegetation Detection Results

Xining was selected for model training. After performance comparison for the designed U-Net network structures, the U-Net model C was finally chosen for vegetation detection. Details of three structures' performance are discussed in following Section 4.1. After model training, the model was applied in two other study areas. Figure 4 shows the results for the three study areas.

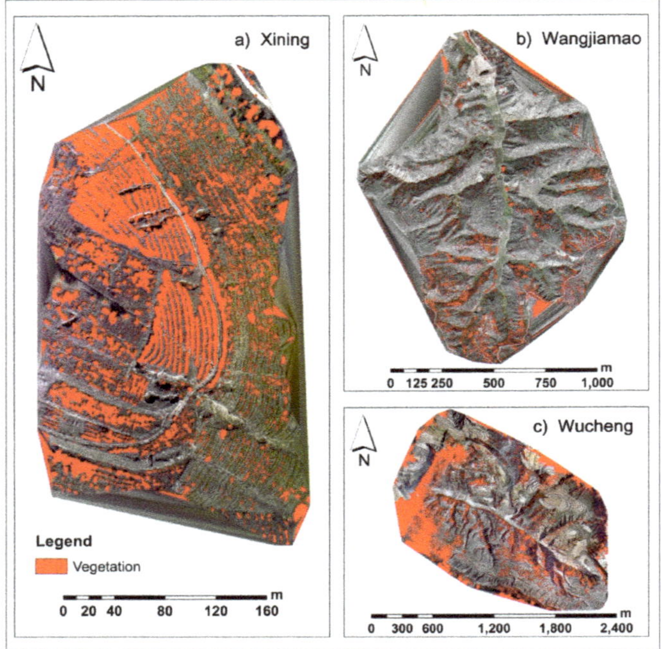

Figure 4. Vegetation detection result. (**a**) Xining; (**b**) Wangjiamao; and (**c**) Wucheng.

Table 4 shows that the confusion matrix of vegetation detection results in three areas with the reference. The detection accuracies were acceptable at 90.9% for Xining, 96.4% for Wangjiamao, and 87.2% for Wucheng. The vegetation detection of Wucheng was not as highly accurate as for the other two areas because the tie points in the southwest corner of Wucheng were relatively insufficient during the automatic image matching. Hence, the accuracy of the original image matching point cloud was reduced.

3.2. Vegetation Identification Results

Identification was conducted in the three study areas, based on the adaptive treatment of discrete and continuous vegetation (Figure 5). The manmade vegetation spatial distribution patterns in Xining and the natural patterns in Wuchenggou and Wangjiawao were successfully identified. Vegetation height estimations ranged from 0.01 to 2.26 m (1.81 m in mean) in Xinning, 0.01 to 7.12 m in Wangjiamao (4.23 m in mean), and 0.66 to 6.38 m in Wucheng (4.21 m in mean), respectively.

Table 4. Confusion matrix of vegetation detection results in three areas for architecture C.

		Detection (In Cells)					
		Xining		Wangjiamao		Wuchenggou	
		Ground	Vegetation	Ground	Vegetation	Ground	Vegetation
Reference	Ground	4,457,886 (62.3%)	225,949 (3.1%)	127,645,071 (90.0%)	2,049,627 (1.4%)	2,095,418 (69.4%)	135,462 (4.5%)
	Vegetation	425,710 (6.0%)	2,039,464 (28.6%)	3,181,941 (2.2%)	9,075,223 (6.4%)	252,464 (8.3%)	535,952 (17.8%)

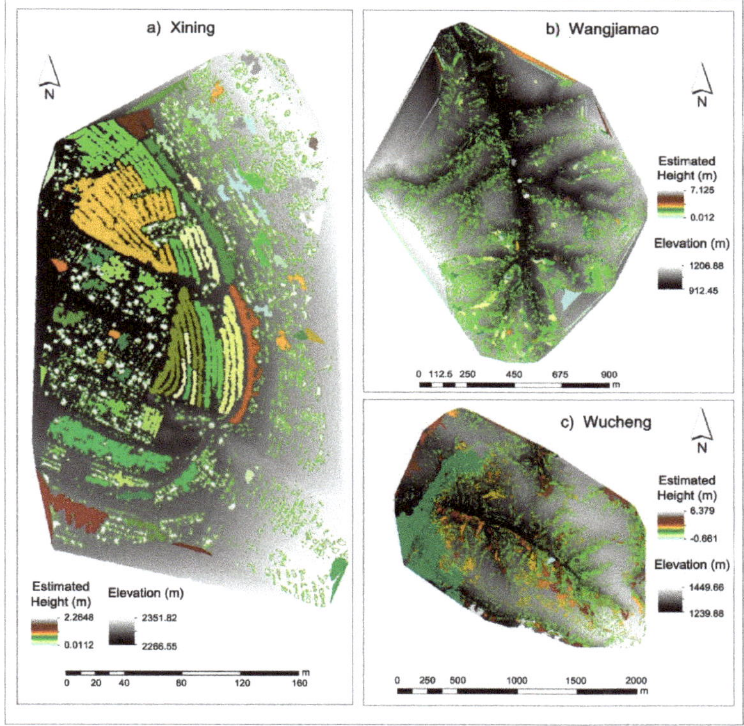

Figure 5. Vegetation identification results. (**a**) Xining; (**b**) Wangjiamao; (**c**) Wucheng. Base map is the digital surface model (DSM), and the estimated vegetation height is colored.

3.3. Terrain Correction Results

After the vegetation identification, terrain correction was done and DTMs with 1 m resolution were then interpolated (Figure 6). The proposed method removed the vegetation points without losing the terrain details and restored the fine DTM. Compared with orthophotos, the terrain reliefs were well presented in the modeling results. The smooth color rendering of DTMs indicated that the vegetation recognition removal was good, and DTM was visually refined.

were well presented in the modeling results. The smooth color rendering of DTMs indicated that the vegetation recognition removal was good, and DTM was visually refined.

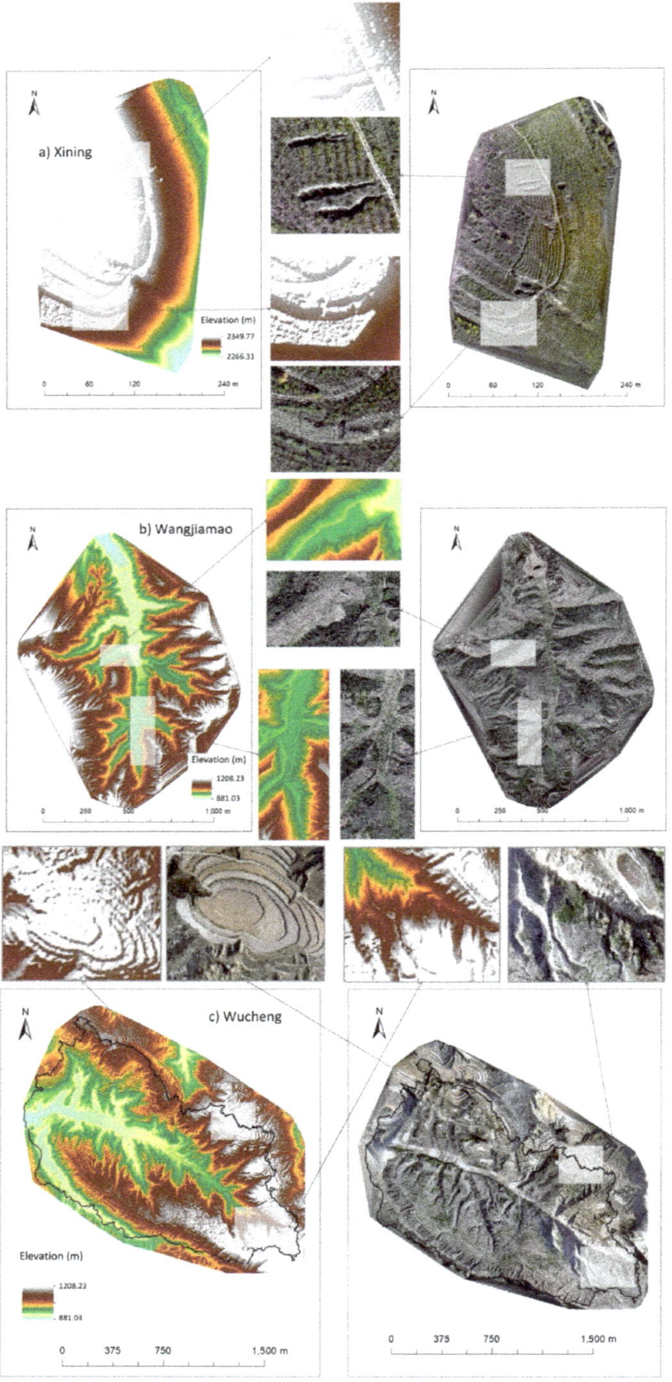

Figure 6. Digital terrain model (DTM) (left) and orthophoto (right) results after terrain corrections. (**a**) Xining; (**b**) Wangjiamao; (**c**) Wucheng. Two detailed windows of each study areas are enlarged.

3.4. Terrain Modeling Result Validation with Field Measurement Data

Ground control points in Xining by field survey were elevated to verify the DTM results (Figure 7).

Figure 7. Elevation uncertainty assessment in Xining.

Table 5 shows the elevation comparison of the CPs. The MSE was 0.024 m, which met the standard of the accurate terrain modeling. Points D and H had the highest prediction accuracy, which were originally ground points. Correctly predicting the vegetation points ensured that the ground elevation values were preserved correctly. Point G failed the accurate elevation, because it was located at a hole even when the vegetation detection was not correct. The terrain correction of the remaining vegetation points was guaranteed.

Table 5. Elevation comparison of CPs.

Sample	Reference/m	Result/m	Error/m
A	2343.246	2343.518	0.272
B	2343.283	2343.424	0.141
C	2339.275	2338.772	−0.497
D	2335.718	2335.739	0.021
E	2328.019	2327.967	−0.948
F	2317.586	2317.699	0.113
G	2331.099	2332.197	1.098
H	2340.806	2340.800	−0.006

4. Discussion

In this section, some extra analyses were conducted to discuss the key to the success of vegetation detection. Hyper-parameter (network structure and epoch) influence analysis was done at first to achieve an optimized parameter setting. The comparison with other two published methods (perceptron and adaptive filtering) was then done for a deeper analyses of the performance of our proposed vegetation detection method.

4.1. U-Net Hyper-Parameter Influence on Vegetation Detection Performance

The performance of the three designed different U-net networks was assessed in terms of training loss, validation accuracy, and training accuracy, to understand the influence of parameter and architecture on vegetation detection.

Figure 8a shows the training loss of the three models with different epoch settings. Model A is simple with a small network layer and capacity. Its training loss reached the local minimum at 48 epochs. The training loss of model B bounced at the 16th and 38th epochs, and was overall faster than that for Model A. The training loss of model C declined smoothly and reached the local minimum at the 45th epoch. Figure 8b shows the training accuracy of the three models with different epoch settings. All three models generally showed an increasing trend. Model A in the 8th epoch to 40th epoch did not meet the saturation. Model B in the 17th and 38th epochs showed a decline in training accuracy. Model C in the 45th epoch achieved the local maximum accuracy. Figure 8c shows the validation accuracy of the three models with different epoch settings. Model A had the lowest verification accuracy. Model B was moderately complex with convolution occurring during pooling, and its verification accuracy was high. However, a substantial decline in the 15th epoch to 0.92, indicated a slightly weakened stability of its performance. Model C was the most stable and accurate with a high accuracy of 0.94 at the 45th epoch (Figure 8c).

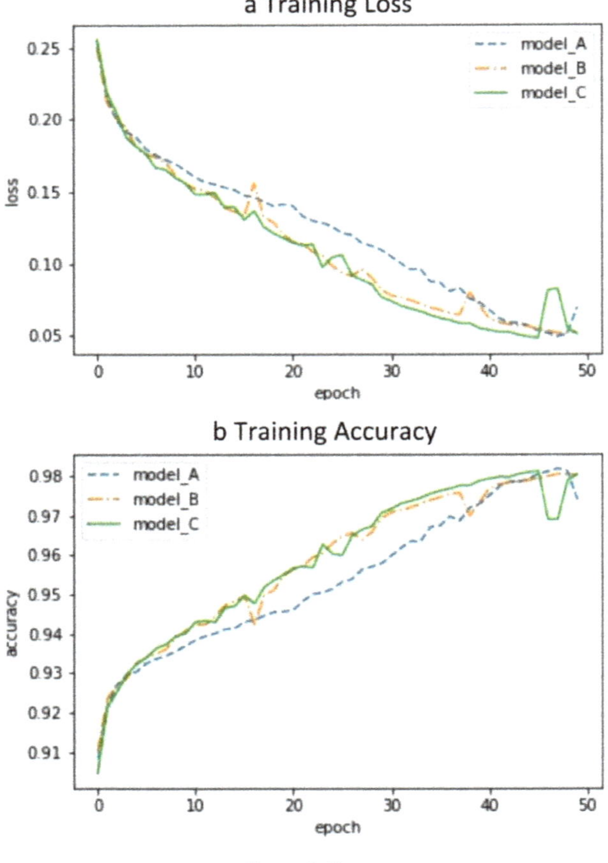

Figure 8. *Cont.*

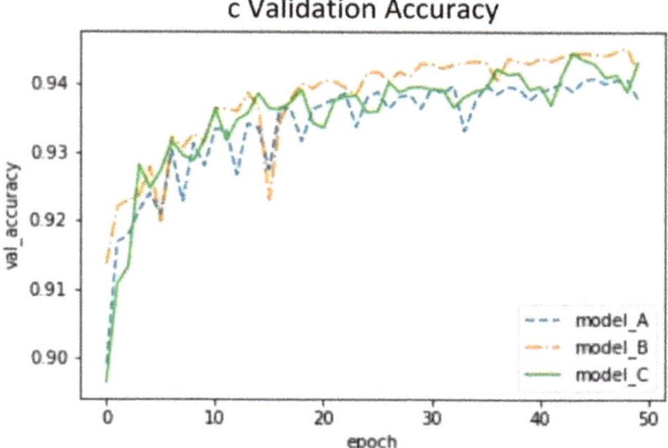

Figure 8. DTM results after terrain corrections. (**a**) training loss, (**b**) training accuracy, and (**c**) validation accuracy of the three different network structures, with different epoch settings.

Model C with an epoch of 45 was selected for vegetation detection, due to its lowest loss function and highest accuracy during training and validation. When the network structure was large, the epoch should be increased appropriately to ensure that the parameters were updated. Merging combined the features of convolution and enhanced the upper sampling of data.

4.2. Comparison of Vegetation Detection Performance with Other Methods

Two other methods, namely, perceptron [59] and adaptive filtering [60] were selected for comparison to additional assessment of vegetation detection. Precision, recall, and F-score values were used for validation. Precision indicated the extent to which the extraction result represented the real target and the error of the model. Prediction was positive when the following values were obtained—true positive (TP) and false positive (FP)—which indicated the extraction of the correct vegetation grid. FP indicated that the ground sample was predicted as a vegetation sample, and FP was a 'false positive' situation. True negative (TN) was achieved when the results predicted for the ground was also a ground sample. Recall represents the extent to which real targets can be extracted and indicates the model's leakage. Predicting vegetation as true (TP) and vegetation samples as ground samples were a false negative (FN), i.e., no vegetation samples were extracted. Precision was the number of samples that were positive relative to the predicted positive, and recall was relative to the number of positive samples in the original sample. The F-value was the reconciliation average of precision and recall. The formulas for precision, recall, and F-values were as follows (Equations (4)–(6).

$$\text{Precision} = \text{TP}/(\text{TP} + \text{FP}) \tag{4}$$

$$\text{Recall} = \text{TP}/(\text{TP} + \text{FN}) \tag{5}$$

$$\text{F} = 2 \times \text{Precision} \times \text{Recall}/(\text{Precision} + \text{Recall}) \tag{6}$$

Figure 9 shows the precision, recall, and F-value of the three methods. Our improved U-Net architecture had the highest values for all three study areas. Particularly, the best identification result was observed in Xining with a precision of 0.91. The performances of last two methods were seriously weaker than that of our improved U-Net architecture. Perceptron lacked the hidden layer and did not introduce random deviation. The final classification result was based on the hyperplane, which could not adapt to the complex terrain, resulting in a low accuracy for vegetation detection. Adaptive filtering was excessive in vegetation recognition, and its results depended on the sketched vegetation range. This phenomenon required the manual sketching of the training area, as a supervised area for vegetation recognition in each study area.

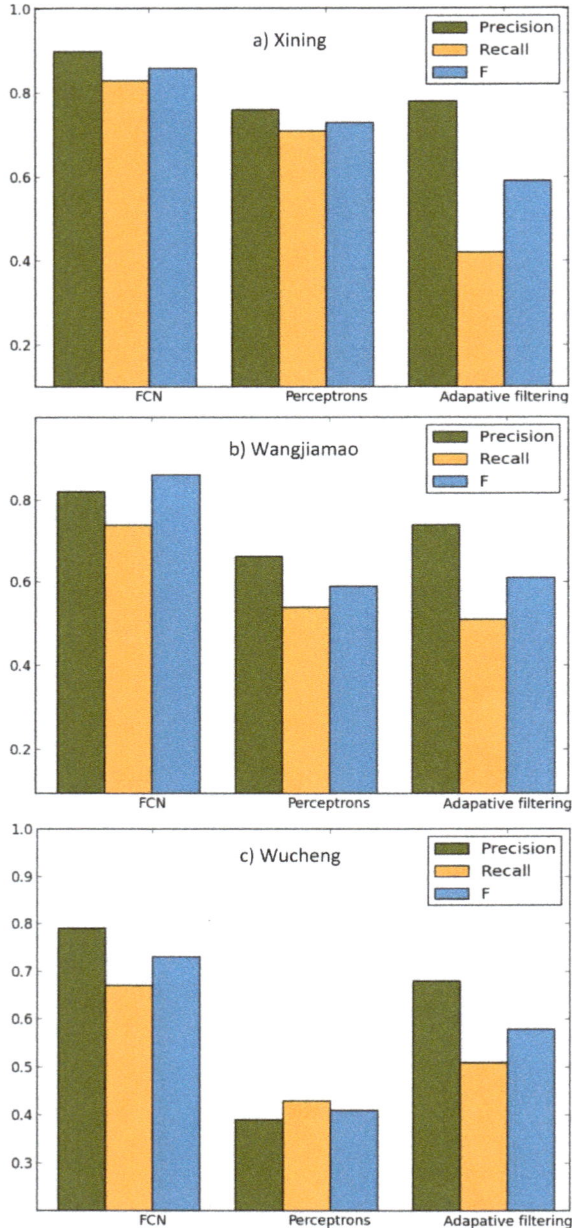

Figure 9. Comparison of accuracy under three methods (FCN by our U-net based method, Perceptrons by Kwak et al., 2007 and Adapative filtering by Hu et al., 2019). (**a**) Xining; (**b**) Wangjiamao; and (**c**) Wucheng. Dark green, orange, and blue bars are Precision, Recall, and F-value respectively.

5. Conclusions

This study proposed a UAV photogrammetric framework for terrain modeling in dense vegetation areas. With the loess plateau of China as the study area, a DL and terrain correction ensemble method was proposed and applied. An improved U-net network for vegetation segmentation was presented. The feature combination of RGB+DSM was used for vegetation detection. According to four-fold cross-verification, the accuracy was 94.97%, and the model had a good generalization ability. The influence of U-Net architecture and parameter epoch setting on vegetation detection performance was also assessed. Comparison with other methods confirmed the better performance of the proposed technique. Fine DTM generation method for terrain modeling was also put forward. The vegetation area was divided into discrete and continuous, and adaptive terrain correction was proposed and realised. DTM accuracy was evaluated with the field measurements. This framework could be applied in dense vegetation, with an advantage of low-cost UAV photogrammetry when laser scanning was limited.

Author Contributions: Conceptualization, J.N. and K.X.; algorithm, J.N. and K.X.; classification analysis, J.N.; process the data, K.X.; writing—original draft preparation, J.N. and K.X.; writing—review and editing, H.D., J.S. and N.P.; supervision, L.X. and G.T.; funding acquisition, L.X. and G.T. All authors have read and agreed to the published version of the manuscript.

Funding: This work was financially supported by the Natural Science Foundation of China, grant numbers 41930102, 41971333, and the Priority Academic Program Development of Jiangsu Higher Education Institutions (No. 164320H116).

Acknowledgments: The authors sincerely thank for the comments from anonymous reviewers and members of the editorial team.

Conflicts of Interest: The authors declare no conflict of interest.

References

1. Kulawardhana, R.W.; Popescu, S.C.; Feagin, R. Airborne lidar remote sensing applications in non-forested short stature environments: A review. *Ann. For. Res.* **2017**, *60*, 173–196. [CrossRef]
2. Galin, E.; Guérin, É.; Peytavie, A.; Cordonnier, G.; Cani, M.; Benes, B.; Gain, J. A Review of Digital Terrain Modeling. *Comput. Graph. Forum* **2019**, *38*, 553–577. [CrossRef]
3. Tmušić, G.; Manfreda, S.; Aasen, H.; James, M.R.; Gonçalves, G.; Ben Dor, E.; Brook, A.; Polinova, M.; Arranz, J.J.; Mészáros, J.; et al. Current Practices in UAS-based Environmental Monitoring. *Remote Sens.* **2020**, *12*, 1001. [CrossRef]
4. Hladik, C.; Alber, M. Accuracy assessment and correction of a LIDAR-derived salt marsh digital elevation model. *Remote. Sens. Environ.* **2012**, *121*, 224–235. [CrossRef]
5. Hladik, C.; Schalles, J.; Alber, M. Salt marsh elevation and habitat mapping using hyperspectral and LIDAR data. *Remote Sens. Environ.* **2013**, *139*, 318–330. [CrossRef]
6. Tsouros, D.C.; Bibi, S.; Sarigiannidis, P.G. A review on UAV-based applications for precision agriculture. *Information* **2019**, *10*, 349. [CrossRef]
7. Szabó, Z.; Tóth, C.A.; Holb, I.; Szabo, S. Aerial Laser Scanning Data as a Source of Terrain Modeling in a Fluvial Environment: Biasing Factors of Terrain Height Accuracy. *Sensors* **2020**, *20*, 2063. [CrossRef]
8. Liu, K.; Ding, H.; Tang, G.; Na, J.; Huang, X.; Xue, Z.; Yang, X.; Li, F. Detection of Catchment-Scale Gully-Affected Areas Using Unmanned Aerial Vehicle (UAV) on the Chinese Loess Plateau. *ISPRS Int. J. Geo-Inf.* **2016**, *5*, 238. [CrossRef]
9. Li, P.; Mu, X.; Holden, J.; Wu, Y.; Irvine, B.; Wang, F.; Gao, P.; Zhao, G.; Sun, W. Comparison of soil erosion models used to study the Chinese Loess Plateau. *Earth-Science Rev.* **2017**, *170*, 17–30. [CrossRef]
10. Toutin, T.; Gray, L. State-of-the-art of elevation extraction from satellite SAR data. *ISPRS J. Photogramm. Remote Sens.* **2000**, *55*, 13–33. [CrossRef]
11. Ludwig, R.; Schneider, P. Validation of digital elevation models from SRTM X-SAR for applications in hydrologic modeling. *ISPRS J. Photogramm. Remote Sens.* **2006**, *60*, 339–358. [CrossRef]
12. Xu, F.; Jin, Y.-Q. Imaging Simulation of Polarimetric SAR for a Comprehensive Terrain Scene Using the Mapping and Projection Algorithm. *IEEE Trans. Geosci. Remote Sens.* **2006**, *44*, 3219–3234. [CrossRef]

13. Sabry, R. Terrain and Surface Modeling Using Polarimetric SAR Data Features. *IEEE Trans. Geosci. Remote Sens.* **2016**, *54*, 1170–1184. [CrossRef]
14. Spinhirne, J. Micro pulse lidar. *IEEE Trans. Geosci. Remote Sens.* **1993**, *31*, 48–55. [CrossRef]
15. Petrie, G.; Toth, C.K. Introduction to Laser Ranging, Profiling, and Scanning. In *Topographic Laser Ranging And Scanning: Principles And Processing*, 2nd ed.; Shan, J., Toth, C.K., Eds.; CRC Press: Boca Raton, FL, USA, 2018; pp. 1–28.
16. Milenković, M.; Ressl, C.; Piermattei, L.; Mandlburger, G.; Pfeifer, N. Roughness Spectra Derived from Multi-Scale LiDAR Point Clouds of a Gravel Surface: A Comparison and Sensitivity Analysis. *ISPRS Int. J. Geo-Information* **2018**, *7*, 69. [CrossRef]
17. Benard, M. Automatic stereophotogrammetry: A method based on feature detection and dynamic programming. *Photogrammetria* **1984**, *39*, 169–181. [CrossRef]
18. Van Zyl, J.J. The Shuttle Radar Topography Mission (SRTM): A breakthrough in remote sensing of topography. *Acta Astronaut.* **2001**, *48*, 559–565. [CrossRef]
19. Rodríguez, E.; Morris, C.S.; Belz, J.E. A Global Assessment of the SRTM Performance. *Photogramm. Eng. Remote Sens.* **2006**, *72*, 249–260. [CrossRef]
20. St-Onge, B.; Vega, C.; Fournier, R.A.; Hu, Y. Mapping canopy height using a combination of digital stereo-photogrammetry and lidar. *Int. J. Remote Sens.* **2008**, *29*, 3343–3364. [CrossRef]
21. Latypov, D. Estimating relative lidar accuracy information from overlapping flight lines. *ISPRS J. Photogramm. Remote. Sens.* **2002**, *56*, 236–245. [CrossRef]
22. Hodgson, M.E.; Bresnahan, P. Accuracy of Airborne Lidar-Derived Elevation. *Photogramm. Eng. Remote Sens.* **2004**, *70*, 331–339. [CrossRef]
23. Salach, A.; Bakuła, K.; Pilarska, M.; Ostrowski, W.; Górski, K.; Kurczyński, Z. Accuracy Assessment of Point Clouds from LiDAR and Dense Image Matching Acquired Using the UAV Platform for DTM Creation. *ISPRS Int. J. Geo-Inf.* **2018**, *7*, 342. [CrossRef]
24. Baltensweiler, A.; Walthert, L.; Ginzler, C.; Sutter, F.; Purves, R.S.; Hanewinkel, M. Terrestrial laser scanning improves digital elevation models and topsoil pH modelling in regions with complex topography and dense vegetation. *Environ. Model. Softw.* **2017**, *95*, 13–21. [CrossRef]
25. Moon, D.; Chung, S.; Kwon, S.; Seo, J.; Shin, J. Comparison and utilization of point cloud generated from photogrammetry and laser scanning: 3D world model for smart heavy equipment planning. *Autom. Constr.* **2019**, *98*, 322–331. [CrossRef]
26. Liu, P.; Chen, A.Y.; Huang, Y.-N.; Han, J.-Y.; Lai, J.-S.; Kang, S.-C.; Wu, T.-H.; Wen, M.-C.; Tsai, M.-H. A review of rotorcraft Unmanned Aerial Vehicle (UAV) developments and applications in civil engineering. *Smart Struct. Syst.* **2014**, *13*, 1065–1094. [CrossRef]
27. Kanellakis, C.; Nikolakopoulos, G. Survey on Computer Vision for UAVs: Current Developments and Trends. *J. Intell. Robot. Syst.* **2017**, *87*, 141–168. [CrossRef]
28. Zhong, Y.; Zhong, Y.; Xu, Y.; Wang, S.; Jia, T.; Hu, X.; Zhao, J.; Wei, L.; Zhang, L. Mini-UAV-Borne Hyperspectral Remote Sensing: From Observation and Processing to Applications. *IEEE Geosci. Remote Sens. Mag.* **2018**, *6*, 46–62. [CrossRef]
29. Tsai, C.-H.; Lin, Y.-C. An accelerated image matching technique for UAV orthoimage registration. *ISPRS J. Photogramm. Remote Sens.* **2017**, *128*, 130–145. [CrossRef]
30. Ziquan, W.; Yifeng, H.; Mengya, L.; Kun, Y.; Yang, Y.; Yi, L.; Sim-Heng, O. A small uav based multi-temporal image registration for dynamic agricultural terrace monitoring. *Remote Sens.* **2017**, *9*, 904.
31. Wan, X.; Liu, J.; Yan, H.; Morgan, G.L. Illumination-invariant image matching for autonomous UAV localisation based on optical sensing. *ISPRS J. Photogramm. Remote Sens.* **2016**, *119*, 198–213. [CrossRef]
32. Jiang, S.; Jiang, W. Hierarchical motion consistency constraint for efficient geometrical verification in UAV stereo image matching. *ISPRS J. Photogramm. Remote Sens.* **2018**, *142*, 222–242. [CrossRef]
33. Zhao, J.; Zhang, X.; Gao, C.; Qiu, X.; Tian, Y.; Zhu, Y.; Cao, W. Rapid Mosaicking of Unmanned Aerial Vehicle (UAV) Images for Crop Growth Monitoring Using the SIFT Algorithm. *Remote Sens.* **2019**, *11*, 1226. [CrossRef]
34. Colomina, I.; Molina, P. Unmanned aerial systems for photogrammetry and remote sensing: A review. *ISPRS J. Photogramm. Remote Sens.* **2014**, *92*, 79–97. [CrossRef]
35. Yu, H.; Li, G.; Zhang, W.; Huang, Q.; Du, D.; Tian, Q.; Sebe, N. The Unmanned Aerial Vehicle Benchmark: Object Detection, Tracking and Baseline. *Int. J. Comput. Vis.* **2019**, *128*, 1141–1159. [CrossRef]

36. Guisado-Pintado, E.; Jackson, D.W.; Rogers, D. 3D mapping efficacy of a drone and terrestrial laser scanner over a temperate beach-dune zone. *Geomorphology* **2019**, *328*, 157–172. [CrossRef]
37. Meng, X.; Shang, N.; Zhang, X.; Li, C.; Zhao, K.; Qiu, X.; Weeks, E. Photogrammetric UAV Mapping of Terrain under Dense Coastal Vegetation: An Object-Oriented Classification Ensemble Algorithm for Classification and Terrain Correction. *Remote Sens.* **2017**, *9*, 1187. [CrossRef]
38. Kolarik, N.E.; Gaughan, A.E.; Stevens, F.R.; Pricope, N.G.; Woodward, K.; Cassidy, L.; Salerno, J.; Hartter, J. A multi-plot assessment of vegetation structure using a micro-unmanned aerial system (UAS) in a semi-arid savanna environment. *ISPRS J. Photogramm. Remote Sens.* **2020**, *164*, 84–96. [CrossRef]
39. Jensen, J.L.R.; Mathews, A.J. Assessment of Image-Based Point Cloud Products to Generate a Bare Earth Surface and Estimate Canopy Heights in a Woodland Ecosystem. *Remote Sens.* **2016**, *8*, 50. [CrossRef]
40. Gruszczyński, W.; Matwij, W.; Ćwiąkała, P. Comparison of low-altitude UAV photogrammetry with terrestrial laser scanning as data-source methods for terrain covered in low vegetation. *ISPRS J. Photogramm. Remote Sens.* **2017**, *126*, 168–179. [CrossRef]
41. Ressl, C.; Brockmann, H.; Mandlburger, G.; Pfeifer, N.; Camillo, R.; Herbert, B.; Gottfried, M.; Norbert, P. Dense Image Matching vs. Airborne Laser Scanning—Comparison of two methods for deriving terrain models. *Photogramm. Fernerkund. Geoinf.* **2016**, *2016*, 57–73. [CrossRef]
42. Manfreda, S.; McCabe, M.F.; Miller, P.E.; Lucas, R.; Pajuelo Madrigal, V.; Mallinis, G.; Ben-Dor, E.; Helman, D.; Estes, L.; Ciraolo, G.; et al. On the Use of Unmanned Aerial Systems for Environmental Monitoring. *Remote Sens.* **2018**, *10*, 641. [CrossRef]
43. Feng, Z.; Yang, Y.; Zhang, Y.; Zhang, P.; Li, Y. Grain-for-green policy and its impacts on grain supply in West China. *Land Use Policy* **2005**, *22*, 301–312. [CrossRef]
44. Delang, C.O.; Yuan, Z. China's Reforestation and Rural Development Programs. In *China's Grain for Green Program*; Springer International Publishing: Cham, Switzerland, 2016; pp. 22–23.
45. DJI Inspire 1. Available online: https://www.dji.com/inspire-1?site=brandsite&from=landing_page (accessed on 28 September 2020).
46. Zenmuse X5. Available online: https://www.dji.com/zenmuse-x5?site=brandsite&from=landing_page (accessed on 28 September 2020).
47. HiPer SR. Available online: https://www.topcon.co.jp/en/positioning/products/pdf/HiPerSR_E.pdf (accessed on 28 September 2020).
48. Pix4D. Available online: https://www.pix4d.com/product/pix4dmapper-photogrammetry-software (accessed on 28 September 2020).
49. Chollet, F. What is deep learning? In *Deep Learning with Python*; Manning Publications: Shelter Island, NY, USA, 2017; pp. 8–9.
50. Zhang, W.; Itoh, K.; Tanida, J.; Ichioka, Y. Parallel distributed processing model with local space-invariant interconnections and its optical architecture. *Appl. Opt.* **1990**, *29*, 4790–4797. [CrossRef]
51. Valueva, M.; Nagornov, N.; Lyakhov, P.; Valuev, G.; Chervyakov, N. Application of the residue number system to reduce hardware costs of the convolutional neural network implementation. *Math. Comput. Simul.* **2020**, *177*, 232–243. [CrossRef]
52. Ronneberger, O.; Fischer, P.; Brox, T. U-net: Convolutional networks for biomedical image segmentation. In *Proceedings of the 18th International Conference on Medical Image Computing and Computer-Assisted Intervention*; Springer: New York, NY, USA, 2015; pp. 234–241.
53. Erdem, F.; Avdan, U. Comparison of Different U-Net Models for Building Extraction from High-Resolution Aerial Imagery. *Int. J. Environ. Geoinf.* **2020**, *7*, 221–227. [CrossRef]
54. Arcgis Pro. Available online: https://www.esri.com/en-us/arcgis/products/arcgis-pro/overview (accessed on 28 September 2020).
55. Long, J.; Shelhamer, E.; Darrell, T. Fully convolutional networks for semantic segmentation. In Proceedings of the IEEE Conference on Computer Vision and Pattern Recognition, Boston, MA, USA, 8–10 June 2015; pp. 3431–3440.
56. Wang, R.; Peethambaran, J.; Chen, D. LiDAR Point Clouds to 3-D Urban Models: A Review. *IEEE J. Sel. Top. Appl. Earth Obs. Remote Sens.* **2018**, *11*, 606–627. [CrossRef]
57. Zlinszky, A.; Boergens, E.; Glira, P.; Pfeifer, N. Airborne Laser Scanning for calibration and validation of inshore satellite altimetry: A proof of concept. *Remote Sens. Environ.* **2017**, *197*, 35–42. [CrossRef]

58. Klápště, P.; Fogl, M.; Barták, V.; Gdulov, K.; Urban, R.; Moudrý, V. Sensitivity analysis of parameters and contrasting performance of ground filtering algorithms with UAV photogrammetry-based and LiDAR point clouds. *Int. J. Digit. Earth* **2020**, 1–23. [CrossRef]
59. Kwak, Y.-T. Segmentation of objects with multi layer perceptron by using informations of window. *J. Korean Data Inf. Sci. Soc.* **2007**, *18*, 1033–1043.
60. Han, H.; Yulin, D.; Qing, Z.; Jie, J.; Xuehu, W.; Li, Z.; Wei, T.; Jun, Y.; Ruofei, Z. Precision global dem generation based on adaptive surface filter and poisson terrain editing. *Acta Geod. Cartogr. Sin.* **2019**, *48*, 374.

Article

UAV Photogrammetry Accuracy Assessment for Corridor Mapping Based on the Number and Distribution of Ground Control Points

Ezequiel Ferrer-González *, Francisco Agüera-Vega, Fernando Carvajal-Ramírez and Patricio Martínez-Carricondo

Department of Engineering, Mediterranean Research Center of Economics and Sustainable Development, (CIMEDES), University of Almería (Agrifood Campus of International Excellence, ceiA3), La Cañada de San Urbano, s/n. 04120 Almería, Spain; faguera@ual.es (F.A.-V.); carvajal@ual.es (F.C.-R.); pmc824@ual.es (P.M.-C.)

* Correspondence: efg520@inlumine.ual.es

Received: 18 June 2020; Accepted: 28 July 2020; Published: 30 July 2020

Abstract: Unmanned aerial vehicle (UAV) photogrammetry has recently emerged as a popular solution to obtain certain products necessary in linear projects, such as orthoimages or digital surface models. This is mainly due to its ability to provide these topographic products in a fast and economical way. In order to guarantee a certain degree of accuracy, it is important to know how many ground control points (GCPs) are necessary and how to distribute them across the study site. The purpose of this work consists of determining the number of GCPs and how to distribute them in a way that yields higher accuracy for a corridor-shaped study area. To do so, several photogrammetric projects have been carried out in which the number of GCPs used in the bundle adjustment and their distribution vary. The different projects were assessed taking into account two different parameters: the root mean square error (RMSE) and the Multiscale Model to Model Cloud Comparison (M3C2). From the different configurations tested, the projects using 9 and 11 GCPs (4.3 and 5.2 GCPs km^{-1}, respectively) distributed alternatively on both sides of the road in an offset or zigzagging pattern, with a pair of GCPs at each end of the road, yielded optimal results regarding fieldwork cost, compared to projects using similar or more GCPs placed according to other distributions.

Keywords: unmanned aerial vehicle (UAV); structure-from-motion (SfM); ground control points (GCP); accuracy assessment; point clouds; corridor mapping

1. Introduction

The availability of high-resolution topographic products, such as orthoimages and digital surface models (DSM), is of increasing importance for many different fields of engineering that require a thorough understanding of topographies. These include, among many others, terrain morphology to perform reliable simulation of soil erosion, flooding phenomena, and assessment of the sediment budget [1–5], landslide mapping and multi-temporal study [6–8], road design [9], road condition surveys for road management [10], precision agriculture [11], or detection of archaeological rests [12]. Unmanned aerial vehicles (UAV) have emerged as a feasible alternative given their lower cost, high temporal and spatial resolution, and flexibility in image acquisition compared to conventional airborne and satellite sensors [13–15]. Most available software applications currently used to process UAV-acquired imagery are based on the structure from motion (SfM) approach. This approach, unlike traditional digital photogrammetry, resolves the collinearity equations without the need for any control point, providing a sparse point cloud in an arbitrary coordinate system and a full camera calibration [16,17]. This is possible due to image matching algorithms that automatically search for similar image objects, called keypoints, through the analysis of the correspondence, similarity, and consistency of the image features [18]. SfM is paired with multi-view stereopsis (MVS) techniques that apply an expanding procedure of the sparse set of matched keypoints in order to obtain a dense point cloud [19].

To georeference the 3D point cloud generated in the photogrammetric process, ground control points (GCPs) are usually employed. These control points can be either permanent ground features or reference targets scattered on the ground before the flight, which must be surveyed to obtain their precise coordinates and ensure that they are clearly identifiable on the raw images. A minimum of three GCPs is necessary to carry out the georeferencing process, although increasing the number of GCPs is highly recommended in order to improve the accuracy of the photogrammetric products. In [20], the influence of the number of GCPs on the DSM and orthoimage accuracies obtained with UAV photogrammetry were studied. A similar conclusion for both horizontal and vertical components was derived: optimal results were reached with 15 GCPs. Furthermore, in [21], different numbers and distributions of GCPs were studied to try to optimize the products obtained by UAV photogrammetry on a surface of 22 ha: it was concluded that more accurate results were reached combining GCPs located around the study area and a stratified distribution inside that area. In [22], the effect of the number and distribution of GCPs on the accuracy of the DSM and orthophoto of a surface of 150 ha was studied. These last two studies reached similar conclusions, proposing 0.5 to 1 GCP ha^{-1} as the optimum concentration of GCPs. In [23], the influence of different variations of GCPs arranged on an area of 2.73 ha on the accuracy of the products of UAV photogrammetry projects was studied. The optimum concentration was 1.8 GCPs ha^{-1} uniformly distributed across the whole surface.

The 3D coordinates of GCPs must be accurate; thus, a suitable survey method, such as GPS or total stations, must be used. Surveying these points is a time-consuming task that can be difficult to carry out depending on the terrain morphology. Alternative to the use of GCPs, differential GPS (DGPS) correction techniques, such as real-time kinematics (RTKs) and post-processing kinematics (PPKs), have been evaluated as methods to provide high-accuracy georeferencing [24–26]. In [25], it was concluded that a UAV RTK/PPK solution can provide highly accurate spatial data (planimetric RMSE = 0.044 m, altimetric RMSE = 0.082 m), compared to data acquired through the use of GCPs. In [27], the repeatability of DSM generation from several blocks acquired with a RTK-enabled drone was studied. Differential corrections were generated by a local master station or a network continuously operating a reference station network. Using identical test fields and flight plans, DSM generation was performed with three block control configurations: GCP only, camera stations only, and with camera stations and one GCP. The results showed that the average DSM accuracy was approximately 2.1 ground sample distance (GSD) with the first and third configurations and 3.7 GSD with the second one.

From the georeferenced dense point cloud, photogrammetric products such as orthoimages and DSM can be obtained. There are several factors that affect the accuracy of these UAV photogrammetry products: the number and distribution of GCPs, flight altitude, studied surface morphology, methodology for camera calibration, image overlap, and the incorporation of oblique images. Agüera et al. [28] carried out a study to determine how flight height, terrain morphology, and number of GCPs influence accuracy. They studied four terrain morphologies (from flat to very rugged) that were approximately square-shaped and had areas between 2 and 4.7 ha, four flight altitudes (50, 80, 100, and 120 m), and three different numbers of GCPs (3, 5, and 10). The results from this work indicated that horizontal accuracy is not influenced by terrain morphology or flight altitude. Furthermore, differences between terrain morphologies were observed only when 5 or 10 GCPs were used. Nevertheless, the number of GCPs influenced the horizontal accuracy: as the number of GCPs increased, the accuracy improved. While both flight altitude and the number of GCPs had a significant influence on vertical accuracy, terrain morphology did not. The lower RMSEs values were reached at a 50 m flight altitude using 10 GCPs (0.053 and 0.049 m for horizontal and vertical components, respectively).

A massive study with 3465 different combinations was conducted by Sanz-Ablanedo et al. [29] to determine the influence of the number and location of GCPs on a 1225 ha coal mining area that was approximately square-shaped. The results demonstrate that the extent to which the accuracy improves as the number of GCPs increases; the accuracy also depends on the location of the GCPs (the RMSE converges slowly to a value approximately double the GSD).

The impact of incorporating oblique images was analyzed by Nesbit et al. [30] to enhance 3D model accuracy in high-relief landscapes, and they concluded that combination datasets including oblique images are preferred over single camera angle datasets. In that research, the study site area was less than 5 ha and was approximately square-shaped. All these recently mentioned studies agree that the accuracy of the DSM and orthoimages obtained through UAV photogrammetry is highly dependent on the number of GCPs used and their distribution across the study area. Furthermore, the accuracy improves as more GCPs are used as long as they are well distributed, although there is a limit, beyond which the accuracy cannot be further improved by increasing the number of GCPs. However, since the fieldwork and associated cost increase with the use of more GCPs, it is necessary to balance the appropriate accuracy with a minimum fieldwork cost.

It is important to keep in mind that all the studies referenced so far were developed on square-shaped terrain or where one dimension is not much larger than another. Thus, it cannot be guaranteed that the conclusions drawn from them can be applied to site studies in which one dimension is much larger than another, as is the case with the so-called linear works in civil engineering (road, linear power distribution, pipelines, or channels).

There is not much research regarding the influence of the number and distribution of GCPs on the accuracy of UAV photogrammetric projects of this type of infrastructure. James and Robson [31] applied SfM and MVS technics to study the erosion of a coastal cliff measuring 50 × 3 m. They used eight GCPs with scale and georeference purposes but did not study the influence of the number or distribution of GCPs on DSM accuracy. Moreover, they did not use check points (CPs) to estimate the accuracy and determine it from the GCPs, which it is not a good methodology for estimating the accuracy or determining if it was affected by the number of GCPs. The title of the work of Zulkipli and Tahar [9] describes the use of UAV-based photogrammetric mapping for road design, but the study site has not one dimension longer than another. They derived a conclusion that could not be generalized to linear work projects. Jaud et al. [32] aimed to assess the extent of the bowl effects on the DSM generated above a linear beach (250 × 25 m) with a restricted distribution of GCPs. The bowl effect or doming deformation is a phenomenon that appears in corridor mapping and is caused by the accumulation of camera calibration errors [33]. To mitigate this effect, two strategies are suggested [33]: densifying GCP distribution or improving the estimation of the exterior orientation of each image. Therefore, using images with geolocation and angular deviations from the terrain reference system included in their EXIF (Exchangeable Image File Format) would limit the geometric distortions [32]. These data are usually included in the images' EXIF of UAV photogrammetry projects because UAVs have GPSs and the camera is mounted on a gimbal that has an inertial measurement unit (IMU) that records the angles to the terrain reference system. Tournadre et al. [34] studied the influence of camera calibration, the inclusion of oblique images, and the number of GCPs on the magnitude of the bowl effect on the UAV photogrammetry project of a corridor of 600 × 15 m. They concluded that those three factors have an effect on DSM accuracy. Regarding GCPs, the results prove that one GCP for each 100 m is optimal for reducing most of the CP reprojection errors to less than one centimeter, but they do not say anything about GCPs distribution. Skarlatos et al. [35] worked on a UAV photogrammetric project on a corridor of 2.2 km × 160 m. They used different numbers of GCPs for bundle adjustment. All combinations had two GCPs at each end of the corridor and from there, they added up to seven GCPs, and in one project, all GCPs measured. Therefore, the minimum distance between GCPs for all combinations was 200 m when all measured points were used as GCPs, which implies that, in this case, the accuracy was not calculated from CPs. Their main conclusion was that, as the number of GCPs increases, accuracy improves.

In view of these studies focused on linear works, it can be concluded that it is necessary to deepen the knowledge of the influence of the number and distribution of GCPs on DSM accuracy in UAV photogrammetry projects on corridors with lengths of several kilometers.

The aim of this study is to determine the number and distribution of GCPs that yields the best balance between accuracy and fieldwork in a linear photogrammetric project, in this case, a road. To achieve this objective, a UAV photogrammetry project was carried out on a road measuring 2.1 km × 190 m. The coordinates of 47 points were measured with a centimeter accuracy GPS. Of these, 18 were used as CPs, and the rest as GCPs. A total of 13 projects were developed, each with a different number and distribution of GCPs. DSM accuracy, derived from these projects, was estimated in two ways: first, by calculating the horizontal and vertical RMSE derived from the 34 CPs, and second, by comparing the 3D point cloud generated by each project with that generated by the project that considered the 47 measured points.

2. Materials and Methods

The methodology used to assess the accuracy of the different photogrammetric projects carried out is summarized in Figure 1.

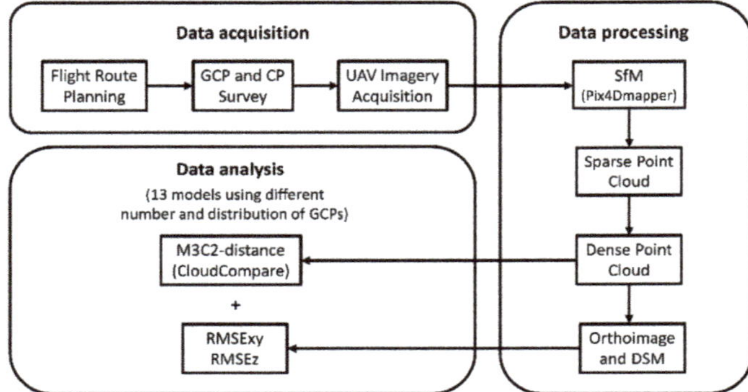

Figure 1. Workflow of image acquisition and processing to assess the influence of number and distribution of ground control points (GCPs) on the accuracy of linear photogrammetric projects.

2.1. Study Site

All coordinates of this study are given in meters and refer to UTM Zone 30N (European Terrestrial Reference System 1989, ETRS89) and the EGM08 geoid model. The study area is located in Roquetas de Mar (Almería), southeast Spain (Figure 2). The southwest and northeast coordinates are 533682, 4065630 and 532371, 4067232, respectively. The study site covers the A-1051-R3 branch road from the A-1051 highway, which measures 2.1 km × 190 m, with 95 m on each side of the road axis, and covers an area of approximately 40 ha. The main feature of the study site is that, in planimetry, one dimension is much larger than the other. The elevation in the studied area varies from 7 to 38 m above mean sea level. Figure 3 shows 3D cloud points corresponding to the northern end of the study site, showing a roundabout and several greenhouses, used for growing horticultural crops.

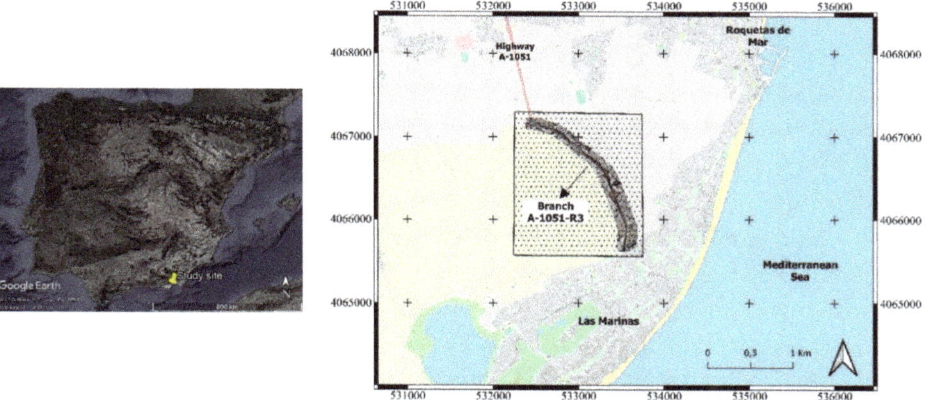

Figure 2. Location of the study site. Coordinates refer to UTM Zone 30N (ETRS89, EGM08 geoid model). The map of the Iberian Peninsula was extracted from Google Earth. The background map on the right side was extracted from OpenStreetMap, and the square inset is an orthophoto of the study site.

Figure 3. 3D cloud point corresponding to the northern end of the study site, showing a roundabout and several greenhouses.

2.2. Data Acquisition

The images used in this study were taken from a rotary wing UAV with four motors. The model employed was the DJI Phantom 4 Pro, which integrates a camera equipped with a one inch and 20 megapixel CMOS sensor, and a f2.8–/f11 wide-angle lens with an equivalent focal length of 24 mm [36].

The whole study area was covered by four different flights, each covering approximately 525 m of the road. Each flight was autonomous, meaning that the UAV followed a previously programmed and loaded path consisting of two passes parallel to the road axis. The flight speed was set at 3 m s^{-1} with images being taken every three seconds in order to achieve an 80% forward overlap. The side overlap was fixed at 60%. The flight altitude was constant at 65 m above ground level, implying that every photo covered a surface of 85.12×63.84 m^2. This resulted in an equivalent ground sample distance (GSD) of 1.75 cm $pixel^{-1}$. A total of 746 images were selected from the four flights to use in the photogrammetric projects.

Prior to the UAV flight, 47 targets were evenly arranged across the study area (Figure 4) to be used as GCPs or CPs. While GCPs help to georeference the project by establishing the coordinates of the model, CPs are used to assess its accuracy. Since the shape of the models adapts to the GCPs, independent CPs are used to assess accuracy by avoiding possible overestimations [25]. These points were surveyed using rover and base GPS receivers, model Trimble R6, working in post-processed kinematic (PPK) mode, and locating the base station within the range of 1 km away to all the measured points. The base station coordinates were previously determined from the geodesic pillar Las Lomas through a fast static process. The base station's 3D coordinates are 533315.482, 4066520.639, and 24.370 m, respectively. For the PPK measurements, according to the manufacturer's specifications, an error of 8 mm + 1 ppm RMS horizontal and 15 mm + 1 ppm RMS vertical can be expected [37].

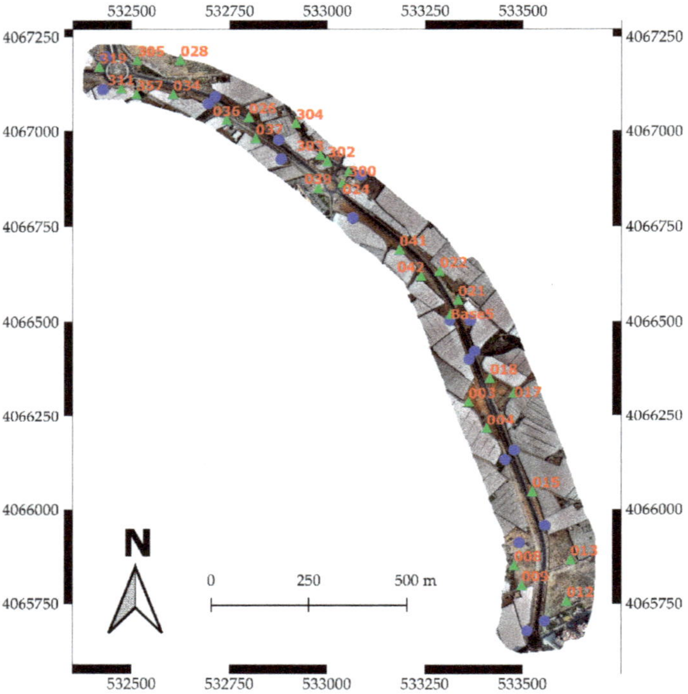

Figure 4. Location of the 47 targets used as GCPs (blue dots) and check points (CPs) (green triangles). The coordinates refer to UTM Zone 30N (ETRS89). The red numbers indicate the CP labels.

2.3. *Image Processing*

The photogrammetric process was carried out using an algorithm based on SfM-MVS techniques. The workflow consists of a three-step process. In the first step, the algorithm searches for common points, usually known as key points, among the uploaded images in order to align them through a matching process. When two different key points from two different images are identical, they become matching points. These matching points, as well as the approximate values of the image position automatically extracted from the EXIF metadata, allow the algorithm to carry out a bundle adjustment and calculate the 3D coordinates of each point. To improve the geolocalization accuracy, the process was supported by both the loading of the GCP coordinates, measured as indicated in the previous section, and the marking of these GCPs in the images. The results obtained from this first step are the exact camera position and orientation for every image, the internal camera calibration parameters, and the 3D coordinates of the sparse point cloud referred to the local coordinated system selected. In the second step, the sparse point cloud is densified through the MVS technique. This technique uses the calculated camera parameters to obtain a higher point cloud density and therefore a more detailed 3D model. The 3D textured mesh is also generated during this second step. In the third step, the DSM can be generated from the densified point cloud, and, in turn, the georeferenced orthomosaic is generated using the DSM. This entire process was carried out by the commercial UAV processing software Pix4Dmapper, version 4.5.6 [38].

2.4. Ground Control Points

Of the 47 targets placed on the ground of the study site, 18 were used as GCPs, while the remaining 29 were used as CPs.

To assess the influence of the number of GCPs and their distribution on the accuracy of the photogrammetric linear projects, 13 different configurations were designed. For this purpose, four GCP distributions, with projects using different numbers of GCPs within each type of distribution, were taken into account for the bundle adjustment. The number of CPs employed for all projects remained constant, independently of the number of GCPs used. The different distributions studied were:

Distribution 1: GCPs were located on both sides of the road and faced each other, as indicated by the red dots in Figure 5. Within this distribution, four projects using 4, 6, 10, and 18 GCPs were performed (Figure 5a–d). The pairs of GCPs were chosen so that they were similarly spaced from one another.

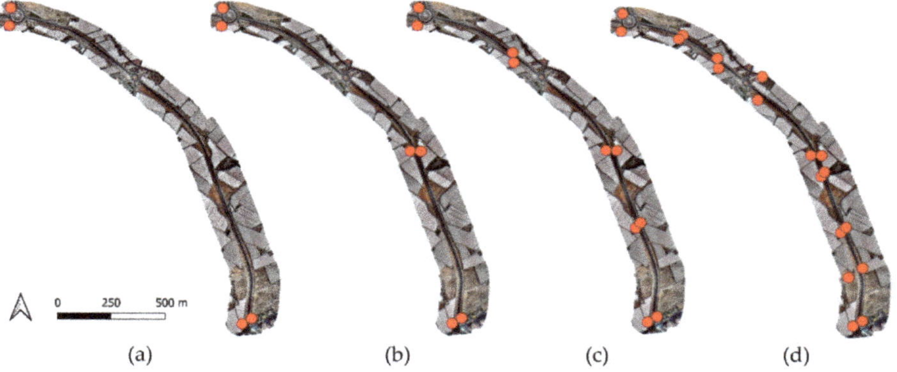

Figure 5. Location of the targets used as GCPs (red dots) for each project within Distribution 1. Four projects using (**a**) 4, (**b**) 6, (**c**) 10, and (**d**) 18 GCPs were carried out.

Distribution 2: GCPs were located on both sides of the road in an offset or zigzagging pattern, as indicated by the red dots in Figure 6. For this configuration, three different projects were carried out using three, five, and nine GCPs (Figure 6a–c).

Distribution 3: GCPs were located on only one side of the road, as indicated by the red dots in Figure 7. Under this distribution, three different projects employing three, five, and nine GCPs were developed (Figure 7a–c).

Distribution 4: as in Distribution 2, GCPs were located on both sides of the road in a zigzagging pattern, but there was an additional pair of GCPs located at each end of the corridor. Under this distribution, three projects were carried out using 7, 9, and 11 GCPs (Figure 8a–c). This configuration can be considered as a combination of Distributions 1 and 2.

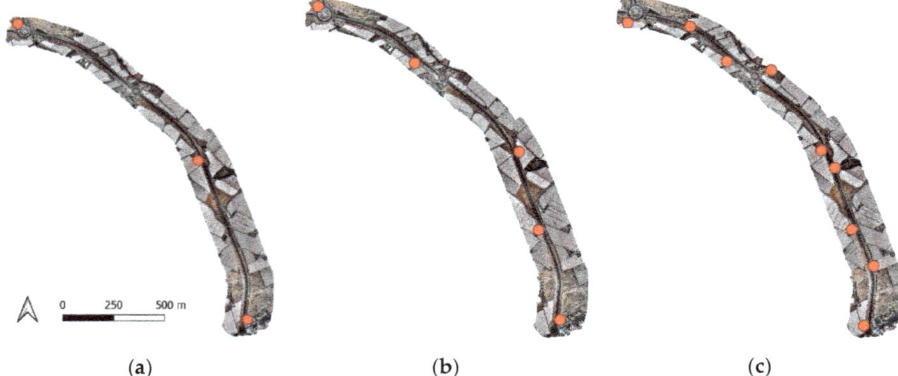

Figure 6. Location of the targets used as GCPs (red dots) for each project within Distribution 2. Three projects using (**a**) 3, (**b**) 5, and (**c**) 9 GCPs.

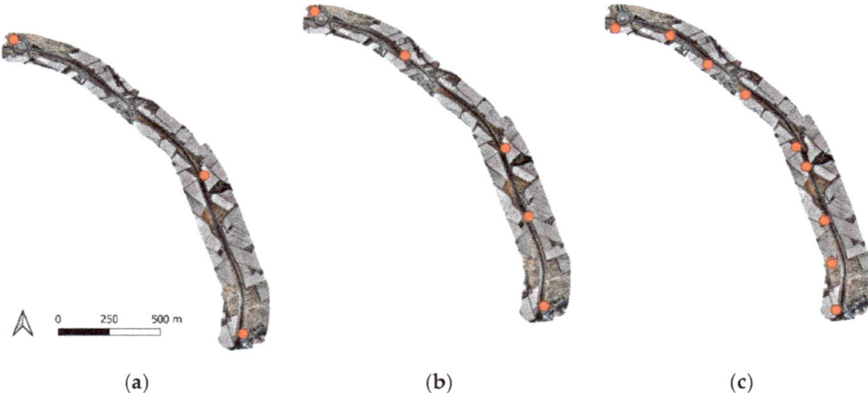

Figure 7. Location of the targets used as GCPs for each project within Distribution 3. Three projects using (**a**) 3, (**b**) 5, and (**c**) 9 GCPs.

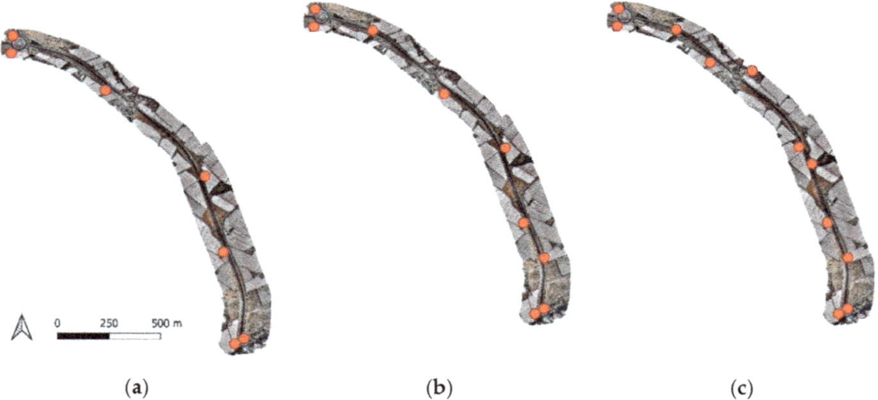

Figure 8. Location of the targets used as GCPs for each project within Distribution 4. Three projects using (**a**) 7, (**b**) 9, and (**c**) 11 GCPs (red dots) were carried out.

2.5. Accuracy Assessment

Two different methods are used to assess the accuracy of photogrammetric products. The first method is the mean square root of square differences between the reconstructed model and the surveyed coordinates of the 29 CPs, known as root mean square error (RMSE), since it can compensate errors with positive and negative values [39]. Differences between the reconstructed model and the surveyed coordinates are called errors, and the effect of each error on the RMSE of each error is proportional to the size of the squared error. Thus, RMSE is sensitive to estimated outlier values because large errors have a big effect on RMSE. Therefore, it is advisable to study the value of the error for each CP to check whether it is an outlier and, if so, to try to find the cause.

The RMSE values for the X component, Y component, XY component, and Z component are estimated as shown in Equations (1)–(4).

$$RMSE_X = \sqrt{\frac{\sum_{i=1}^{n} (X_{Oi} - X_{GPSi})^2}{n}} \quad (1)$$

$$RMSE_Y = \sqrt{\frac{\sum_{i=1}^{n} (Y_{Oi} - Y_{GPSi})^2}{n}} \quad (2)$$

$$RMSE_{XY} = \sqrt{\frac{\sum_{i=1}^{n} [(X_{Oi} - X_{GPSi})^2 + (Y_{Oi} - Y_{GPSi})]^2}{n}} \quad (3)$$

$$RMSE_Z = \sqrt{\frac{\sum_{i=1}^{n} (Z_{Oi} - Z_{GPSi})^2}{n}}. \quad (4)$$

where:

n is the number of CPs;

X_{Oi}, Y_{Oi}, and Z_{Oi} are the X, Y, and Z coordinates estimated by the model for the ith CP, respectively;

X_{GPSi}, Y_{GPSi}, and Z_{GPSi} are the X, Y, and Z coordinates measured by GPS for the ith CP, respectively.

The second method used in this study to assess the accuracy of the point clouds obtained through a UAV paired with SfM-MVS techniques consists of the freely available Multiscale Model to Model Cloud Comparison (M3C2) plugin offered by the CloudCompare software [40]. For the comparison of the different point clouds, a reference cloud was computed using the 47 surveyed points as GCPs, assuming that it is the most accurate and precise that can be achieved with the available data.

The M3C2 algorithm calculates the local differences between the reference cloud and the compared point cloud relative to local surface normal orientation. The algorithm does this through two different steps [41]:

1. A user-defined diameter of the spherical neighborhood in the reference point cloud is used to compute the local normal orientations. This user-defined diameter is known as the normal scale;
2. The normal orientation calculated is then used to project a cylinder, with a user-defined diameter called the projection scale, inside which equivalent points in the compared point cloud are searched for. From the points intercepted within the cylinder in each cloud, the average position along the normal direction is calculated for both clouds. The local distance between the two clouds is then given based on the distance between these averaged positions.

To ensure that the normal orientation is unaffected by point cloud roughness, the normal scale was set as 25 times the average local roughness calculated for the reference point cloud by CloudCompare [41]. To compare point clouds, the M3C2 distances between the reference point cloud and the point clouds generated for the 13 projects were calculated. Mean and standard deviation values calculated from the M3C2 distance were then used to assess the accuracy and the precision, respectively, of each point cloud. The mapping of the errors and their distribution curve allowed us to determine the influence of the GCP distribution on the M3C2 difference spatial distribution and to identify possible patterns for the spatial distribution of the errors.

3. Results

3.1. Accuracy Based on RMSE

For all four types of distribution considered in this study, the planimetric accuracy ($RMSE_{XY}$) decreases as the number of GCPs increases (Figure 9). For Distribution 1, in which GCPs were placed on both sides of the road facing each other, $RMSE_{XY}$ values ranged from 0.061 using 4 GCPs to 0.027 m using 18 GCPs. For Distribution 2, in which GCPs were located on both sides of the road and offset from one another, $RMSE_{XY}$ values ranged from 0.076 using three GCPs to 0.026 m using nine GCPs. For Distribution 3, in which GCPs were located on only one side of the road, $RMSE_{XY}$ ranged from 0.084 using three GCPs to 0.029 m using nine GCPs. Finally, for Distribution 4, in which GCPs were placed according to a combination of Distributions 1 and 2, the planimetric error ranged from 0.031 using 7 GCPs to 0.028 m using 11 GCPs. Our results show that an increase in the number of GCPs used in the bundle adjustment leads to an increase in planimetric accuracy, independent of the spatial distribution. Only five GCPs (approximately 2.4 GCPs km^{-1}) were necessary to achieve an $RMSE_{XY}$ less than two times the GSD of the project, and no fewer than nine GCPs (4.3 GCPs km^{-1}) were required to achieve $RMSE_{XY}$ values less than 0.03 m. The improvement in planimetric accuracy when 18 GCPs (8.6 GCPs km^{-1}) were used was less than 0.01 m compared to the accuracy obtained with five GCPs. The increase in accuracy became insignificant when more than nine GCPs were used.

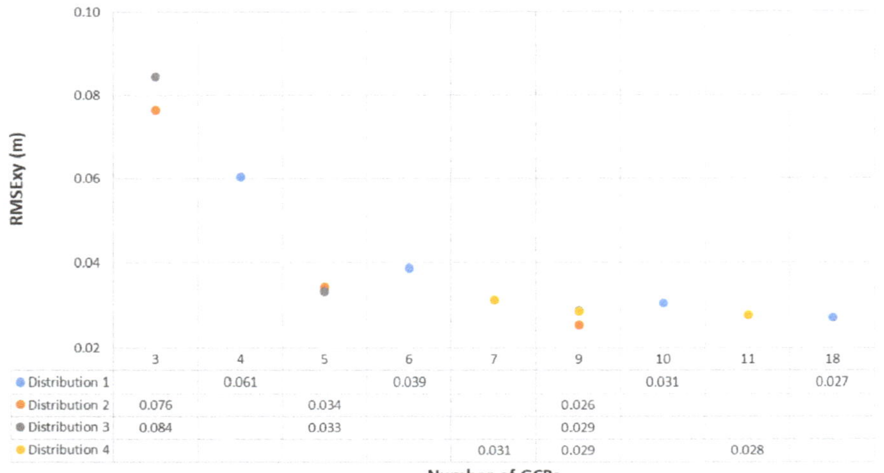

Figure 9. RMSExy values (in meters) obtained according to the number of GCPs used in the bundle adjustment. Each distribution is represented by a different color.

Our results also show that GCP distribution influences planimetric accuracy. Distributions 2 and 3 yielded very similar results. Furthermore, with an identical number of GCPs, Distributions 2 and 3 achieved better accuracy values than Distribution 1. Distribution 4 also improves upon the results obtained by Distribution 1, yielding better or similar accuracy values with fewer GCPs. The lowest RMSExy value was obtained with nine GCPs in a Distribution 2 configuration.

For all the photogrammetric projects performed, the values obtained for vertical accuracy ($RMSE_Z$) are higher than those obtained for $RMSE_{xy}$. As with planimetric accuracy, the $RMSE_z$ also decreases as the number of GCPs used for the bundle adjustment increases for the four types of distribution

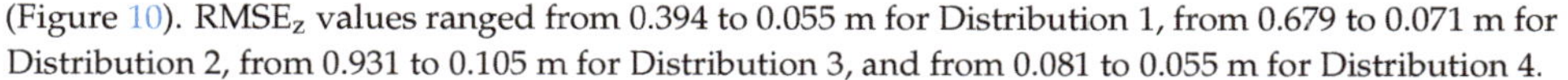

(Figure 10). $RMSE_z$ values ranged from 0.394 to 0.055 m for Distribution 1, from 0.679 to 0.071 m for Distribution 2, from 0.931 to 0.105 m for Distribution 3, and from 0.081 to 0.055 m for Distribution 4.

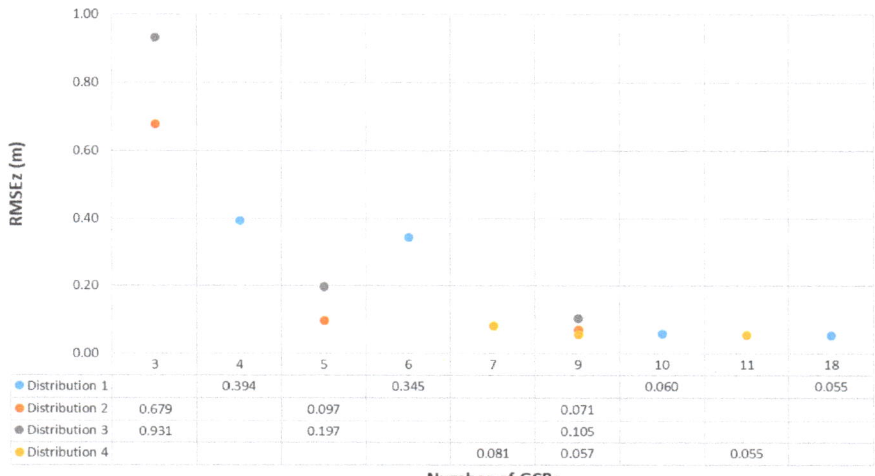

Figure 10. RMSEz values in meters obtained according to the number of GCPs used in the bundle adjustment. Each distribution is represented by a different color.

Independent of the distribution, at least seven GCPs (3.3 GCPs km^{-1}) are necessary to achieve $RMSE_z$ values significantly less than 0.1 m, and nine or more are required to obtain values less than 0.06 m. Three projects were close to the recommended $RMSE_z$ value of three times the GSD of the project (0.053 m) [35]. For all distributions, the accuracy improves along with the number of GCPs included. For Distribution 1, the difference between the use of 10 and 18 GCPs is less significant considering the large increase in the number of GCPs. The accuracies obtained for projects in Distribution 1 using 10 and 18 GCPs and in Distribution 4 using 9 and 11 GCPs are very similar, ranging from 0.055 to 0.06 m.

Regarding vertical accuracy, the influence of GCP distribution is more evident than for planimetric accuracy. While Distributions 2 and 3 yielded similar results in terms of horizontal accuracy, when it comes to height accuracy, Distribution 2 achieves lower $RMSE_z$ values, with a difference of up to 0.03 m when nine GCPs are used.

Distribution 1 significantly improves the vertical accuracy obtained for Distributions 2 and 3. Using just 11 GCPs, Distribution 4 achieves an $RMSE_z$ value similar to the accuracy yielded with 18 GCPs in Distribution 1 (0.055 m). Furthermore, with the use of nine GCPs in Distribution 4, a very close value (0.057 m) to Distribution 1 with 18 GCPs (0.055m) was obtained.

Taking into account both horizontal and vertical accuracy, the configurations with the lowest total RMSEs are Distribution 1 with 18 GCPs (8.6 GCPs km^{-1}, $RMSE_{XY}$ = 0.027 m, $RMSE_Z$ = 0.055 m), and Distribution 4 with 11 GCPs (5.2 GCPs km^{-1}, $RMSE_{xy}$ = 0.028 m, $RMSE_z$ = 0.055 m). The configuration with nine GCPs (4.3 GCPs km^{-1}) in Distribution 4 yielded a total RMSE value of 0.064 m. For the remaining configurations, the total RMSE is 0.07 m in the case of 10 GCPs (4.8 GCP km^{-1}) in Distribution 1, while the other distributions resulted in better values.

3.2. Accuracy Based on M3C2-Distances

For every distribution, point clouds become more accurate and more precise as the number of GCPs increases. In most cases, with nine GCPs or more, mean difference values are around 0.02 m or less and standard deviation values are less than 0.07 m. Distribution 3 yielded the worst results with nine GCPs: 0.053 m (Figure 11).

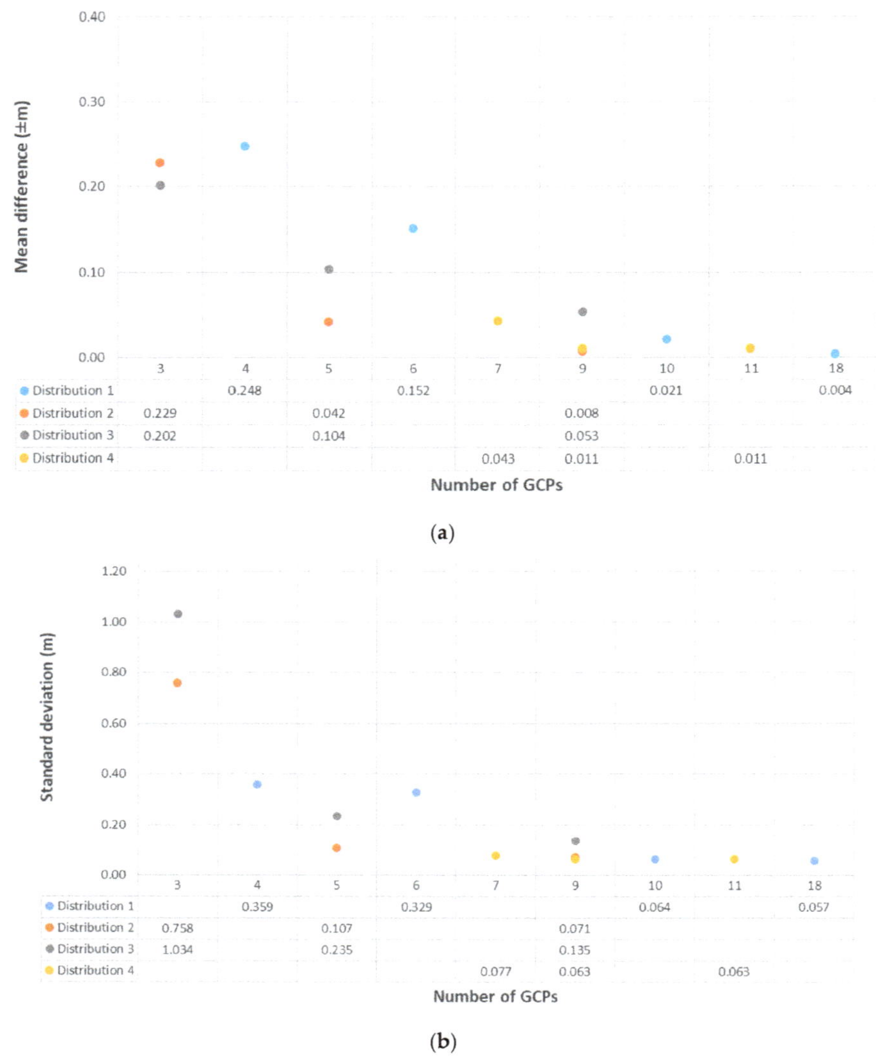

Figure 11. Multiscale Model to Model Cloud Comparison (M3C2) distance measurements between the reference cloud and the clouds obtained from the different photogrammetric projects carried out. (**a**) Mean difference (accuracy); (**b**) standard deviation (precision).

Further, for the majority of the projects carried out, neither accurate nor precise point clouds were achieved with fewer than seven GCPs, regardless of the type of distribution employed. Distribution 3 presents higher standard deviations and mean values, which means lower precision and accuracy than the other distributions. For projects with nine GCPs, the one placed according to Distribution 2, yielded better accuracy but lower precision than those with 9 and 11 GCPs in Distribution 4. Distribution 1 with 18 GCPs resulted in the lowest standard deviation and smallest mean difference than any other configuration, although similar values can be achieved with 9 or 11 GCPs placed according to Distribution 4.

In order to better understand how the distribution of GCPs impacts the accuracy of the projects, the spatial distributions of the M3C2-calculated distance between the reference cloud and the clouds from the photogrammetric projects were analyzed. For this purpose, only projects that used nine or more GCPs were considered (Figures 12 and 13).

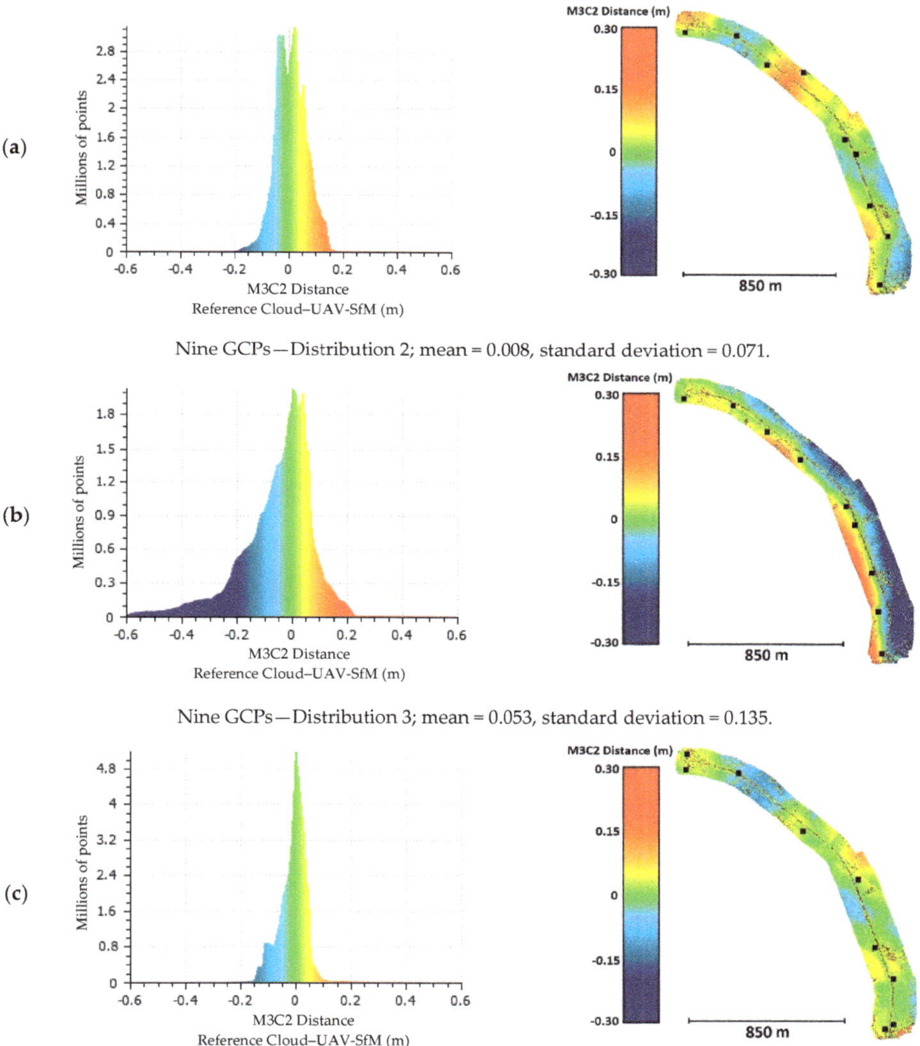

Nine GCPs—Distribution 4; mean = −0.011, standard deviation = 0.063.

Figure 12. Distribution of errors for the M3C2-calculated distance between the reference cloud and the clouds obtained from the photogrammetric projects that used nine GCPs. The black squares represent the locations of the GCPs. (**a**) Distribution 2, (**b**) Distribution 3, and (**c**) Distribution 4.

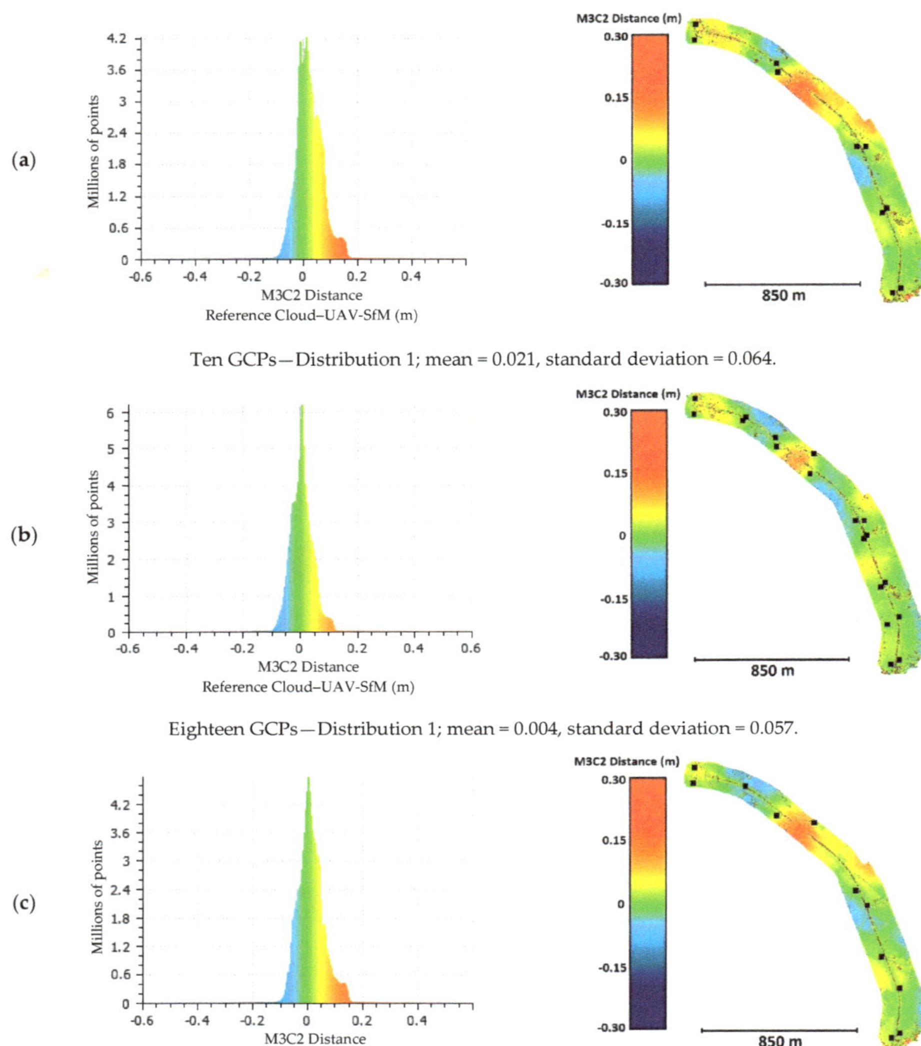

Figure 13. Distribution of errors for the M3C2-calculated distance between the reference cloud and the clouds obtained from the photogrammetric projects that used more than nine GCPs. The black squares represent the locations of the GCPs. (**a**) Project with 10 GCPs in Distribution 1, (**b**) project with 18 GCPs in Distribution 1, and (**c**) project with 11 GCPs in Distribution 4.

Although the mean difference was lower for Distribution 2, the precision improved considerably when a pair of GCPs was placed at each end of the corridor (Distribution 4, Figure 12). Distribution 2 (mean = −0.008 m, standard deviation = 0.071 m) and Distribution 4 (mean = −0.011 m, standard deviation = 0.063 m) achieved much better results for both accuracy and precision than Distribution 3, which yielded a higher mean (0.053 m) and standard deviation (0.135 m).

Figure 13 shows the spatial distributions of the M3C2-calculated distance between the reference cloud and the clouds generated from the photogrammetric projects in which more than nine GCPs were used to carry out the bundle adjustment. Although the project with 18 GCPs yielded the lowest values for both accuracy (0.004 m) and precision (0.057 m), close values can be obtained with just 9 (accuracy = −0.0121 m, precision = 0.0643 m, Figure 12) or 11 GCPs (accuracy = 0.011 m, precision = 0.063 m) placed according to Distribution 4. This distribution has better values of accuracy and precision with both 9 (accuracy = −0.011 m, precision = 0.063 m) and 11 (accuracy = 0.011 m, precision = 0.063 m) GCPs than those of Distribution 1 with 10 GCPs (accuracy = 0.021 m, precision = 0.064 m). This could be caused by the better distribution of GCPs in Distribution 4 than in Distribution 1, where the pairs of GCPs are very close.

4. Discussion

In the literature, there is little research that focuses on studying the effect of the number and distribution of GCPs on the accuracy of UAV photogrammetric projects on corridors. Most of the studies are focused on surfaces where one dimension is not much larger than the other. Skarlatos et al. [35], on a corridor measuring 2.2 km × 160 m, used a GCP distribution similar to our Distribution 4, with two points at each end of the corridor and others (one, two, and three points) along the corridor. Therefore, their project with seven GCPs is equivalent to our Distribution 4 with seven GCPs, which represents 3.3 GCPs km^{-1}. In this situation, Skarlatos et al. reported an $RMSE_{xy}$ = 0.130 m and an $RMSE_z$ = 0.170 m, while our results were $RMSE_{xy}$ = 0.031 m and $RMSE_Z$ = 0.081 m. The main difference between Skarlatos et al.'s study and our own is the GSD: 0.040 m for their images and 0.0175 m for our images. If we consider the GSD, Skarlatos et al. achieved horizontal and vertical accuracies of approximately three and four times the GSD. In our work, the planimetric accuracy was better (less than two times the GSD of the project), but the vertical was similar (in the range of four times the GSD). These accuracies can be improved by adding more GCPs, independently of their distribution. When Skarlatos et al. used all 16 measured points as GCPs, they report an $RMSE_{xy}$ = 0.070 m and an $RMSE_z$ = 0.130 m, which are higher than those found in our work for Distribution 4 with 11 GCPs ($RMSE_{xy}$ = 0.028 m and a $RMSE_z$ = 0.055 m). If we again consider using the GSD to compare the results, the values are similar: 1.75 and 1.6 GSD for horizontal accuracy and 3.25 and 3.14 GSD for vertical accuracy. In any case, it should be noted that the accuracy values of Skarlatos et al. when 16 GCPs were considered were calculated from the GCPs themselves.

Tahar [22] evaluated different numbers of GCPs in a UAV photogrammetric block. Although Tahar did not indicate the linear dimension of the study area, the text refers a road to this. Several combinations of numbers (from four to nine) and distributions of GCPs were tested to study their influence on the achieved accuracy. The best RMSEs calculated in that study were reached using nine GCPs: 0.48 m for the horizontal component, and 0.78 m for the vertical component, which are larger than any value found in any of our projects. In that work, the GSD is not reported, so it is not possible to make a comparison using this value.

Zulkipli and Tahar [9] focused on using UAV as a tool to capture data of the ground for road design. Considering that the study site was not a corridor and the results are not comparable to ours, they obtained RMSE values of 0.155, 0.228, and 0.479 m for X, Y, and Z, respectively, with six GCPs and a fly height of 148 m (the GSD value is not reported). These values mean that, although the height accuracy is close to the one presented in the present study for six GCPs and Distribution 1, the planimetric accuracy is much higher for the same number of GCPs since an increase in the number of GCPs is necessary to improve the accuracy of photogrammetric projects, as the authors concluded. Nevertheless, since the fly height of the present study is 65 m, while in Zulkipli and Tahar it was 148 m, their GSD was likely larger than ours, and it is important to note that, as the flight height (and, in turn, the GSD) increases, the accuracy deteriorates [27]. One of our main findings is that the project using more GCPs was not the most accurate. It is very important to consider not only the number of GCPs but also their distribution across the study area.

Tournadre et al. [34] aimed to present a method to assure precise accuracy in UAV photogrammetric projects of linear works and to minimize the number of GCPs required. Their study was developed on a corridor of 600 × 15 m. The influence of camera calibration, the inclusion of oblique images, and the number of GCPs on the magnitude of the bowl effect in the UAV photogrammetric project was studied. They concluded that one GCP for each 100 m (six GCPs in the studied corridor) is the optimal distribution to reduce most of the CPs reprojection errors to less than one centimeter, they do not mention GCP distribution. Our accuracies with six GCPs are better than the accuracy given by Tournadre et al., but they do not report the project GSD.

Several studies have already proven that the distribution of the GCPs affects the accuracy of the projects, and a good geometrical distribution of GCPs will lead to better accuracies [21,29]. In terms of the distribution of GCPs, since the results of the projects using GCPs on only one side of the road were the worst in both the RMSE and M3C2 distance values, we found that, to improve the accuracy in corridor-shaped projects, it is necessary to place GCPs on both sides of the road. The distribution in which the GCPs are placed alternately on each side of the road and separated by an offset distance presented results similar to those of the distribution in which GCPs are set out in pairs along the road. However, the best results were yielded by a combination of both, in which the GCPs were set out in an offset pattern but with the addition of a pair of GCPs at each end of the road, yielding better results with just 11 GCPs (5.2 GCPs km^{-1}) than another distribution using 18 GCPs (8.6 GCPs km^{-1}). This configuration yielded values less than two and three times the GSD of the project for both horizontal and vertical accuracy.

In view of Figures 12 and 13, it can be deduced that there are no significant errors in the clouds of the projects represented and that these are not concentrated in certain areas. An exception is Distribution 3 with nine GCPs (Figure 12), where values of approximately 0.3 m are reached in the southeast area. This is related to the RMSE values found for the CPs located in the same area (points 12, 13, and 17, Figure 4). In the other representations of the error distribution in Figures 12 and 13, the values of the errors observed in an area are in agreement with the RMSE values calculated for the CPs located in that same area.

The results derived from both methodologies used to assess the accuracy are coherent. Similar results were obtained through these two different approaches, thus strengthening the conclusions of the work carried out. Furthermore, when the distribution of errors for the M3C2-calculated distance between the reference cloud and the clouds was obtained from the different photogrammetric projects, no bowl effect was observed, even when the number of GCPs was small.

5. Conclusions

This study was performed to assess how the number of GCPs and their distribution impact the accuracy of UAV photogrammetry projects in a corridor-shaped study site. For that purpose, several projects with different configurations were carried out on a 2.1 km road, where 47 points were surveyed to be used either as GCPs or CPs. To assess accuracy, RMSE values from the georeferencing process and the M3C2 distance from the point clouds comparison were used.

For all the distributions studied, both horizontal and vertical accuracy improved as the number of GCPs used in the bundle adjustment increased, and planimetric accuracy was always better than vertical accuracy. Independent of the chosen distribution, no fewer than seven GCPs (3.3 GCPs km^{-1}) must be used to reach values of $RMSE_{xy} \leq 0.031$ m and $RMSE_z \leq 0.081$ m. The best results were achieved for those distributions where the GCPs were placed on both sides of the road. Placing GCPs alternatively on each side of the road and separating them by an offset distance, with a pair of GCPs placed at each end of the corridor, proved to yield the best results.

Considering the results, configurations with 9 or 11 GCPs (4.3 and 5.2 GCPs km^{-1}, respectively) placed on both sides of the road in an offset pattern, with a pair of GCPs at each end, yielded the best results in terms of balancing the accuracy and fieldwork, with RMSE mean values of 0.029 and 0.028 m for horizontal and 0.057 and 0.055 m for vertical accuracy, respectively. Similar results in terms of

RMSE values (0.027 for horizontal and 0.055 m for vertical) and slightly better results in terms of M3C2 distance (mean difference and standard deviation) were achieved with 18 GCPs (8.6 GCPs km^{-1}) set out in pairs along the corridor. Since every GCP must be surveyed using high-accuracy technology, the use of 9 or 11 GCPs, with the offset distribution mentioned previously, is recommended in study areas similar to that assessed in this study, since it can significantly reduce both the fieldwork and survey duration without a loss in accuracy, compared to the use of a higher number of GCPs placed according to other distributions.

To determine if the conclusions derived from this study are generally applicable, it would be necessary to carry out related studies in corridors with different terrain morphologies.

Author Contributions: This research was conceptualized and designed by F.A.-V., F.C.-R., P.M.-C. and E.F.-G.; field work and data acquisition was coordinated by F.A.-V., F.C.-R. and P.M.-C.; data processing was carried out by E.F.-G.; data analysis was performed by E.F.-G., F.A.-V., F.C.-R. and P.M.-C.; original manuscript preparation was conducted by E.F.-G.; manuscript was reviewed and edited by F.A.-V., F.C.-R. and P.M.-C.; all authors provided helpful discussion. All authors have read and agreed to the published version of the manuscript.

Funding: This research received no external funding.

Conflicts of Interest: The authors declare no conflict of interest.

References

1. Mancini, F.; Dubbini, M.; Gattelli, M.; Stecchi, F.; Fabbri, S.; Gabbianelli, G. Using Unmanned Aerial Vehicles (UAV) for High-Resolution Reconstruction of Topography: The Structure from Motion Approach on Coastal Environments. *Remote Sens.* **2013**, *5*, 6880–6898. [CrossRef]
2. Mourato, S.; Fernandez, P.; Pereira, L.; Moreira, M. Improving a DSM Obtained by Unmanned Aerial Vehicles for Flood Modelling. *IOP Conf. Ser. Earth Environ. Sci.* **2017**, *95*. [CrossRef]
3. Rodrigo-Comino, J.; Seeger, M.; Iserloh, T.; Senciales González, J.M.; Ruiz-Sinoga, J.D.; Ries, J.B. Rainfall-simulated quantification of initial soil erosion processes in sloping and poorly maintained terraced vineyards—Key issues for sustainable management systems. *Sci. Total Environ.* **2019**, *660*, 1047–1057. [CrossRef] [PubMed]
4. Campbell, S.; Simmons, R.; Rickson, J.; Waine, T.; Simms, D. Using Near-Surface Photogrammetry Assessment of Surface Roughness (NSPAS) to assess the effectiveness of erosion control treatments applied to slope forming materials from a mine site in West Africa. *Geomorphology* **2018**, *322*, 188–195. [CrossRef]
5. Gong, C.; Lei, S.; Bian, Z.; Liu, Y.; Zhang, Z.; Cheng, W. Analysis of the Development of an Erosion Gully in an Open-Pit Coal Mine Dump During a Winter Freeze-Thaw Cycle by Using Low-Cost UAVs. *Remote Sens.* **2019**, *11*, 1356. [CrossRef]
6. Rossi, G.; Tanteri, L.; Tofani, V.; Vannocci, P.; Moretti, S.; Casagli, N. Multitemporal UAV surveys for landslide mapping and characterization. *Landslides* **2018**, *15*, 1045–1052. [CrossRef]
7. Menegoni, N.; Giordan, D.; Perotti, C. Reliability and uncertainties of the analysis of an unstable rock slope performed on RPAS digital outcrop models: The case of the gallivaggio landslide (Western Alps, Italy). *Remote Sens.* **2020**, *12*, 1635. [CrossRef]
8. Fernández, T.; Pérez, J.; Cardenal, J.; Gómez, J.; Colomo, C.; Delgado, J. Analysis of Landslide Evolution Affecting Olive Groves Using UAV and Photogrammetric Techniques. *Remote Sens.* **2016**, *8*, 837. [CrossRef]
9. Zulkipli, M.A.; Tahar, K.N. Multirotor UAV-Based Photogrammetric Mapping for Road Design. *Int. J. Opt.* **2018**, *2018*. [CrossRef]
10. Tan, Y.; Li, Y. UAV photogrammetry-based 3D road distress detection. *ISPRS Int. J. Geo-Inf.* **2019**, *8*, 409. [CrossRef]
11. Tsouros, D.C.; Bibi, S.; Sarigiannidis, P.G. A review on UAV-based applications for precision agriculture. *Information* **2019**, *10*, 349. [CrossRef]
12. Agudo, P.; Pajas, J.; Pérez-Cabello, F.; Redón, J.; Lebrón, B. The Potential of Drones and Sensors to Enhance Detection of Archaeological Cropmarks: A Comparative Study Between Multi-Spectral and Thermal Imagery. *Drones* **2018**, *2*, 29. [CrossRef]

13. Westoby, M.J.; Brasington, J.; Glasser, N.F.; Hambrey, M.J.; Reynolds, J.M. 'Structure-from-Motion' photogrammetry: A low-cost, effective tool for geoscience applications. *Geomorphology* **2012**, *179*, 300–314. [CrossRef]
14. Harwin, S.; Lucieer, A. Assessing the Accuracy of Georeferenced Point Clouds Produced via Multi-View Stereopsis from Unmanned Aerial Vehicle (UAV) Imagery. *Remote Sens.* **2012**, *4*, 1573–1599. [CrossRef]
15. Hugenholtz, C.H.; Whitehead, K.; Brown, O.W.; Barchyn, T.E.; Moorman, B.J.; LeClair, A.; Riddell, K.; Hamilton, T. Geomorphological mapping with a small unmanned aircraft system (sUAS): Feature detection and accuracy assessment of a photogrammetrically-derived digital terrain model. *Geomorphology* **2013**, *194*, 16–24. [CrossRef]
16. Fonstad, M.A.; Dietrich, J.T.; Courville, B.C.; Jensen, J.L.; Carbonneau, P.E. Topographic structure from motion: A new development in photogrammetric measurement. *Earth Surf. Process. Landf.* **2013**, *38*, 421–430. [CrossRef]
17. Snavely, N.; Seitz, S.M.; Szeliski, R. Modeling the World from Internet Photo Collections. *Int. J. Comput. Vis.* **2008**, *80*, 189–210. [CrossRef]
18. Ao, T.; Liu, X.; Ren, Y.; Luo, R.; Xi, J. An Approach to Scene Matching Algorithm for UAV Autonomous Navigation. In Proceedings of the 2018 Chinese Control and Decision Conference (CCDC), Shenyang, China, 9–11 June 2018; pp. 996–1001.
19. Furukawa, Y.; Ponce, J. Accurate, Dense, and Robust Multi-View Stereopsis. *IEEE Trans. Softw. Eng.* **2010**, *32*, 1362–1376. [CrossRef]
20. Agüera-Vega, F.; Carvajal-Ramírez, F.; Martínez-Carricondo, P. Assessment of photogrammetric mapping accuracy based on variation ground control points number using unmanned aerial vehicle. *Meas. J. Int. Meas. Confed.* **2017**, *98*, 221–227. [CrossRef]
21. Martínez-Carricondo, P.; Agüera-Vega, F.; Carvajal-Ramírez, F.; Mesas-Carrascosa, F.J.; García-Ferrer, A.; Pérez-Porras, F.J. Assessment of UAV-photogrammetric mapping accuracy based on variation of ground control points. *Int. J. Appl. Earth Obs. Geoinf.* **2018**, *72*, 1–10. [CrossRef]
22. Tahar, K.N. An Evaluation on Different Number of Ground Control Points in Unmanned Aerial Vehicle Photogrammetric Block. *Int. Arch. Photogramm. Remote Sens. Spat. Inf. Sci. ISPRS Arch.* **2013**, *XL-2/W2*, 27–29. [CrossRef]
23. Reshetyuk, Y.; Mårtensson, S.G. Generation of Highly Accurate Digital Elevation Models with Unmanned Aerial Vehicles. *Photogramm. Rec.* **2016**, *31*, 143–165. [CrossRef]
24. Li, T.; Zhang, H.; Gao, Z.; Niu, X.; El-sheimy, N. Tight Fusion of a Monocular Camera, MEMS-IMU, and Single-Frequency Multi-GNSS RTK for Precise Navigation in GNSS-Challenged Environments. *Remote Sens.* **2019**, *11*, 610. [CrossRef]
25. Tomaštík, J.; Mokroš, M.; Surový, P.; Grznárová, A.; Merganič, J. UAV RTK/PPK method-An optimal solution for mapping inaccessible forested areas? *Remote Sens.* **2019**, *11*, 721. [CrossRef]
26. Kim, J.; Song, J.; No, H.; Han, D.; Kim, D.; Park, B.; Kee, C. Accuracy improvement of DGPS for low-cost single-frequency receiver using modified Flächen Korrektur parameter correction. *ISPRS Int. J. Geo-Inf.* **2017**, *6*, 222. [CrossRef]
27. Forlani, G.; Dall'Asta, E.; Diotri, F.; di Cella, U.M.; Roncella, R.; Santise, M. Quality assessment of DSMs produced from UAV flights georeferenced with on-board RTK positioning. *Remote Sens.* **2018**, *10*, 311. [CrossRef]
28. Agüera-Vega, F.; Carvajal-Ramírez, F.; Martínez-Carricondo, P. Accuracy of digital surface models and orthophotos derived from unmanned aerial vehicle photogrammetry. *J. Surv. Eng.* **2017**, *143*, 1–10. [CrossRef]
29. Sanz-Ablanedo, E.; Chandler, J.H.; Rodríguez-Pérez, J.R.; Ordóñez, C. Accuracy of Unmanned Aerial Vehicle (UAV) and SfM photogrammetry survey as a function of the number and location of ground control points used. *Remote Sens.* **2018**, *10*, 606. [CrossRef]
30. Nesbit, P.R.; Hugenholtz, C.H. Enhancing UAV-SfM 3D model accuracy in high-relief landscapes by incorporating oblique images. *Remote Sens.* **2019**, *11*, 239. [CrossRef]
31. James, M.R.; Robson, S. Straightforward reconstruction of 3D surfaces and topography with a camera: Accuracy and geoscience application. *J. Geophys. Res. Earth Surf.* **2012**, *117*, 1–18. [CrossRef]
32. Jaud, M.; Passot, S.; Le Bivic, R.; Delacourt, C.; Grandjean, P.; Le Dantec, N. Assessing the accuracy of high resolution digital surface models computed by PhotoScan® and MicMac® in sub-optimal survey conditions. *Remote Sens.* **2016**, *8*, 465. [CrossRef]

33. James, M.R.; Robson, S. Mitigating systematic error in topographic models derived from UAV and ground-based image networks. *Earth Surf. Process. Landf.* **2014**, *39*, 1413–1420. [CrossRef]
34. Tournadre, V.; Pierrot-Deseilligny, M.; Faure, P.H. UAV linear photogrammetry. *Int. Arch. Photogramm. Remote Sens. Spat. Inf. Sci. ISPRS Arch.* **2015**, *40*, 327–333. [CrossRef]
35. Skarlatos, D.; Procopiou, E.; Stavrou, G.; Gregoriou, M. Accuracy assessment of minimum control points for UAV photography and georeferencing. *First Int. Conf. Remote Sens. Geoinf. Environ.* **2013**, *8795*, 879514. [CrossRef]
36. DJI Phantom 4 Pro. Available online: https://dl.djicdn.com/downloads/phantom_4_pro/20170719/Phantom_4_Pro_Pro_Plus_User_Manual_ES.pdf (accessed on 18 July 2020).
37. Trimble Trimble R6. Available online: http://www.orient-mediterranee.com/IMG/pdf/R8-R6-R4-5800M3_UserGuide.pdf (accessed on 18 July 2020).
38. Pix4Dmapper Pix4D. Available online: https://www.pix4d.com/product/pix4dmapper-photogrammetry-software (accessed on 18 July 2020).
39. Congalton, R.G. Thematic and Positional Accuracy Assessment of Digital Remotely Sensed Data. In Proceedings of the Seventh Annual Forest Inventory and Analysis Symposium, Portland, ME, USA, 3–6 October 2005.
40. CloudCompare v2.10.2. Available online: https://www.danielgm.net/cc/ (accessed on 18 July 2020).
41. Lague, D.; Brodu, N.; Leroux, J. Accurate 3D comparison of complex topography with terrestrial laser scanner: Application to the Rangitikei canyon (N-Z). *ISPRS J. Photogramm. Remote Sens.* **2013**, *82*, 10–26. [CrossRef]

Technical Note

UAV + BIM: Incorporation of Photogrammetric Techniques in Architectural Projects with Building Information Modeling Versus Classical Work Processes

Carlos Rizo-Maestre [1,*], Ángel González-Avilés [1], Antonio Galiano-Garrigós [1], María Dolores Andújar-Montoya [2] and Juan Antonio Puchol-García [3]

1 Department of Architectural Constructions, University of Alicante, 03690 San Vicente del Raspeig, Alicante, Spain; angelb@ua.es (Á.G.-A.); antonio.galiano@ua.es (A.G.-G.)
2 Building Sciences and Urbanism Department, University of Alicante, 03690 San Vicente del Raspeig, Alicante, Spain; lola.andujar@ua.es
3 Department of Computer Science and Artificial Intelligence, University of Alicante, 03690 San Vicente del Raspeig, Alicante, Spain; puchol@ua.es
* Correspondence: carlosrm@ua.es

Received: 29 May 2020; Accepted: 16 July 2020; Published: 20 July 2020

Abstract: The current computer technology facilitates the processing of large volumes of information in architectural design teams, in parallel with recent advances in-flight automation in unmanned aerial vehicles (UAVs) along with lower costs, facilitates their use to capture aerial photographs and obtain orthophotographs and 3D models of relief and terrain textures. With these technologies, 3D models can be produced that allow different geometric configurations of the distribution of construction elements on the ground to be analyzed. This article presents the process of implementation in a terrain integrated into the early stages of architectural design. A methodology is proposed that covers the detailed capture of terrain, the relationship with the architectural design environment, and its implementation on the plot. As a novelty, an inverse perspective to the remaining disciplines is presented, from the inside of the object to the outside. The proposed methodology for the use of UAVs integrates terrain capture, generation of the 3D mesh, superimposition of environmental realities and architectural design using building information modeling (BIM) technologies. In addition, it represents the beginning of a line of research on the implementation of the plot and the layout of foundations using UAVs. The results obtained in the study carried out in three different projects comparing traditional technologies with the integration of UAVs + BIM show a clear improvement in the second option. The use of new technologies applied to the execution and control of work not only improves accuracy but also reduces errors and saves time, which undoubtedly indicates significant savings in costs and deviations in the project.

Keywords: photogrammetry; orthophotography; construction planning; sustainable construction; urbanism; BIM; building maintenance; UAV; unmanned aerial vehicle

---

1. Introduction

The use of unmanned aerial vehicles (UAVs) for photogrammetry has been driven by three aspects: the improvement of their performance both in-flight stability and in increasing the quality of photographic capture, and the developments in the field of graphic computing with structure from motion algorithms (SfM) Eisenbeiss and Sauerbier [1], García-Pulido et al. [2] Irschara et al. [3]. Today's UAVs, costing less than $1000, can fly over planned routes and carry out photo capture plans at predefined heights with photo resolution above 20 megapixels Colomina et al. [4] Rodriguez-Navarro et al. [5]

García-Pulido et al. [2]. With the improvement of image resolution, it is still important to improve the accuracy of point correspondence calculations between different photographs Agüera-Vega et al. [6], Dai and Lu [7] in addition to improving computational processes Fraser and Edmundson [8]. The current status of UAVs and their photogrammetry allows 2D and 3D models to be obtained from photographs taken with certain restrictions Everaerts [9] and are based on classic stereo viewing techniques Hiep et al. [10]. The field of photogrammetry UAV application has evolved in all disciplines. This article presents a design methodology based on two starting points: obtaining 2D and 3D terrain models with acceptable quality and precision in construction and the availability of integrated systems in the design and execution phases of buildings to be integrated into BIM (building information modeling) tools. There are three different situations: the possibility of using it as an instrument of measurement and representation (data collection), a tool for the architectural design phase (decision making) and finally, as a tool for reconsideration (implementation).

1.1. From Outside-to-Inside

The work with UAVs in buildings is mainly from an outside view to the inside (of the buildings), that is, as an element of remote observation, from a mapping point of view to obtain a catalog of what exists. This technology is increasingly being used to support inspection tasks in industrial and civil applications. Usually, the end user completes the procedure once the flight mission is over, and the video transmission and joint data collected by the UAV are examined. At this point, it can be integrated with the BIM methodology.

Applied in disciplines closer to architecture such as archeology, digital cataloging improves conservation, archeological research and local tourism, as demonstrated by the Delphi4Delphi International Project on Cyberarchaeology in Greece. Archeology has an advantage over architecture in the application of UAVs in the initial cataloging phase, and the first moral and technical debates arise from the fact that it involves a very high cost of labor and concerns arise as to whether a midrange scanner can capture enough details about rock art Opitz et al. [11]. The use of UAVs is rapidly advancing in almost all disciplines where there are objects to be observed, cataloged or recorded. The creative use of this instrument shows that there is a wide field of application in all of these disciplines, which leads us to investigate their inclusion in the field of architecture and propose new uses.

The origin of unmanned aerial vehicles (UAVs), commonly known as "drones", can be placed chronologically at the beginning of the 20th century with mainly military purposes Dalamagkidis et al. [12], and many civilian uses have emerged in this decade. The use of UAVs has been extended to search and rescue missions Kim et al. [13], surveillance, transport systems Sánchez-Bou and López-Pujol [14], high-resolution map production, fire detection Wing et al. [15], crop monitoring or fumigation Faiçal et al. [16], forest inventory studies, and the propagation of trees by studying conifer tree cover Ivosevic et al. [17]. UAVs have also been applied in geothermal energy Harvey et al. [18], biodiversity and biology Bohmann et al. [19]. One of the most frequent uses of UAVs is their application to the calculation of volumes. These techniques are complemented by others such as the evaluation of the accuracy of ice measurement with terrestrial laser scanner Gasinec et al. [20].

Applied to earthworks in civil works, it can reach accuracies of up to 2.5 cm per pixel, which allows us to record the furrows of tractor wheels Vergouw et al. [21]. Another field of application for UAVs is the study of wildlife conservation by remote control; their use allows information to be collected from places that are difficult to access, minimizing disturbance and allowing biodiversity to be predicted Ivošević et al. [22]. In the field of application of automatic object detection, its use allows a step to be taken between manual operation and the complete automation of traditional inspection procedures Vaquero-Melchor et al. [23].

1.2. From Fieldwork to Virtual Design: BIM as a Tool towards Sustainability

In disciplines such as architecture and within the set of activities that we refer to as outside-to-inside, the most frequent area of application is an intervention in heritage. Europe is at the forefront of digital cultural heritage. In addition to topography, soil measurements, inspections, photography, construction site monitoring or surveillance, 3D optical documentation of both buildings and urban elements (old or modern) is also available Arias et al. [24]. Applications include the control of historic buildings Püschel et al. [25] and the detection of defects in structures or building facades Aydin [26] Remondino et al. [27]. For its application in construction, activities with UAVs such as weaving Mirjan et al. [28], painting surfaces Vempati [29], collecting and laying bricks Augugliaro [30] or spraying mortar Chaltiel et al. [31] have been underway for years.

The search for efficiency within the construction industry has led to savings in all areas to minimize production costs. The new tools for generating virtual scenarios, provided by the UAVs among other systems (Figure 1), facilitate the study of the constructive possibilities of a terrain; that is, the possible studies of landscape integration, orientation and composition of future buildings. Based on the mesh generated on the plot, together with the necessary tools, the final building results can be approached with a computer model Garrigós and Kouider [32].

Figure 1. Sample of the topographic survey generated by a UAV in the illustrated forest of the University of Alicante.

Poor management of information in previous phases of the construction process (design, orientation, distribution, etc.) turns buildings into energy loss systems Gaujena et al. [33]. There are researchers devoted exclusively to interpreting the relationship between the shape of a building and its energy consumption demand for both temperate and cold climates Domínguez et al. [34]. The projection forms generated by the new software allow for more accurate representations that help generate more efficient buildings in all fields and stages Cho et al. [35]. The common objective of this process is to achieve the highest efficiency buildings Pellegrino et al. [36] Larsen et al. [37]. Researchers worldwide are developing strategies to address and integrate these techniques into construction processes Wang et al. [38]. There are even BIM tools that allow us to detect whether the modeling of a building is accurate according to the construction conditions Lou et al. [39].

The demand for sustainable buildings with a low environmental impact is increasing. Building information modeling (BIM) gives the building construction process greater control over the energy performance of buildings from the earliest stages. Studies have measured the performance of different programs, such as Virtual Environment^{TM}, Ecotect^{TM} or Green Building Studio^{TM} Azhar and Brown [40]. Construction management programs also extend to the modeling of the complete life cycle of buildings (LCAs) Bruce-Hyrkäs et al. [41]. IT developments help to ensure that efficient construction measures are provided even in the early stages of projects Jalaei and Jrade [42]. Buildings are becoming more sustainable due to these processes and generate a lower environmental impact Turner et al. [43]. Building designs managed from BIM software can help sustainability directly or indirectly, saving in construction processes, optimizing orientations, spaces, piping or even providing information on energy costs according to different project strategies Krygiel and Nies [44]. The construction design should start by analyzing the evaluation of the thermal load loss of the building according to its location Peruzzi et al. [45] Pérez-Lombard et al. [46]. In Italy, this process is defined by the UNE EN ISO 13790 standard.

Research in these fields is also applied to buildings Vollaro et al. [47]. There are branches of research that apply techniques for recognizing the properties of buildings that have already been built Ham and Golparvar-Fard [48]. Evaluation tools increasingly condition interventions in existing buildings Motawa and Carter [49]. Simulations of the energy demand of a building allow the interpretation of thermal needs according to the location conditions Kim et al. [50], and even according to the materials used and their life cycle from their construction to their use stage Martínez-Rocamora et al. [51]. There are studies that combine these work processes with case studies and compare the different typologies and constructions Scheuer et al. [52] Raji et al. [53]. These processes also help to find the optimal conditions for passive buildings with almost no energy consumption Mahdavi and Doppelbauer [54] or tall buildings Raji et al. [53].

The carbon footprint is the maximum element of control of whether a generated process is sustainable. Buildings can be measured to confirm the carbon footprint they generate during their lifetime Qin et al. [55], even during their end-of-life stage for subsequent recycling Thormark [56]. Regulations are becoming increasingly stringent in the field of construction, even if the measurement systems for these processes are not yet optimally developed Bribián et al. [57] Tronchin and Fabbri [58]. Therefore, technology is being applied to develop more efficient buildings and to monitor those already implemented Serrano and Álvarez [59] Evangelisti et al. [60]. In addition, work is underway to introduce these advances into the regulatory and control processes for building energy systems Evangelisti [61]. These jobs can lead to savings that could be of social interest Macias et al. [62].

2. Objectives

2.1. Bringing the Outside-to-Inside Methodology Closer to the User

All the examples of UAV applications we have discussed are from an outside-to-inside view. Our proposal for application to the architectural discipline is proposed in two phases: the topography phase and the stakeout phase:

1. **Topography**: Its application not only consists of the collection and cataloging of data but also if the novelty of the inside to outside perspective is added in this chronological stage of the design implementation process, it can serve as a real scenario of the architecture and be used as a projection tool.
2. **Stakeout**: Through the application of UAVs, the novelty of serving as a tool for staking out foundations and arming is proposed.

2.2. UAV Environment Acquisition. Photogrammetric 3D Reconstruction

The management of information prior to the development of a building has different stages, from the fieldwork where the environmental information is accumulated (the work with UAVs creates great advantages) to the creation of the final images of the future project. Figure 2 shows the four stages of work for the creation of virtual images in architecture: photogrammetry, 3D design, rendering and postproduction.

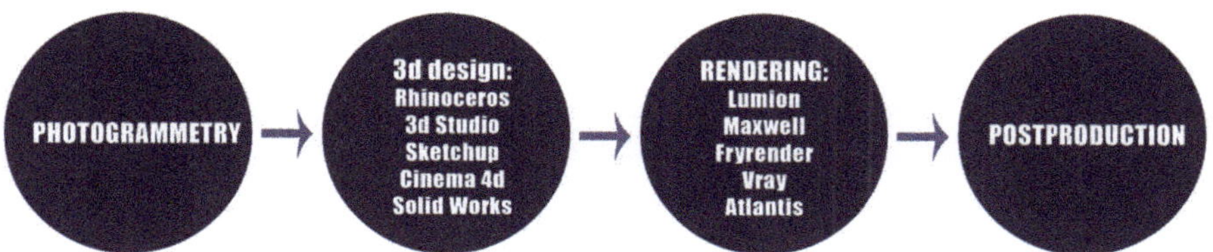

Figure 2. Image showing the different stages of work for the creation of virtual building images.

The technology of capturing environments with UAVs applied to this design phase makes it possible to realize the way in which an architectural space is inhabited and the relationship with the place from within the architecture; something that is difficult to achieve with other tools such as models. A model allows the previous study from outside-to-inside to be a very powerful instrument of architectural analysis Carrión et al. [63].

The use of the UAV in buildings is used to obtain all the fundamental aspects of a place, location or urban environment by using the UAVs for the 3D reconstruction associated with its cadastral reference as a tool to project any object.

The technology used by UAVs to reconstruct 3D surfaces from photographs is a mathematical problem that has been explained by Koch Koch [64]; it uses the correspondence of different images taken from different positions, which combined with interpolation and triangulation allows the height map needed to create a 3D mesh of the terrain to be reconstructed.

In the case presented in this study, a UAV was used to systematically scan the terrain following a predetermined path, in which one of the starting conditions of this route is the high overlap between the programmed images. Thus, the system uses multiple photographs of the terrain, which are paired to form a stereo pair. In other words, by capturing the same area from different positions, it is possible to make out the correspondence of points between them which, together with the necessary trigonometric information, produces a model with the 3D surface reconstructed in a way faithful to reality.

The programmed flight carried out by the UAV has a trajectory and height of flight fixed in each instant that allows us to interpret the data of coating or overlap between the different images where each one of the photographs was taken with respect to the contiguous photograph. The stages of this process are as follows:

- Programming the route with the premises
- Obtaining images
- Image adjustment and correlation.
- Correspondence of the same point in several images.
- Dimensioning by means of epipolar geometry.

3. Method

3.1. Photogrammetry with UAV

The interpretation of the data obtained by the UAV is a process based on a trigonometric problem. In a photograph, there is a main point where the perpendicular axis of the photograph intersects the plane of the ground. The main distance is a feature of the camera that corresponds to the distance between the lens and the negative plane of the photograph.

In addition to the data that must be recognized in the photograph, the scale at which the work is performed is also important, since it is directly proportional to the height at which the photograph was taken. Obtaining the 3D reconstruction of the terrain is considered in two situations represented in Figure 3 Otero et al. [65].

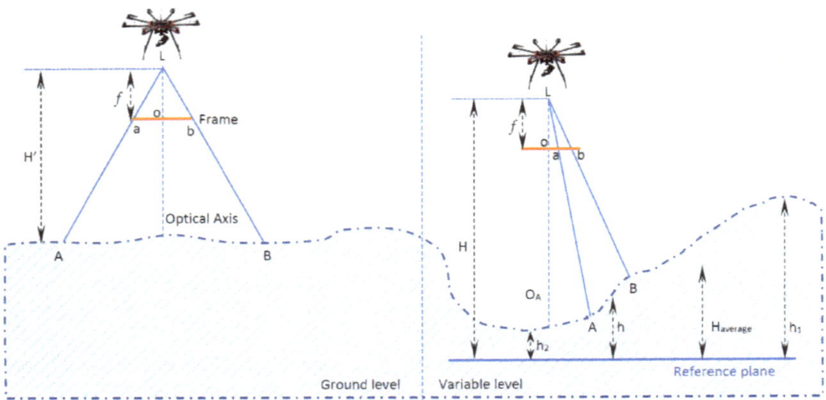

Figure 3. Image showing the difference between the type of photograph studied. (**Left**) flat terrain. (**Right**) variable terrain.

When the topography is practically horizontal, it is considered to be flat terrain, as it is easy to obtain the working scale with the following equation directly proportional to the focal length of the camera and inversely proportional to the height of flight:

$$\frac{ab}{AB} = \frac{f}{H} = \frac{1}{E} = Scale\ of\ the\ photograph,$$

where E is the scale module.

Unlike flat topography, when the working height at the different points is uneven, it is considered a variable terrain. Therefore, to calculate the working scale, the height must be recalculated at each of the points since the terrain approaches and moves away from the camera lens according to the main working point. With an image taken at a height h, you have the following relationship:

$$\frac{1}{E} = \frac{f}{H - h_m} \qquad h_m = \frac{h_1 + h_2}{2}$$

In the departure orders that are established before the flight, some indispensable parameters are set: the flight height and the route and the image overlap (Figure 4). The flight altitude is obtained by GPS from the takeoff point, which is considered to be the starting point, at altitude 0. The flight path marks the different areas through which the UAV will pass and the moment when it will take an image, making an overlap or coating between one and the other. The distance between the main points of consecutive pictures allows us to fix a longitudinal coating between consecutive pictures and a lateral overlap between adjacent pictures, usually taken in opposite directions.

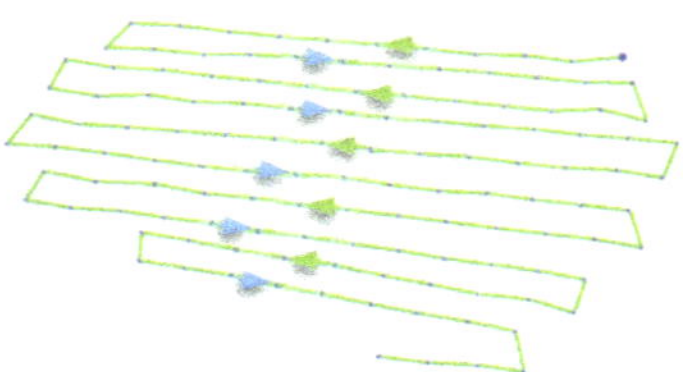

Figure 4. Image showing the outline of a route where the course and position in which the photographs will be taken are shown.

If we consider that the starting parameters are the following: frame side (s), frame scale (E) and longitudinal overlap (p), the air base, which is the distance in the air between two consecutive photographs (Figure 5), is calculated according to the following relation:

$$s * (1 - p\%)\ E_v.$$

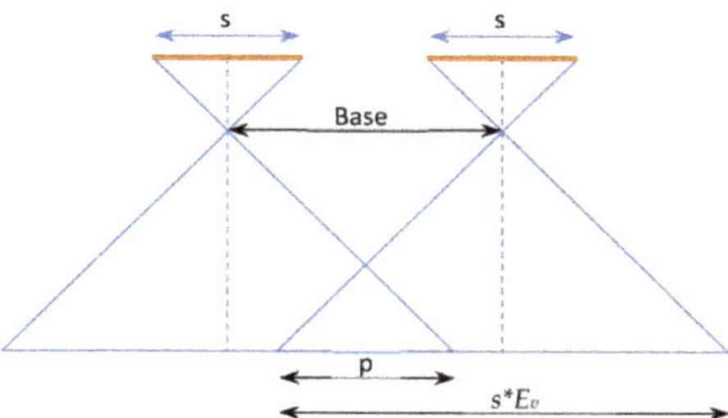

Figure 5. Diagram showing the relationship between two points where the photographs are taken.

Thus, the distance of the photographs taken on the route can also be defined with the following expression:

$$s * (1 - q\%)\ E_v,$$

where q is the transverse coating.

3.2. Stereoscopic Correspondence

The main problem encountered in terrain reconnaissance using UAVs is identifying specific points in different images. This has to do with the overlap of the images; the more overlap, the easier it is to find points of correspondence in different images. The way to solve this situation is the same way that it is used by human beings. We measure distances from the images captured by both eyes that come together in the brain. Applying this concept to the field of photogrammetry means solving a problem of homologous ray geometry. This process consists of four stages Vergouw et al. [21], Sánchez and Sobrino [66]:

- Internal orientation. Determination of the perspective beam from the data known as the focal length and other parameters mentioned.
- Relative guidance. Determination of the position of one beam in relation to the other.
- Absolute orientation. Location and scaling according to a system of terrain coordinates
- Restitution. Identification and correspondence of homologous rays and therefore of the height of the points on the ground.

In a predetermined flight, the height is calculated by the GPS data, considering the takeoff height, which is considered a height of 0 m. The identification of the homologous points is made by means of epipolar geometry that is based on a stereo pair of images; that is, once the correspondence of the points has been identified, the intersection of the projected lines is made to calculate the 3D coordinate (Figure 6).

The distance z' from point P to the UAV camera is determined by the following trigonometric equations, where b is the air base and f is the focal length of the camera:

$$z' = \frac{bf}{(x_i{}' - x_d{}')}$$

$$\frac{x}{z'} = \frac{x_i'}{f} \qquad \frac{x-b}{z'} = \frac{x_d'}{f}$$

$$z = H - z'.$$

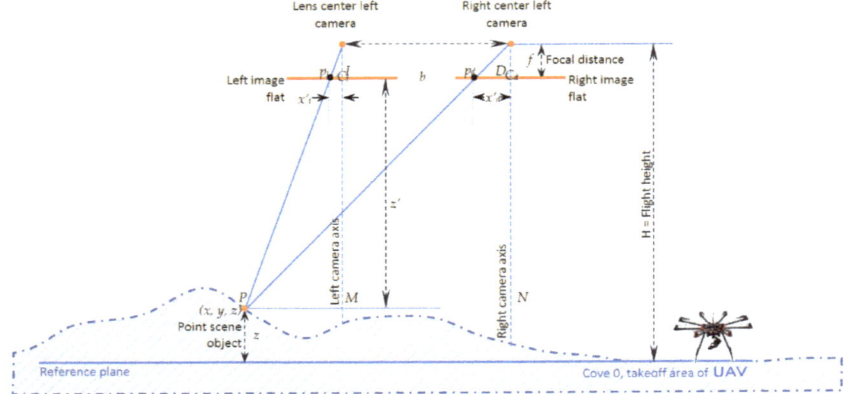

Figure 6. Representation of how to take data on a predetermined flight. The height is calculated with the GPS data, considering the takeoff height, which is considered to be 0 m.

3.3. Description of the Photogrammetric Techniques Used

The UAV used in this article is the DJI Phantom 3 Advanced quadcopter, which has a GPS-based navigation system and a barometric system for altitude control. The UAV imaging system consists of a 12-megapixel camera with an f/2.8 lens and a 94-degree FOV mounted on a three-axis gimbal.

The photogrammetric techniques used are aimed at obtaining the 2D model (orthophotography of the terrain) and the 3D model (terrain relief). Two software systems were used for the tests: Dronedeploy and Pix4d. Both are goal-oriented, although each brings positive aspects (Figure 7).

We start by defining the area to be captured, first establishing the height above ground level at the UAV launch point; in the example used as a guide to the method, this was performed with a flight height of 90 m. At higher altitudes, fewer photographs will be needed, in exchange for losing resolution. Next, the outer perimeter of the capture is defined, and the overlap values of the photographs are set (lateral overlap is set at 65% and frontal overlap is set at 85%). In addition to the flight height, these two values influence both the quality and the number of photographs that will finally be taken (Figure 8).

Figure 9 shows the overlaps between photographs in a contour where the measurement is to be made of 2.69 hectares of surface. The image shows the points from which the 77 photographs required for this example are taken with the preset parameters. The UAV camera is oriented vertically to the ground.

From the 77 photographs, the orthophotography of the terrain is obtained, which is superimposed on the Google Maps satellite image in Figure 10. The orthophoto allows different operations to be carried out that serve to study the area and determine the different options for creating and locating buildings: measuring with precision, determining the elements present in the terrain, carrying out different simulations of plot occupation, sunshine, shadows cast, etc.

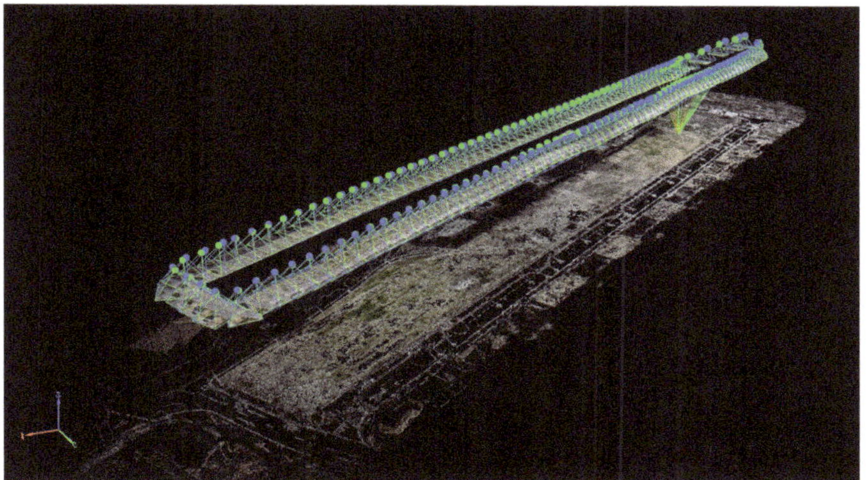

Figure 7. Unmanned aerial vehicle (UAV) flight path and image generation.

Figure 8. Image capture path preparation process. (**Left**) contour of the plot. (**Right**) route and shooting points.

Figure 9. Overlaps between the different photographs on the selected plot.

The triangulation of each point allows the terrain elevation map to be obtained, as shown in Figure 11, in which a gradient between blue (level 0) and red can be seen. The points of zero height are represented in dark blue and the height of 12 meters, maximum present in the region, of the adjacent buildings in red.

This technology allows the option of representing the terrain in 3D, that is, with the high-resolution orthophoto obtained and the difference in height of each point, a solid representation is generated. This solid representation is represented by a relief and texture, as shown in Figure 12, and serves as a starting point for working on the geometry of the building, its location within the plot and its relationship with the environment.

Figure 10. Image of the final orthophoto superimposed on the Google Maps satellite image.

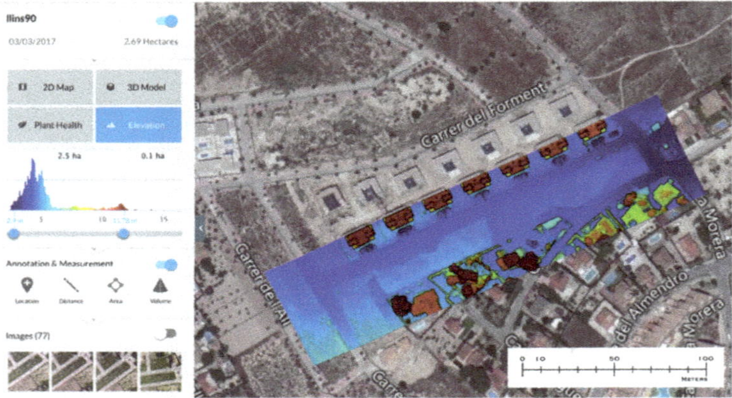

Figure 11. Image of the program Pix4D that represents the difference in heights in each point of the region in a blue-red gradient.

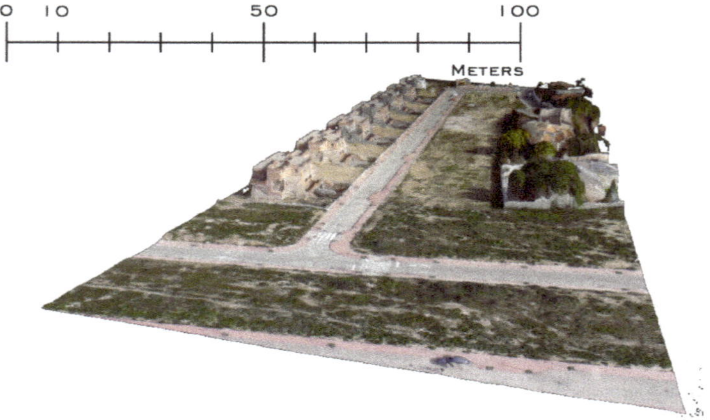

Figure 12. Image of the solid generated by the UAV with the information obtained from photographs and heights.

3.4. The BIM Software Interconnection. Connection of BIM Modules

The construction process from the beginning of the idea to the execution of the project is linked to the method of the architect in charge of the work. Computer software advances have made it possible to move from making all calculations and drawings by hand to being able to make them with a computer. The first software tools to help in this process were simple 2D line drawing tools and calculation systems for both structures and costs (time and economy). These programs have been developed over time to the second-generation tools, the BIM software.

BIM software programs, or building information modeling, are computer tools designed to digitally model and manage the integral information of a building for all phases of its life cycle in the form of a database Gu and London [67]. For this, 2D lines are not drawn as in CAD or computer-aided design programs, as the management of 3D files is direct; that is, working from one of the views or construction phases implies an update in the derivatives.

In BIM programs, all building information is managed Succar et al. [68], not only its design plans but also the definition of facilities and structure, which imply other indirect variations such as costs. In a large majority of the interventions analyzed, the same common denominator is found; there is no complete level of interoperability between data collection programs and BIM software Achille et al. [69]. However, the research developed allows such interoperability, which is a great advantage. From the point of view of the relationship with the environment, the method in which work is performed on an urban scale and on the scale of the architectural object is also fundamental. Some experts argue that the goal is to integrate GIS and BIM Scianna et al. [70]. The use of XML-based formats would allow the standardization of information and consult metadata of both scales on the same platform.

3.5. Integration of the Projected Building with Building Information Modeling (BIM)

The tools for generating topographies described by means of UAVs have made it possible to obtain detailed information about a certain area where the building is planned to be constructed. These data are treated with both 2D and 3D software tools, that is, they work on both the X–Y and Z planes. This is also reflected in the type of program used by BIM for the interconnection of all parts of the construction. In the introduction, we mentioned the advantage that archaeology has over UAV architecture for the data collection phase, which also applies to BIM integration.

Today, the accuracy of 3D photo surveying allows information to be quickly recorded and loaded into BIM systems. That the technological advances in data collection and its incorporation into BIM mark a before and after in the conception and design of a building is currently an objective reality. However, it is necessary to go deeper into the relationship with the environment in which the building is located with the application of these technologies to the execution on site.

4. Evaluation of the Results Obtained in the Experience. Classic Method against UAV + BIM

At present, the process to follow after the purchase of a plot is to carry out a topographical and geotechnical study as initial tasks to carry out an execution project of a building. The main difference between the two is that the second is mandatory by law when the building to be executed must comply with the Spanish Technical Building Code (CTE) 699 [71]. However, as we mentioned, the first can be crucial for the development of the proposal and even more for the control of the execution in its initial phase of rethinking, as we will see in the following case study. Good architecture is traversed, crossed, both inside and outside Corbusier et al. [72], and we have to learn to appreciate the effective confluences of the exterior and interior Neutral et al. [73]. In this sense, it is essential to strengthen the precision of the 3D models for a better understanding of the architecture and its environment. For this reason, the paths and 360° images are parts of the projects and must integrate the architecture's environment not as a static image but as an augmented reality.

The 2D topographic study of a plot of land is between 350 and 475 m^2, as in the case study. The practical case we present of this work system, adapted to the region studied by the UAV, allows us to analyze from outside-to-inside the different options of volumetry and fit in the studied plot and the relation from the inside with the environment. Figure 13 shows volumetric work done in a 3D software program and the adaptation of this program to the solid representation generated by the UAV.

Figure 14 shows the process carried out for the integration of the building within the plot. This work is very useful within the first stages of study of certain buildings since it allows for different tests to see in real time a faithful approximation of the final construction.

To clarify the progress in time and cost reduction, three projects were analyzed according to the protocol followed in each of them. The three methods of approaching the implementation in the plot were the traditional method of data collection with 2D computational work, the traditional method with BIM and UAV + BIM.

Figure 13. (**Left**) volumetry generated in 3D software. (**Right**) integration of the volumetry generated in the solid generated with the UAV flight.

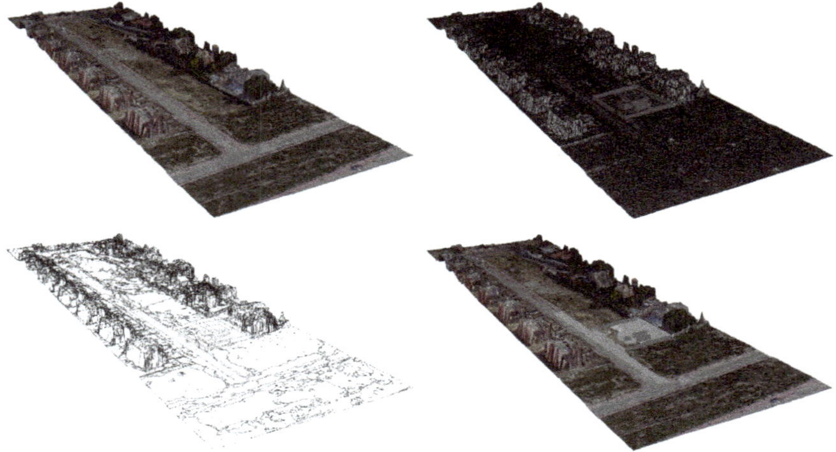

Figure 14. Image showing the integration process of a volumetry generated in 3D software over the solid representation generated by the UAV flight. The process described shows the proposed technique used. The photogrammetric data of the plot are obtained, the new work is adjusted and the two processes are integrated (UAV + building information modeling (BIM)).

Table 1 shows the synthesis of the results obtained in two situations compared to the architectural design, the design with 2D tools or with 3D software (BIM methodology). There are two methods of taking data from the plot; the traditional method with a total station and by means of DRON. In the early phase studied in the three assumptions, it was possible to quantify the time savings in the data collection phase for restatement. The accuracy factor is directly proportional to the number of points and, therefore, to the execution time of the data collection, so that in the face of the need for more time for greater precision of a technician in topography, the use of UAVs requires a maximum of 1 h for an area of 500 m^2. The incorporation of UAVs in the early phase of implementation is proposed as a qualitative leap compared to traditional techniques. The implementation of this technique in the topography phase, the flight and its subsequent incorporation into BIM programs allows the complete database to be available for all subsequent operations (Figure 15).

Table 1. Comparison between the three assumptions in the early project phase.

Early Project Phase	**Assumption 1 Traditional/2D**	**Assumption 2 Traditional/3D BIM**	**Assumption 3 UAV/3D BIM**
Accuracy	50%	75%	100%
Time (plot 500 m^2)	4 h +1 h	4 h +1 h	1 h +1 h
Interconnection	70%	85%	100%
Changes	NOT	NOT	YES
360°	NOT	YES	YES
Tour	NOT	NOT	YES

Currently, the common methodology for staking out a plot is performed by means of plaster on the ground and at a later stage, tile on concrete for cleaning. The accuracy of these tools common in the construction world is far from the levels of accuracy of design tools. For this reason, UAV technology has been applied both to the creation of the topographical plan of the plot and to the laying out of the foundations, giving higher levels of precision than traditional methods. In Table 2, the precision and time data for the three scenarios studied were collected. The result is an improvement in accuracy during stakeout and assembly of 100%. In addition, although the process in scenario 3 took longer than scenarios 1 and 2 using the traditional method of restatement, it required only one person. The implementation with traditional methods required two people to correctly establish the main axes and foundation limits with respect to the boundaries of the plot, something that the use of UAVs and georeferences did not require.

Table 2. Comparison between the three assumptions in the stakeout/armament phase.

Armed Stakeout	**Assumption 1 Traditional/2D**	**Assumption 2 Traditional/3D BIM**	**Assumption 3 UAV/3D BIM**
Accuracy	mm/cm	mm/cm	mm/mm
Time 100 m^2 slab	1 h	1 h	0.7 h
Accuracy	70%	85%	100%
Staff	2	2	1 (UAV pilot)
Control	under	medium	high

Figure 15. Image showing the process of building stakeout before excavation with gypsum, slab perimeter stakeout on polyethylene film with spray coupled to DRON; marking pillars with a traditional shape strip and reinforcement of the foundation slab with dotting of slab rebar by UAV and spray. The different processes show the difficulties in marking the lines executed with a computer in the field.

5. Conclusions

The benefits of the use of UAVs in the early stages of architectural design and the connection with BIM programs have been demonstrated, making them a tremendously effective working tool. The suitability of the 3D photogrammetric survey has been proven against traditional techniques. The 3D technique allows the information to be quickly recorded and loaded into BIM systems.

The triangulation of each point of the survey allows us to obtain the complete terrain elevation map and a flight with 360° visibility that allows greater control and detail of the relationship of the architectural project with the environment in all directions.

The results obtained in the study carried out in three different projects comparing traditional technologies with UAV + BIM integration show a clear improvement in the second option. In the early phases of work, both the accuracy and the time spent show an improvement in UAV + BIM technology, as well as allowing changes in the project phase. In the phase of setting out the assembly or structure, greater precision, and less time and need for personnel, as well as more control over changes are achieved.

Finally, the use of new technologies applied to the execution and control of work not only improves accuracy but also reduces errors and saves time, which undoubtedly means a significant saving in costs and deviations in the project.

Author Contributions: The work presented here was developed in collaboration amongst all authors. All authors have contributed to, seen and approved the manuscript. All authors have read and agreed to the published version of the manuscript.

Funding: This work has been supported by the Ministerio de Ciencia, Innovación y Universidades (Spain), project RTI2018-096219-B-I00. Project cofinanced with FEDER funds.

Acknowledgments: The authors of this paper thank the University of Alicante for the grant that allowed part of this research.

Conflicts of Interest: The authors declare no conflict of interest.

References

1. Eisenbeiss, H.; Sauerbier, M. Investigation of uav systems and flight modes for photogrammetric applications. *Photogramm. Rec.* **2011**, *26*, 400–421. [CrossRef]
2. García-Pulido, L.J.; Jaramillo, J.R.; Dorado, M.I.A. Heritage survey and scientific analysis of the watchtowers that defended the last Islamic Kingdom in the iberian peninsula (Thirteen To Fifeteenth Century). *ISPRS Int. Arch. Photogramm. Remote Sens. Spat. Inf. Sci.* **2017**, *XLII-2/W5*, 259–265.
3. Irschara, A.; Kaufmann, V.; Klopschitz, M.; Bischof, H.; Leberl, F. Towards Fully Automatic Photogrammetric Reconstruction Using Digital Images Taken from UAVs. *Int. Arch. Photogramm. Remote Sens. Spat. Inf. Sci.* **2010**, *38*, 65–70.
4. Colomina, I.; Blázquez, M.; Molina, P.; Parés, M.E.; Wis, M. Towards a new paradigm for high-resolution low-cost photogrammetryand remote sensing. *Int. Arch. Photogramm. Remote Sens. Spat. Inf. Sci.* **2008**, *XXXVII*, 1201–1206.
5. Rodriguez-Navarro, P.; Gil-Piqueras, T.; Verdiani, G. Drones for Architectural Surveying. Their Use in Documenting Towers of the Valencian Coast. Availble online: https://www.researchgate.net/publication/320066302_Drones_for_Architectural_Surveying_Their_Use_in_Documenting_Towers_of_the_Valencian_Coast (accessed on 12 January 2020).
6. Agüera-Vega, F.; Carvajal-Ramírez, F.; Martínez-Carricondo, P. Accuracy of Digital Surface Models and Orthophotos Derived from Unmanned Aerial Vehicle Photogrammetry. *J. Surv. Eng.* **2017**, *143*, 04016025. [CrossRef]
7. Dai, F.; Lu, M. Assessing the Accuracy of Applying Photogrammetry to Take Geometric Measurements on Building Products. *J. Constr. Eng. Manag.* **2010**, *136*, 242–250. [CrossRef]
8. Fraser, C.S.; Edmundson, K.L. Design and implementation of a computational processing system for off-line digital close-range photogrammetry. *ISPRS J. Photogram. Remote Sens.* **2000**, *55*, 94–104. [CrossRef]
9. Everaerts, J. The use of unmanned aerial vehicles (UAVs) for remote sensing and mapping. *Int. Arch. Photogramm. Remote Sens. Spat. Inf. Sci.* **2008**, *37*, 1187–1192.
10. Hiep, V.H.; Keriven, R.; Labatut, P.; Pons, J.P. Towards high-resolution large-scale multi-view stereo. In Proceedings of the 2009 IEEE Conference on Computer Vision and Pattern Recognition, Miami, FL, USA, 20–25 June 2009; pp. 1430–1437.
11. Opitz, R.; Simon, K.; Barnes, A.; Fisher, K.; Lippiello, L. Close-Range Photogrammetry vs. 3D Scanning: Comparing Data Capture, Processing and Model Generation in the Field and the Lab. Available online: https://www.google.com.hk/url?sa=t&rct=j&q=&esrc=s&source=web&cd=&ved=2ahUKEwj685XmstjqAhUhGaYKHbY6CQAQFjABegQIAhAB&url=https%3A%2F%2Fncptt.nps.gov%2Fwp-content%2Fuploads%2Fsimon.pdf&usg=AOvVaw2sB1-arFK-LCViuMGoKuox (accessed on 12 January 2020).
12. Dalamagkidis, K.; Valavanis, K.P.; Piegl, L.A. *On Integrating Unmanned Aircraft Systems into the National Airspace System*; Springer: Dordrecht, The Netherlands, 2012.
13. Kim, K.; Hyun, J.; Myung, H. Development of aerial image transmitting sensor platform for disaster site surveillance. In Proceedings of the 2017 17th International Conference on Control, Automation and Systems, Jeju, South Korea, 18–21 October 2017; pp. 794–796.
14. Sánchez-Bou, C.; López-Pujol, J. The coming revolution: The use of drones in plant conservation. *Collect. Bot.* **2014**, *33*, 7.
15. Wing, M.G.; Burnett, J.D.; Sessions, J. Remote Sensing and Unmanned Aerial System Technology for Monitoring and Quantifying Forest Fire Impacts. *Int. J. Remote Sens. Appl.* **2014**, *4*, 18. [CrossRef]

16. Faiçal, B.S.; Costa, F.G.; Pessin, G.; Ueyama, J.; Freitas, H.; Colombo, A.; Fini, P.H.; Villas, L.; Osório, F.S.; Vargas, P.A.; et al. The use of unmanned aerial vehicles and wireless sensor networks for spraying pesticides. *J. Syst. Archit.* **2014**, *60*, 393–404. [CrossRef]
17. Ivosevic, B.; Han, Y.G.; Kwon, O. Calculating coniferous tree coverage using unmanned aerial vehicle photogrammetry. *J. Ecol. Environ.* **2017**, *41*, 10. [CrossRef]
18. Harvey, M.C.; Harvey, C.C.; Rowland, J.; Luketina, K.M. Drones in geothermal exploration: Thermal infrared imagery, aerial photos and digital elevation models. In Proceedings of the 6th African Rift Geothermal Conference, Addis Ababa, Ethiopia, 2–4 November 2016.
19. Bohmann, K.; Schnell, I.B.; Gilbert, M.T.P. When bugs reveal biodiversity. *Mol. Ecol.* **2013**, *22*, 909–911. [CrossRef] [PubMed]
20. Gasinec, J.; Gašincová, S.; Černota, P.; Staňková, H. Uses of terrestrial laser sanning in mnitoring of ground ice within dobšinská ice cave. *J. Pol. Miner. Eng. Soc.* **2012**, *30*, 31–42.
21. Vergouw, B.; Nagel, H.; Bondt, G.; Custers, B. Drone Technology: Types, Payloads, Applications, Frequency Spectrum Issues and Future Developments In *The Future of Drone Use*; TMC Asser Press: The Hague, The Netherlands, 2016.
22. Ivošević, B.; Han, Y.G.; Cho, Y.; Kwon, O. The use of conservation drones in ecology and wildlife research. *J. Ecol. Environ.* **2015**, *38*, 113–118. [CrossRef]
23. Vaquero-Melchor, D.; Campaña, I.; Bernardos, A.M.; Bergesio, L.; Besada, J.A. A distributed drone-oriented architecture for in-flight object detection. In *International Conference on Hybrid Artificial Intelligence Systems*; Springer: Cham, The Netherlands, 2018.
24. Arias, P.; Herráez, J.; Lorenzo, H.; Ordóñez, C. Control of structural problems in cultural heritage monuments using close-range photogrammetry and computer methods. *Comput. Struct.* **2005**, *83*, 1754–1766. [CrossRef]
25. Püschel, H.; Sauerbier, M.; Eisenbeiss, H. A 3D model of Castle Landenberg (CH) from combined photogrammetric processing of terrestrial and UAV-based images. *Int. Arch. Photogramm. Remote Sens. Spat. Inf. Sci.* **2008**, *37*, 93–98.
26. Aydin, C.C. Designing building façades for the urban rebuilt environment with integration of digital close-range photogrammetry and geographical information systems. *Autom. Constr.* **2014**, *43*, 38–48. [CrossRef]
27. Remondino, F.; Barazzetti, L.; Nex, F.; Scaioni, M.; Sarazzi, D. UAV photogrammetry for mapping and 3d modeling—Current status and future perspectives. *ISPRS Int. Arch. Photogramm. Remote Sens. Spat. Inf. Sci.* **2012**, *XXXVIII-1/C22*, 25–31. [CrossRef]
28. Mirjan, A.; Gramazio, F.; Kohler, M.; Augugliaro, F.; Andrea, R.D. Architectural fabrication of tensile structures with flying machines. In *Green Design, Materials and Manufacturing Processes*; Taylor & Francis Group: London, UK, 2013.
29. Vempati, A.S. PaintCopter: An Autonomous UAV for Spray Painting on Three-Dimensional Surfaces. *IEEE Robot. Autom. Lett* **2018**, *3*, 2862–2869. [CrossRef]
30. Augugliaro, F. The flight assembled architecture installation: Cooperative contruction with flying machines. *IEEE Control Syst.* **2014**, *34*, 46–64.
31. Chaltiel, S.; Bravo, M.; Chronis, A. Digital fabricationwith Virtual and Augmented Reality for Monolithic Shells. Available online: https://www.researchgate.net/publication/320839954_Digital_fabrication_with_Virtual_and_Augmented_Reality_for_Monolithic_Shells (accessed on 12 January 2020).
32. Garrigós, A.; Kouider, T. The Virtual Interactive Relationship Between BIM Project Teams: Effective Communication to aid Collaboration in the Design Process. In *Healthy Buildings, Innovation, Desing & Technology: Conference Proceedings of the 6th International Congress of Architectural Technology*; Universidad de Alicante: San Vicente del Raspeig, Spain, 2016; pp. 12–14.
33. Gaujena, B.; Borodinecs, A.; Zemitis, J.; Prozuments, A. Influence of Building Envelope Thermal Mass on Heating Design Temperature. *IOP Conf. Ser. Mater. Sci. Eng.* **2015**, *96*, 012031. [CrossRef]
34. Domínguez, S.; Sendra, J.J.; León, A.L.; Esquivias, P.M. Towards Energy Demand Reduction in Social Housing Buildings: Envelope System Optimization Strategies. *Energies* **2012**, *5*, 2263–2287. [CrossRef]
35. Cho, Y.K.; Ham, Y.; Golpavar-Fard, M. 3D as-is building energy modeling and diagnostics: A review of the state-of-the-art. *Adv. Eng. Inform.* **2015**, *29*, 184–195. [CrossRef]

36. Pellegrino, J.; Fanney, A.H.; Persily, A.K.; Domanski, P.A.; Healy, W.M.; Bushby, S.T. *Measurement Science Roadmap for Net-Zero Energy Buildings*; National Institute of Standards and Technology: Gaithersburg, MD, USA, 2010; p. 1660.
37. Larsen, K.E.; Lattke, F.; Ott, S.; Winter, S. Surveying and digital workflow in energy performance retrofit projects using prefabricated elements. *Autom. Constr.* **2011**, *20*, 999–1011. [CrossRef]
38. Wang, C.; Cho, Y.K.; Gai, M. As-Is 3D Thermal Modeling for Existing Building Envelopes Using a Hybrid LIDAR System. *J. Comput. Civ. Eng.* **2013**, *27*, 645–656. [CrossRef]
39. Lou, J.; Xu, J.; Wang, K. Study on Construction Quality Control of Urban Complex Project Based on BIM. *Procedia Eng.* **2017**, *174*, 668–676. [CrossRef]
40. Azhar, S.; Brown, J. BIM for Sustainability Analyses. *Int. J. Constr. Educ. Res.* **2009**, *5*, 276–292. [CrossRef]
41. Bruce-Hyrkäs, T.; Pasanen, P.; Castro, R. Overview of Whole Building Life-Cycle Assessment for Green Building Certification and Ecodesign through Industry Surveys and Interviews. *Procedia CIRP* **2018**, *69*, 178–183. [CrossRef]
42. Jalaei, F.; Jrade, A. Integrating Building Information Modeling (BIM) and energy analysis tools with green building certification system to conceptually design sustainable buildings. *J. Inf. Technol. Constr.* **2014**, *19*, 494–519.
43. Turner, W.J.; Kinnane, O.; Basu, B. Demand-side Characterization of the Smart City for Energy Modelling. *Energy Procedia* **2014**, *62*, 160–169. [CrossRef]
44. Krygiel, E.; Nies, B. *Green BIM: Successful Sustainable Design with Building Information Modeling*; John Wiley & Sons: Hoboken, NJ, USA, 2008.
45. Peruzzi, L.; Salata, F.; de Lieto Vollaro, A.; de Lieto Vollaro, R. The reliability of technological systems with high energy efficiency in residential buildings. *Energy Build.* **2014**, *68*, 19–24. [CrossRef]
46. Pérez-Lombard, L.; Ortiz, J.; Pout, C. A review on buildings energy consumption information. *Energy Build.* **2008**, *40*, 394–398. [CrossRef]
47. Vollaro, R.D.L.; Guattari, C.; Evangelisti, L.; Battista, G.; Carnielo, E.; Gori, P. Building energy performance analysis: A case study. *Energy Build.* **2015**, *87*, 87–94. [CrossRef]
48. Ham, Y.; Golparvar-Fard, M. Mapping actual thermal properties to building elements in gbXML-based BIM for reliable building energy performance modeling. *Autom. Constr.* **2015**, *49*, 214–224. [CrossRef]
49. Motawa, I.; Carter, K. Sustainable BIM-based Evaluation of Buildings. *Procedia Soc. Behav. Sci.* **2013**, *74*, 419–428. [CrossRef]
50. Kim, J.B.; Jeong, W.; Clayton, M.J.; Haberl, J.S.; Yan, W. Developing a physical BIM library for building thermal energy simulation. *Autom. Constr.* **2015**, *50*, 16–28. [CrossRef]
51. Martínez-Rocamora, A.; Solís-Guzmán, J.; Marrero, M. LCA databases focused on construction materials: A review. *Renew. Sustain. Energy Rev.* **2016**, *58*, 565–573. [CrossRef]
52. Scheuer, C.; Keoleian, G.A.; Reppe, P. Life cycle energy and environmental performance of a new university building: modeling challenges and design implications. *Energy Build.* **2003**, *35*, 1049–1064. [CrossRef]
53. Raji, B.; Tenpierik, M.; van den Dobbelsteen, A. Early-Stage Design Considerations for the Energy-Efficiency of High-Rise Office Buildings. *Sustainability* **2017**, *9*, 623. [CrossRef]
54. Mahdavi, A.; Doppelbauer, E.M. A performance comparison of passive and low-energy buildings. *Energy Build.* **2010**, *42*, 1314–1319. [CrossRef]
55. Qin, Z.; Zhai, G.; Wu, X.; Yu, Y.; Zhang, Z. Carbon footprint evaluation of coal-to-methanol chain with the hierarchical attribution management and life cycle assessment. *Energy Convers. Manag.* **2016**, *124*, 168–179. [CrossRef]
56. Thormark, C. A low energy building in a life cycle—its embodied energy, energy need for operation and recycling potential. *Build. Environ.* **2002**, *37*, 429–435. [CrossRef]
57. Bribián, I.Z.; Usón, A.A.; Scarpellini, S. Life cycle assessment in buildings: State-of-the-art and simplified LCA methodology as a complement for building certification. *Build. Environ.* **2009**, *44*, 2510–2520. [CrossRef]
58. Tronchin, L.; Fabbri, K. Energy performance building evaluation in Mediterranean countries: Comparison between software simulations and operating rating simulation. *Energy Build.* **2008**, *40*, 1176–1187. [CrossRef]
59. Serrano, A.R.; Álvarez, S.P. Life Cycle Assessment in Building: A Case Study on the Energy and Emissions Impact Related to the Choice of Housing Typologies and Construction Process in Spain. *Sustainability* **2016**, *8*, 287. [CrossRef]

60. Evangelisti, L.; Battista, G.; Guattari, C.; Basilicata, C.; de Lieto Vollaro, R. Analysis of Two Models for Evaluating the Energy Performance of Different Buildings. *Sustainability* **2014**, *6*, 5311–5321. [CrossRef]
61. Evangelisti, L. Influence of the Thermal Inertia in the European Simplified Procedures for the Assessment of Buildings' Energy Performance. *Sustainability* **2014**, *6*, 4514–4524. [CrossRef]
62. Macias, J.; Iturburu, L.; Rodriguez, C.; Agdas, D.; Boero, A.; Soriano, G. Embodied and operational energy assessment of different construction methods employed on social interest dwellings in Ecuador. *Energy Build.* **2017**, *151*, 107–120. [CrossRef]
63. Carrión, M.T.P.; Prieto, J.I.; Boza, R.E.; Tomás, R.; Cardona, M.G.S.; Díaz, M.D.C. Las Maquetas Como Material Didáctico Para la enseñAnza y Aprendizaje de la Lectura e Interpretación de Planos en la Ingeniería. Available online: https://docplayer.es/18845885-Las-maquetas-como-material-didactico-para-la-ensenanza-y-aprendizaje-de-la-lectura-e-interpretacion-de-planos-en-la-ingenieria.html (accessed on 12 January 2020).
64. Koch, R. *3-D Modeling of Human Heads from Stereoscopic Image Sequences*; Springer: Berlin/Heidelberg, Germany, 1996; pp. 169–178.
65. Otero, I.; Ezquerra, A.; Solano, R.; Martín, L.; Bachelor, I. FOTOGRAMETRÍA. Available online: http://ocw.upm.es/pluginfile.php/1068/mod_label/intro/fotogrametria_cap_libro.pdf (accessed on 12 January 2020).
66. Sánchez, J.A.; Sobrino. *Introduction to Photogrammetry*; ETSI Roads, Canals and Ports; The Ohio State University: Columbus, OH, USA, 2006.
67. Gu, N.; London, K. Understanding and facilitating BIM adoption in the AEC industry. *Autom. Constr.* **2010**, *19*, 988–999. [CrossRef]
68. Succar, B.; Sher, W.; Williams, A. Measuring BIM performance: Five metrics. *Archit. Eng. Des. Manag.* **2012**, *8*, 120–142. [CrossRef]
69. Achille, C.; Lombardini, N.; Tommasi, C. BIM and cultural heritage: Compatibility tests in an archaeological site. *Trans. Built Environ.* **2015**, *149*, 593–604.
70. Scianna, A.; Gristina, S.; Paliaga, S. Experimental BIM applications in archaeology: A work-flow. In *Euro-Mediterranean Conference*; Springer: Cham, The Netherlands, 2014.
71. Real Decreto 314/2006, de 17 de Marzo, por el que se Aprueba el Código Técnico de la Edificación. Available online: https://www.boe.es/buscar/act.php?id=BOE-A-2006-5515 (accessed on 12 January 2020).
72. Corbusier, L.; C, I.; Architecture, M. *To Studenten: The "Charte d'Athènes"*; Rowohlt: Reinbek, Germany, 1962.
73. Neutra, R. *Gestaltete Umwelt: Erfahrungen und Forderungen Eines Architekten*; Verlag der Kunst: Dresden, Germany, 1976

Article

Structure from Motion of Multi-Angle RPAS Imagery Complements Larger-Scale Airborne Lidar Data for Cost-Effective Snow Monitoring in Mountain Forests

Patrick D. Broxton [1,*] and Willem J. D. van Leeuwen [1,2]

1 School of Natural Resources and the Environment, University of Arizona, Tucson, AZ 85721, USA; leeuw@arizona.edu

2 School of Geography and Development, University of Arizona, Tucson, AZ 85721, USA

* Correspondence: broxtopd@arizona.edu

Received: 25 May 2020; Accepted: 16 July 2020; Published: 18 July 2020

Abstract: Snowmelt from mountain forests is critically important for water resources and hydropower generation. More than 75% of surface water supply originates as snowmelt in mountainous regions, such as the western U.S. Remote sensing has the potential to measure snowpack in these areas accurately. In this research, we combine light detection and ranging (lidar) from crewed aircraft (currently, the most reliable way of measuring snow depth in mountain forests) and structure from motion (SfM) remotely piloted aircraft systems (RPAS) for cost-effective multi-temporal monitoring of snowpack in mountain forests. In sparsely forested areas, both technologies give similar snow depth maps, with a comparable agreement with ground-based snow depth observations (RMSE ~10 cm). In densely forested areas, airborne lidar is better able to represent snow depth than RPAS-SfM (RMSE ~10 cm vs ~10–20 cm). In addition, we find the relationship between RPAS-SfM and previous lidar snow depth data can be used to estimate snow depth conditions outside of relatively small RPAS-SfM monitoring plots, with RMSE's between these observed and estimated snow depths on the order of 10–15 cm for the larger lidar coverages. This suggests that when a single airborne lidar snow survey exists, RPAS-SfM may provide useful multi-temporal snow monitoring that can estimate basin-scale snowpack, at a much lower cost than multiple airborne lidar surveys. Doing so requires a pre-existing mid-winter or peak-snowpack airborne lidar snow survey, and subsequent well-designed paired SfM and field snow surveys that accurately capture substantial snow depth variability.

Keywords: snow; remotely piloted aircraft systems; structure from motion; lidar; forests

---

1. Introduction

Snowpack in mountain forests is a major source of water for reservoirs that provide water and hydropower for many urban and agricultural communities [1–3]. Mountain snowpacks are affected by many climatic, topographic and ecological variables, and are sensitive to forest disturbance such as thinning, prescribed fires, wildfire, and tree die-off [4–13]. It is important to monitor how snowpacks in these areas respond to changing environmental conditions in order to understand and forecast available water resources for both natural and human consumption.

Characterization of snowpack in these mountain forests is challenging because of the large amount of small-scale spatial variability of the snowpack, due to topographic heterogeneity and variable forest structure [14–18]. This extreme heterogeneity makes it difficult to monitor snowpack with point snow measurements [19]. Remote sensing of snowpack is better able to characterize this heterogeneity, though it suffers in forested environments because trees can interfere with the remote sensing of snowpack below the canopy [20]. Furthermore, remote sensing of snowpack is hindered by clouds and can be affected by sub-grid landscape heterogeneity [21,22]. Therefore, monitoring and assessing the snowpack response to environmental changes requires multi-scale and multi-sensor monitoring that takes advantage of field data in combination with multiple observing platforms (e.g., remotely piloted aircraft systems (RPAS), airplanes and satellites) and payloads (e.g., visible, multispectral and lidar sensors) [21,23,24]. These snow monitoring platforms and payloads all have advantages and disadvantages that relate to spatial and spectral resolutions, geolocation, extent, cloud cover, wind, reliability, timeliness, terrain accessibility, resources and training, and government regulations [25,26].

Of the remote sensing techniques for measuring snowpacks in forested environments, lidar from crewed aircraft (hereafter, referred to as 'airborne lidar', or simply 'lidar') is one of the most successful because it covers relatively large areas, has very high sampling rates, and has the ability to penetrate the canopy through canopy gaps [27–30]. However, there are many costs associated with these lidar acquisitions, and they can be time-consuming to process [31,32], making it hard to use them for multi-temporal snow monitoring. While some groups, such as NASA's Airborne Snow Observatory [29] have deployed multitemporal airborne lidar surveys, such acquisitions are costly and are not available for most sites.

Three-dimensional modelling created from multi-angle imagery from RPAS, such as quadcopter platforms with gimballed digital cameras, offers an exciting addition to airborne lidar for snow monitoring in heterogeneous forests [33,34]. This imagery, which is collected at multiple points along the RPAS's flight trajectory, is stitched together to create a 3-D representation of the landscape using a process called structure from motion (SfM; which uses points on multiple overlapping images to triangulate and bundle-adjust the locations of image pixels [35]). This process is able to achieve good accuracy for snow depth and forest structure assessments [34,36–39]. In certain respects, SfM from RPAS imagery (hereafter, referred to as 'RPAS-SfM' or simply 'SfM') offers advantages over airborne lidar. For example, RPAS have become relatively inexpensive and easy to operate, allowing them to be economically deployed with minimal labor [40]. However, there are still challenges associated with RPAS monitoring of snow depth [37,41]. These include variable field site conditions (which might include a dense canopy that obscures the snowpack), survey design, SFM processing software (which is computationally expensive), and validation data. In addition, RPAS-SfM is generally limited to smaller areas, due to physical limitations to RPAS flight extent (such as battery life, which limits flight times), and there are sometimes logistical challenges with RPAS flights such as obtaining permission before flights.

Due to their strengths for monitoring snowpack, both airborne lidar and RPAS-SfM technologies have been extensively used for snowpack monitoring (e.g., References [20,27,29,30,34,36,37,40,42–50]). However, these studies do not include a strategy for combining smaller-area RPAS-SfM measurements with larger-area airborne lidar measurements for multitemporal snowpack monitoring, especially at basin-scales. This combination is natural because the strengths of each technology are complementary to the other. For example, airborne lidar provides consistent accuracy and larger spatial extents, which are more appropriate for hydrological applications, such as water supply monitoring and Land Surface Model evaluation. However, the cost (in terms of both money and processing time) can be quite high. Meanwhile, RPAS-SfM allows for increased temporal frequency and on-demand data acquisitions with rapid processing times at a much lower monetary cost over limited spatial domains.

In this research, we show how RPAS-SfM and airborne lidar can be used in a complementary fashion to attain regular snowpack monitoring data that can lead to seasonal large area snow depth and snow water equivalent (SWE) fields in complex forested environments. We investigate (1) how well SfM is able to characterize snowpack for a variety of forest cover and topographic conditions using a rich dataset of field-based and airborne lidar measurements; (2) how can multi-temporal SfM data be combined with existing airborne lidar snow data to achieve multi-temporal estimates of snow amount and distribution outside of spatially-limited SfM plots; and (3), what is the accuracy of these estimates at different times of the year and under different forest and terrain conditions. We use extensive field data collected from 11 surveys from 2017–2020 for four study plots along Arizona's Mogollon Rim, representing the diverse forest and snowpack conditions in the region. All surveys have multi-angle aerial photography acquisitions from a RPAS paired with precisely geolocated ground-based snow surveys, while seven of these surveys are coincident with acquisitions of high point density airborne lidar. The combination of the RPAS surveys with the airborne lidar and ground surveys provides an unprecedented opportunity not only to evaluate the performance of the SfM remote sensing under a variety of forest and snowpack conditions, but also to explore opportunities to combine all of these measurements for cost-effective, high quality multi-temporal snowpack monitoring in mountain forests.

2. Materials and Methods

2.1. Field Data

Data in this study come from four intensively studied snow-research plots near the Mogollon Rim, a mid- to high-elevation (~2000–3000 m) forested region in central Arizona (AZ; Figure 1). Two "montane" plots, located ~50 km east of Show Low, AZ, are in close proximity to each other at a high-elevation (~2800 m) field site. These plots have low to moderate relief slopes and contain a mixture of burned (from a 2011 wildfire [51]) and live mixed conifer forests and montane grasslands. One plot (the "montane meadow" plot) contains large snow variability, due to heterogeneous forest conditions (i.e., tree stands separated by meadows). In contrast, the other (the "montane valley" plot) contains large snow variability due mostly to heterogeneous topography (i.e., north- vs. south-facing slopes). A "dense forest" plot, located ~20 km north of Payson, AZ, has gentle slopes and is densely forested, primarily with Ponderosa Pine (pinus ponderosa) trees. Finally, ~10 km to the north of the dense forest plot, is another plot (called the "thinning comparison" plot) that is located along with the transition between recently thinned and unthinned ponderosa pine forest patches near Clints Well, AZ. The dense forest and thinning comparison plots have shallower snowpack than the montane plots. These four plots span a range of forest densities, tree heights, and topographic conditions (Table 1), allowing us to evaluate the performance of snow measurement with SfM and airborne lidar under a variety of forest conditions.

Table 1. Topographic, canopy cover/height and sampling information for each SfM survey.

Survey	Area [1] (ha)	Slope [2] (%)	Canopy Cover (%)	Canopy Height (m)			# Sample Locations [3]	Lidar
				Mean	Median	Max		
Montane Meadow (1 February 2017)	5.9	7.4	12.9	3.4	0.1	33.4	94	Y
Montane Meadow (5 March 2018)	12.2	8.1	11.3	3.1	0.1	34.0	58	N
Montane Meadow (4 March 2019)	9.5	7.9	8.9	2.5	0.1	34.0	108	Y
Montane Meadow (4 March 2020)	11.6	8.1	10.3	2.9	0.1	34.0	38	N
Montane Valley (5 March 2018)	6.8	8.4	16.2	4.3	0.1	34.5	37	N
Montane Valley (4 March 2019)	6.9	8.2	12.8	3.8	0.1	34.7	37	Y
Montane Valley (4 March 2020)	7.4	7.9	13.1	3.9	0.1	34.7	22	N
Dense Forest (1 February 2017)	4.0	3.9	47.0	11.2	12.2	37.3	70	Y
Dense Forest (7 March 2017)	5.1	3.6	49.1	11.5	12.9	29.7	70	Y
Dense Forest (4 March 2019)	9.5	4.4	44.8	10.8	11.6	30.4	79	Y
Thinning Comparison (4 March 2019)	17.0	3.1	32.7	9.0	8.4	41.5	105	Y

[1] Area refers to the size of the SfM-generated snow maps. [2] Slope, canopy cover, and canopy height are computed from 2019 airborne lidar data [52]; [3] Sample locations refers to the number of individual locations where snow depth is sampled.

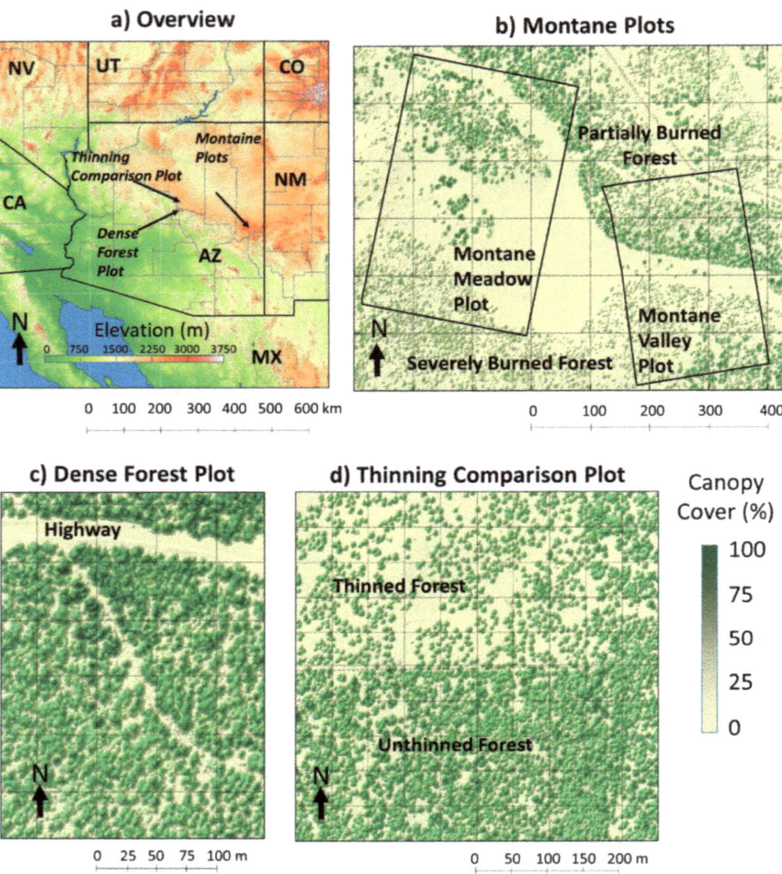

Figure 1. Locations of the four study plots (**a**—overview, **b**—montane meadow and montane valley plots, **c**—dense forest plot, **d**—thinning comparison plot). Canopy cover data are derived from 2019 airborne lidar data [52]. At these field sites, we collected rich snowpack datasets, including a variety of airborne and ground-measured snowpack data that are ideal for creating and evaluating high-quality SfM models. Snow surveys for these plots consisted of multi-angle RPAS imagery, airborne lidar, and ground measurements of snow depth and SWE. In total, there were 11 snow surveys (Table 1), with the most surveys occurring at the montane plots (as this area is important for snowpack monitoring for local water supply managers), with fewer surveys occurring at the dense forest and thinning comparison plots. All surveys included RPAS flights to acquire aerial imagery (using a DJI Phantom 3 Pro drone with a 12 megapixel camera—for 2017 surveys—and a DJI Phantom 4 Pro drone with a 20 megapixel camera—for 2018–2020 surveys) to generate snow-on 3-D models of the study plots using SfM (see Section 3). All RPAS flights occurred on clear, sunny days between 10:00 AM and 2:00 PM to minimize the length of shadows. For the flights in 2017, the RPAS was controlled manually with a combination of nadir and oblique (~30% nadir) imagery along back and forth flight lines spanning each domain. For the flights in 2018-2020, flight planning software called Altizure® was used to collect the imagery in a regular flight pattern (using back and forth flight lines) with a set of nadir imagery, as well as two sets of oblique images (with 75% overlap for all imagery). All flights occurred at ~75 m above ground level with horizontal velocities of ~5–7 m/s resulting in ~4 cm and ~2.1 cm spatial resolution on the ground for the Phantom 3 and Phantom 4 systems, respectively. The cameras used automatic exposure settings.

These surveys also included ground measurements of snow depth and snow density. A majority of these measurements were collected by more than 25 employees and volunteers from five public and private partner entities (see Acknowledgements) along carefully pre-marked snow survey transects with precisely geolocated (using a differential GPS) endpoints to enable the precise geolocation of snow measurements in order to relate them to the airborne data. Following the sampling methodology described in Reference [53], five snow depth samples were taken every 5 m along the transects in a star pattern to ensure spatial representativeness of the snow depth samples, and snow density was sampled every 10–20 m along the transects. There were also some additional measurements along transects (also positioned with a differential GPS) where snow depth is monitored using remote cameras for temporal snowpack monitoring (however, this study only uses the snow depth data, as well as manual measurements of snow density along these transects for the survey dates in Table 1). Not all dates had identical ground surveys for each study plot, depending on logistical considerations (e.g., available workers). Overall, individual surveys included ground measurements from 22–108 individual locations (Table 1). Note that this translates to >3000 measurements as there were five snow depth measurements at each manual sample location. Further details about the snow survey methodology can be found in Reference [54].

2.2. Generation of Lidar Snow Depth, Density, and SWE Maps

For seven of the surveys, we generated state-of-the-art airborne lidar-based maps of snow depth, snow density, and SWE, as they included airborne lidar data. These maps were created for the two ~100 km^2 lidar footprint areas (which each have three separate lidar acquisitions), that encompass the study plots. The two montane plots are within a "high-elevation" lidar footprint, and the dense forest and thinning comparison plots are within a "mid-elevation" lidar footprint. These maps were created using high point density (~10–15 points/m^2) snow-on airborne lidar data (which are differenced from similar snow-off lidar data) and are bias-corrected with the field-measured snow depths. All lidar data was flown by Quantum Spatial Inc. [52,55–58].

The snow depth maps were then combined with snow density maps generated using artificial neural network (ANN) machine learning of the field-measured snow density measurements, using various lidar-derived physiographic attributes as predictors. The methodology for creating these maps (and their validation) is detailed in Reference [54], but is briefly summarized here. We use an ensemble of ANNs with relatively simple network structure (1 hidden layer with ten neurons) with Levenberg-Marquardt (L-M) backpropagation (Marquardt, 1963) to model the field snow density measurements using lidar-derived metrics as predictor variables. The predictor variables include physiographic variables (elevation, slope, northness—or sin(slope) × sin(aspect), canopy height, and canopy closure), other GIS-derived quantities using the lidar data (skyview factor, below canopy solar forcing index—or the ratio of incident solar radiation reaching the ground surface over a given period to that hitting a flat surface with no obstructions), and lidar snow depth (from 1 February 2017). These snow density maps are multiplied by the lidar snow depth maps to generate maps of SWE.

2.3. Generation of SfM Snow Depth Maps

Snow-on point clouds were generated from the RPAS imagery using Agisoft Metashape software. After the point clouds were created, the ground points were isolated by using a combination of Metashape's built-in ground filtering algorithm (for a first cut separation of the point cloud) and a Cloth Simulation Filter algorithm [59] implemented in R (to remove remaining debris from the ground point cloud).

Next, we used an automated post-processing workflow to accurately georeference the point clouds using available snow-off point cloud data. In this study, we use snow-off lidar point cloud data (from summer 2014) for this purpose (though we also tried, with similar results, applying the workflow using snow-off SfM data that we collected—see Section 3.1 below). This workflow consists of two general steps. First, the snow-on SfM point clouds were co-registered with the snow-off point cloud canopy layer by removing any vertical offsets between the snow-on and snow-off point clouds and then using an Iterative Closest Point algorithm implemented in CloudCompare software to match the canopy layers. This successfully adjusted the snow-on point clouds to match the canopy elements in the snow-off point clouds, even when there was sometimes substantial vegetation change between the snow-on and snow-off data (Figure 2). Next, warping and tilting of the point clouds were removed using low-order polynomial filters (a 1st order polynomial correction to correct tilt followed by a 2nd order polynomial correction to correct warping) by comparing them to the lidar ground surface plus a first guess of snow depth, based on the ground survey data. This first guess was generated using the same ANN methodology as the creation of snow density maps from field snow density measurements in Reference [54] and described in Section 2.2, except that field measured snow depths, rather than field measured snow densities, are used as the predicted variable.

Figure 2. Example showing the snow-on SfM point cloud in the center of the montane meadow plot (mostly white, green, and brown colored dots represent snow, evergreen canopy, and woody material) in relation to the lidar point cloud (red dots) for an area in the montane meadow plot. Note that in some areas, there is a difference between the lidar (from summer 2014) and SfM (from winter 2017) point clouds, due to post-wildfire effects (lost branches, fallen snags and trees), but for the snags and trees that did not change during this period, the SFM and lidar point clouds are very close to one another.

This processing workflow was automated using CloudCompare's command line mode (for the ICP algorithm), as well as the US Forest Service's FUSION/LDV [60] software and a python script to model and correct warping and tilting in the snow-on SfM point cloud. The ANN machine learning methodology is implemented in Matlab® software, as described in Reference [54].

2.4. Using SfM Snowpack Data to Supplement Lidar Snowpack Data

We used these SfM snow depth maps to advance our understanding of snow distributions for our research sites in a number of ways. First, following the methodology in Reference [54], we combined them with maps of snow density generated using ANN modeling of snow density measurements to generate maps of SWE for our study plots. This enabled us to characterize differences in terms of snow depth, snow density, and SWE for different seasons. This is important because while the directly measured variable using lidar and SfM is snow depth, SWE is the snowpack variable that is of most interest in hydrological applications [61]. Although the distribution of SWE is broadly similar to that of snow depth, it is important to consider the effects of variable snow densities, which can vary substantially in both space and time [54,62].

Next, we investigated the relationships between snow depths for the larger mid- and high-elevation lidar footprints on different dates to see how well they could be used to extrapolate snow depths measured for the SfM study plots to the larger lidar footprints. In particular, we compared the relationships between snow depths from the mid-winter (on 1 February 2017) and late winter (on 7 March 2017 and 4 March 2019) lidar snow surveys with those from the mid-winter lidar snow survey and the late-winter SfM data. To ensure that only the best quality SfM data was used to construct these relationships, we only considered SfM data for open areas without an overhead canopy for these comparisons. Of particular interest was (1) the strength of the relationships between the lidar data on multiple dates (which indicated how well they could be used to predict snow depths for different times), and (2) how well the SfM-data over the small study plots could capture those relationships. Even though the late winter surveys occurred at the same time of year, the 7 March 2017 survey reflected more evolved snowpack conditions: For the mid-elevation lidar coverage, it reflected mid-ablation to nearly snow-free conditions (depending on elevation), and for the high-elevation lidar coverage, it reflected peak-SWE to mid-ablation conditions, depending on elevation. On the other hand, the 4 March 2019 survey reflected conditions just prior to Peak SWE for the high-elevation lidar coverage, and mid-ablation conditions for the mid-elevation lidar coverage. The 1 February survey reflected mid-winter snowpack conditions in both areas.

Finally, we tested how well snow depths could be estimated from the SfM data for areas outside the SfM plots (but within the lidar coverages) on these different dates using the relationships between the late-winter SfM data and the mid-winter (1 February 2017) lidar data. These relationships were modeled using a third order polynomial fit (which was chosen based on the data analysis described above), and the regression parameters were applied to the 1 February 2017 lidar data to generate the predictions. These simulated maps were then compared to the actual lidar observations on 7 March 2017 and 4 March 2019.

3. Results

3.1. Comparison Between SfM, Field-Based, and Lidar Snow Depths

Overall, there is good agreement between the lidar and the SfM snow depth maps in the montane plots, which generally have sparse canopies. Figure 3a–i shows that most major snow depth features are consistently represented in both maps, such as the areas with deep snow (which usually correspond to north-facing aspects), and the extremely shallow snow on south-facing sides of large tree stands. Overall, there are high squared correlation coefficient (R^2) values (0.78 to 0.89), relatively low Root Mean Squared Error (RMSE) values (8.5 to 9.4 cm) and small average differences (SfM average depths are 1.0 to 3.5 cm shallower) between the SfM and lidar data (Table 2). The coefficient of variation (CV) differences between the maps are also relatively low (0.01–0.04). At the same time, the agreement between the SfM and field measured snow depths is comparable to that between the lidar and field measured snow depths for the montane plots (Table 3, Figure S1). Although R^2 values between the SfM data and ground measurements (0.60–0.90) are a little lower than those for the lidar data (0.68–0.94), SfM RMSEs (8.6–13.4 cm) are similar to lidar RMSEs (8.5–12.4 cm). Both the SfM and lidar data are relatively unbiased with respect to ground observations for the montane plots.

There is generally a lower correspondence between the lidar and SfM snow depth maps for the dense forest plots. Major features can clearly be seen in both sets of maps (e.g., the road on the northern end of the plot as well as the snowbanks on either side of the road), though other details in the forest are less consistent (Figure 3j–r). For example, the SfM maps show more pockets of shallow snow (especially on 1 February 2017). Compared to the montane plots, the dense forest plot has lower R^2 values (0.23 to 0.57), higher RMSEs (12.4 to 24.0 cm), and higher CV differences (0.09 to 0.26; with the SfM data having higher CVs than the lidar data; Table 2). At the same time, though, average depth differences are only a little larger than those observed in the montane plots (3.8 to 6.3 cm). Compared to the ground observations, the lidar generally performs better than the SfM data, though the performance of the SfM data is also highly variable (Table 3, Figure S1) for the dense forest plots. RMSEs between SfM and field measured snow depths range from 8.8 and 17.2 cm (vs. 8.3 to 10.3 cm for the lidar data), and R^2 values are 0.26 to 0.29 (vs 0.33 to 0.62 for the lidar data). However, the SfM data has lower biases, compared to the ground observations, than the lidar data at the dense forest plot (−0.4 to 2.1 cm vs 0.8 to 5.8 cm).

Table 2. Agreement statistics between lidar and SfM snow depth data for the paired airborne lidar–SfM surveys. Statistics are for the outlined areas for each plot in Figure 3.

	Lidar Depth (cm)	SfM Depth (cm)	Lidar CV	SfM CV	R^2	RMSE (cm)
Montane Meadow (1 February 2017)	65.1	62.1	0.26	0.30	0.78	9.2
Montane Meadow (4 March 2019)	75.3	71.8	0.25	0.26	0.80	9.4
Montane Valley (4 March 2019)	63.9	62.9	0.40	0.39	0.89	8.5
Dense Forest (1 February 2017)	54.3	48.0	0.29	0.55	0.23	24.0
Dense Forest (7 March 2017)	30.7	26.9	0.61	0.70	0.34	17.6
Dense Forest (4 March 2019)	34.9	29.6	0.45	0.55	0.57	12.4
Thinning Comparison (4 March 2019)	9.5	10.7	0.90	0.80	0.41	7.4

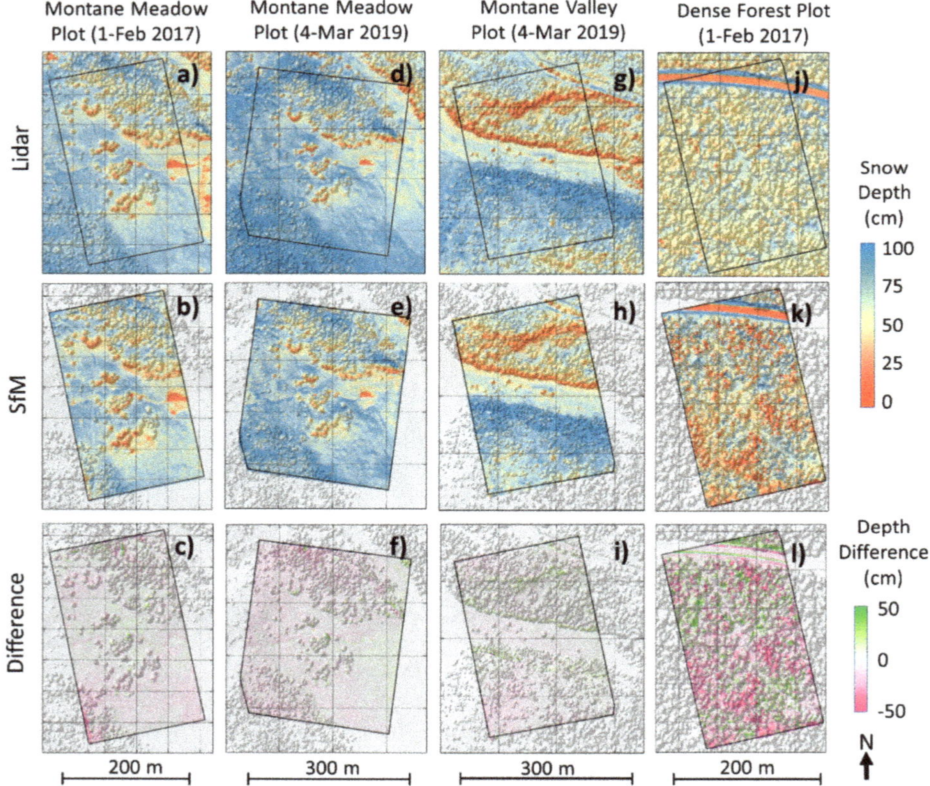

Figure 3. *Cont.*

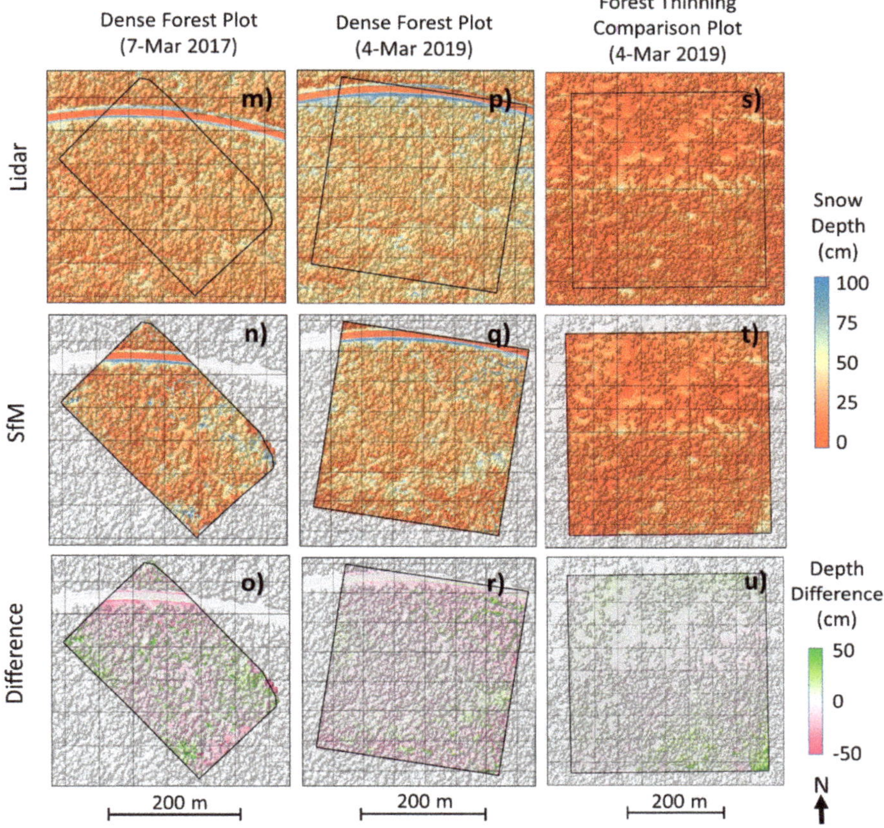

Figure 3. Maps showing lidar snow depths, SfM snow depths, and their differences (SfM-lidar) for the paired airborne lidar–RPAS SfM surveys for the montane meadow plot on 1 February 2017 (**a**–**c**) and 4 March 2019 (**d**–**f**); the montane valley plot on 4 March 2019 (**g**–**i**); the dense forest on 1 February 2017 (**j**–**l**), 7 March 2017 (**m**–**o**), and 4 March 2019 (**p**–**r**); and thinning comparison plot on 4 March 2019 (**s**–**u**). A canopy hillshade effect is added to show the locations of trees.

Table 3. Agreement statistics between lidar/SfM and field measured snow depth data for the paired airborne lidar–surveys.

	SfM R^2	SfM RMSE (cm)	SfM Bias (cm)	Lidar R^2	Lidar RMSE (cm)	Lidar Bias (cm)
Montane Meadow (1 February 2017)	0.82	8.6	0.7	0.79	9.2	1.6
Montane Meadow (4 March 2019)	0.60	13.4	−0.7	0.68	12.4	2.4
Montane Valley (4 March 2019)	0.90	9.7	2.7	0.94	8.5	4.1
Dense Forest (1 February 2017)	0.29	17.2	2.1	0.33	10.3	5.2
Dense Forest (7 March 2017)	0.26	16.3	1.8	0.40	9.3	0.8
Dense Forest (4 March 2019)	0.27	8.8	−0.4	0.62	8.3	5.8
Thinning Comparison (4 March 2019)	0.37	7.9	2.0	0.46	7.1	1.9

For the thinning comparison plot (Figure 3s–u), there is a higher correspondence between the lidar and SfM maps than for the dense forest plot. The thinning comparison plot also has very little snow (~10 cm on average) during the March 2019 survey (the only snow survey performed at this site), yet both technologies show similar snow patterns. Overall, RMSE (7.4 cm) between the lidar and SfM data is lower than for the other surveys, R^2 (0.41) is higher than some of the dense forest plots surveys but lower than those in the montane plots (with the opposite true for CV differences), and the average snow depth difference (1.2 cm) is small compared to the other surveys. Compared to the ground survey data, the performance of the lidar and SfM data is similar (with R^2 values of 0.37 and 0.46 and RMSEs of 7.1–7.9 cm), again with a relatively small bias (1.9 cm). Overall, the thinning comparison plot has lower vegetation density than the dense forest plot, but higher vegetation density than the montane plots (Table 1).

The accuracy of these SfM data depends, in part, on the quality of the first guess snow depth maps (which are shown in Figure S2 Supplemental Materials). In general, these first guess maps are comparable, in magnitude to the SfM and lidar maps, shown in Figure 3, but they typically do not reflect as much variability as the SfM and lidar maps do. The overall agreement between theses maps is usually lower than that between the SfM and lidar maps (with the exception of the 1 February 2017 dense forest plot, R^2 values range from 0.09–0.33 lower, and RMSE values range from 0.4–3.6 cm higher). Note that for the forest thinning comparison plot, the snow depths were so shallow that a spatially variable first guess map was not needed. The first guess snow depth map showed virtually no spatial variability, due to the small snow depth signal compared to the uncertainty of these maps (Figure S2g). Nevertheless, it was included here to maintain methodological consistency with all other surveys.

Finally, note that the same methodology used here can be used to generate SfM snow depth maps that based on snow-off SfM point clouds instead of snow-off lidar point clouds (Figure S3). The agreement between the SfM and lidar snow depth maps is a little lower for these maps (compare Table 2 and Table S1). This discrepancy may be related to inconsistent ground filtering algorithms used for the lidar data (which was performed by QSI) and SfM data (see Section 2.3).

3.2. SWE Monitoring Using SfM

Due to the hydrological importance of quantifying water content of the snowpack, the snow density and SWE maps for each plot prepared using the methodology of Broxton et al. [54] (see Section 2.3) were useful for multi-temporal monitoring of snowpack. In this study, they were especially useful for the montane plots because this area is very important for predicting local water supplies. For this area, there were snow surveys for four subsequent water years near the time of peak SWE (in early March). Figure 4 shows spatial differences between snow depth, snow density, and SWE for this area from 2017–2020 (note that in March 2017, there was airborne lidar, but no SfM data in the area).

While the 2019 maps depict snowpack conditions just prior to peak SWE, the 2017 and 2020 maps reflect snowpack conditions that are further along in the melt season (note the relatively large areas with no snow on south-facing slopes), and the 2018 maps depict historically low snow conditions (as 2018 was a very dry and warm year). Despite the differences between snowpack conditions between 2018 and the other years, the 2018 data shows relatively deeper snowpack in the same areas as in the other sets of maps (e.g., on north-facing slopes in the montane valley plot and on the shaded sides of the montane meadow plot). There is a relatively high correlation between the 2017, 2019, and 2020 surveys (R^2 between the 2017 and 2019 survey data are 0.71 and 0.69 for depth and SWE, respectively, and the R^2 between the 2017 and 2020 surveys are 0.67 and 0.60 for depth and SWE, respectively). The R^2 between the 2017 and 2018 survey is lower, (R^2 = 0.36 for snow depth and 0.39 for SWE). Overall, the 2019 survey had the most snow, but snow variability (as measured by the CV of snow depth and SWE) was higher during 2017 (when there was more advanced snowmelt) and 2018 (when the snowpack was shallow; Table 4) surveys. Snow density was also lowest in 2018 (due to shallow snowpack) and highest in 2020 (as the survey occurred after a long period of warm, dry weather).

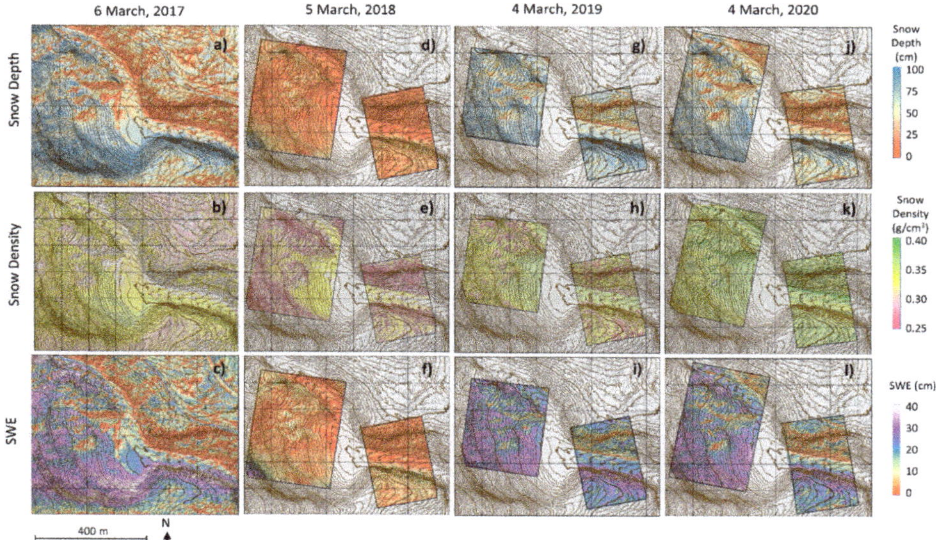

Figure 4. Maps of snow depth, snow density, and SWE for the montane plots near-maximum snow accumulation on 6 March 2017 (**a–c**), 5 March 2018 (**d–f**), 4 March 2018 (**g–i**), and 4 March 2020 (**j–l**). A canopy hillshade effect is added to show the locations of trees and contours (contour interval = 2 m) are added to help interpret the topography of the area.

Table 4. Averages and coefficient of variation (CV) for the snow depth, snow density, and SWE maps in Figure 5 (for areas that have data for all years). For 7 March 2017, statistics are derived from the airborne lidar survey (as no SfM was flown for this area on this date), and for the other dates, they are derived from the SfM surveys.

Date	Average Depth (cm)	Depth CV (-)	Average Density (g/cm^3)	Density CV	Average SWE (cm)	SWE CV
7 March 2017	52.0	0.63	0.32	0.05	17.0	0.65
5 March 2018	14.2	0.85	0.30	0.07	4.2	0.83
4 March 2019	69.3	0.34	0.33	0.05	22.7	0.35
4 March 2020	56.7	0.40	0.36	0.04	20.2	0.38

3.3. *Using Airborne Lidar to Extend the SfM Data*

Figure 4 shows that while there are differences between the snowpack distributions for different years (e.g., between the peak-SWE conditions on 4 March 2019 vs the post-peak SWE conditions on 6 March 2017), there are also similarities (e.g., areas with deeper snow in one scene tend to have deeper snow in other scenes). To further understand the consistency between the snowpack at different times across much larger (~100 km^2) lidar coverages, we plotted the relationship between snow depth for three sets of lidar coverages (at different times in the snow season) for the mid- and high- elevation lidar domains (Figure 5). The mid-winter (1 February 2017) lidar data is used as the abscissa, and the late-winter (7 March 2017 and 4 March 2019) lidar data are used as the ordinates in these plots. Figure 5 (red dots) shows that the relationships between the snow depth data change between different dates. For example, while the relationship between 1 February 2017 and 4 March 2019 lidar data for the high-elevation lidar coverage is linear, the relationship between the 1 February 2017 and 7 March 2017 lidar data for the high-elevation lidar box is nonlinear. The former represents a comparison between mid-winter and conditions just prior to peak SWE while the latter represents a comparison between mid-winter and primarily post-peak SWE conditions. Despite this changing shape, the strength of these relationships is consistent in both cases (with R^2 values of ~0.9). For the mid-elevation lidar coverage, these relationships (which both represent a comparison of mid-winter conditions and mostly mid-ablation conditions) are nonlinear for both pairs of data, and are somewhat weaker, but still consistent (R^2 = ~0.77 in both cases). This suggests there a high degree of intra-seasonal and inter-annual relatability between snow depth patterns for our study sites at different times. Note that these relationships are similar if all lidar pixels are considered (thick red dashed lines in Figure 5) vs. if only open pixels (where SfM performs well) are considered (thin red dashed lines in Figure 5).

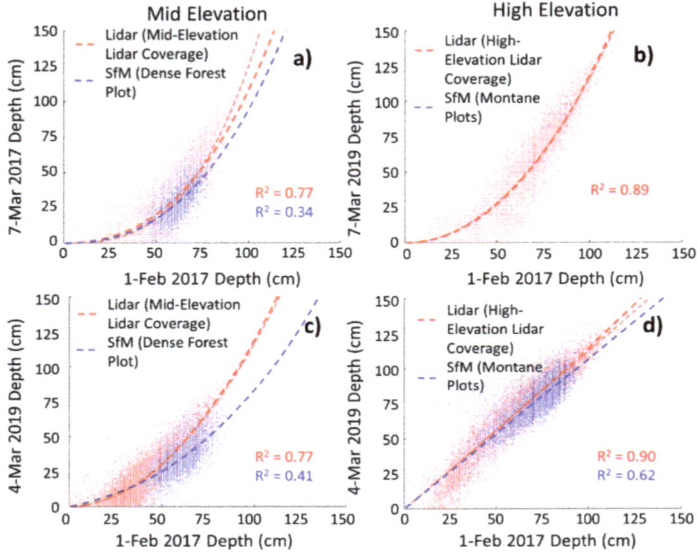

Figure 5. Relationships between 1 February 2017 lidar snow depth and 7 March and 4 March 2019 lidar (red dots, for the entire lidar coverages) and SfM (blue dots, for the SfM survey plots, based on open pixels) snow depth data for the mid-elevation (**a**,**c**) and high-elevation (**b**,**d**) lidar coverages. The dashed lines show the third order polynomial regressions for each data series (for the lidar regression, additional lines have been added to show the similarity of the comparisons when using only open pixels (thin red dashed lines) and all pixels (thick red dashed lines). The open area SfM regressions (blue lines) are used to simulate the 4 March 2019 snow depths for the airborne lidar coverages in Figures 6 and 7.

Next, we tested how well the SfM data within the small study plots can capture these relationships. Also plotted in Figure 5 are the relationships between the 1 February 2017 lidar snow depth and the 7 March 2017 and 4 March 2019 SfM data for the dense vegetation plot (for the mid-elevation panels) and for the montane plots (for the high-elevation panels), considering only open pixels (where the SfM data perform well). Note that there was no SfM survey for the montane plots on 7 March 2017. The data follow similar trends to those found for the larger lidar coverages (just a little flatter; Figure 5), though with lower R^2 values. This may be due to the lower quality of the SfM data in some environments (e.g., the dense forest plot), as well as the smaller dynamic range represented in these small plots compared with the overall lidar coverages (note that the dense forest plot, in particular, is flat with fairly uniform forest cover across most of the plot).

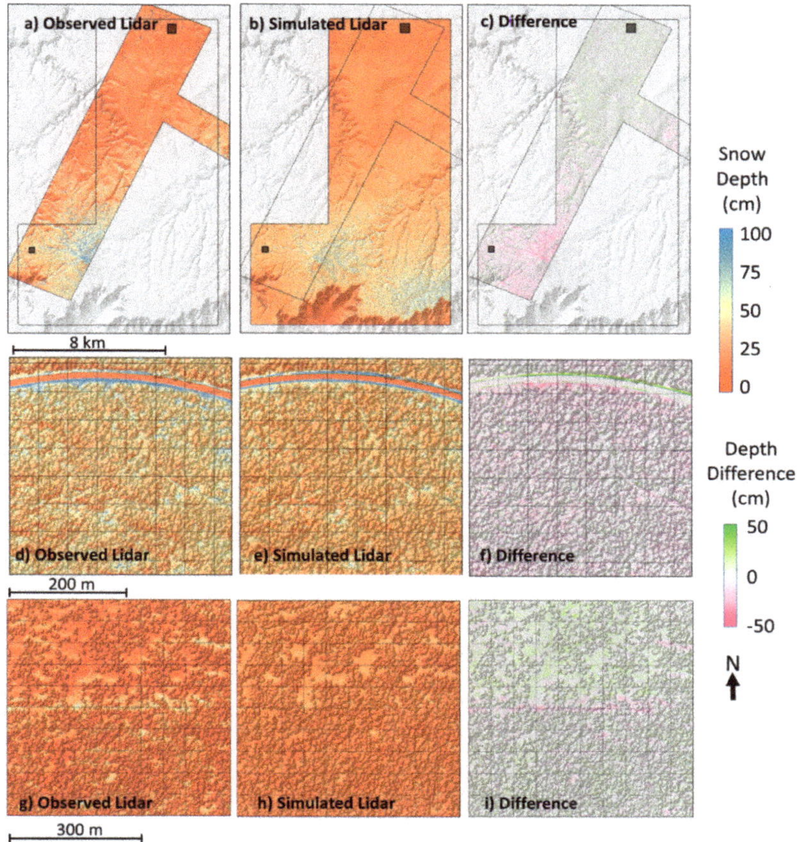

Figure 6. Observed lidar-based snow depths for 4 March 2019, simulated data (using the regression between the 4 March 2019 SfM data and 1 February 2017 lidar data and shown in Figure 5c), and the difference (simulated-observed) for the mid-elevation lidar coverage (**a–c**) and the area around the dense forest (**d–f**) and thinning-comparison (**g–i**) plots. Note that the simulated depths are for the area covered by the 1 February 2017 lidar, while the March 2019 lidar has a different footprint. The small dark boxes in (**a–c**) show the area covered by the inset maps (**d–i**) corresponding to the plots in Figure 1.

The strength of the relationships between the lidar snow depth data on different dates, as well as the ability of the SfM data to capture these trends suggests that the SfM data can be used to predict snow depth distributions over the larger airborne lidar coverages, given that a lidar snow depth map already exists. To evaluate how well this extrapolation works, we used the relationships between the 1 February 2017 lidar, 7 March 2017 and 4 March 2019 SfM data, shown in Figure 5, applied to the 1 February 2017 lidar map, as described in Section 2.4. Figures 6 and 7, which show observed and simulated snow depth data for the mid- and high-elevation lidar coverages on 4 March 2019, illustrate that these extrapolations are fairly successful at estimating the observed lidar snow depth data. This can be seen for both the entire ~100 km^2 lidar coverages, as well as for the < 1 km^2 study plots. Note, however, that there can be significant differences between the observed and simulated snow depth patterns in areas that have undergone substantial vegetation changes (such as the thinning comparison plot in Figure 6g–i, which underwent mechanical thinning between the 2017 and 2019 snow surveys).

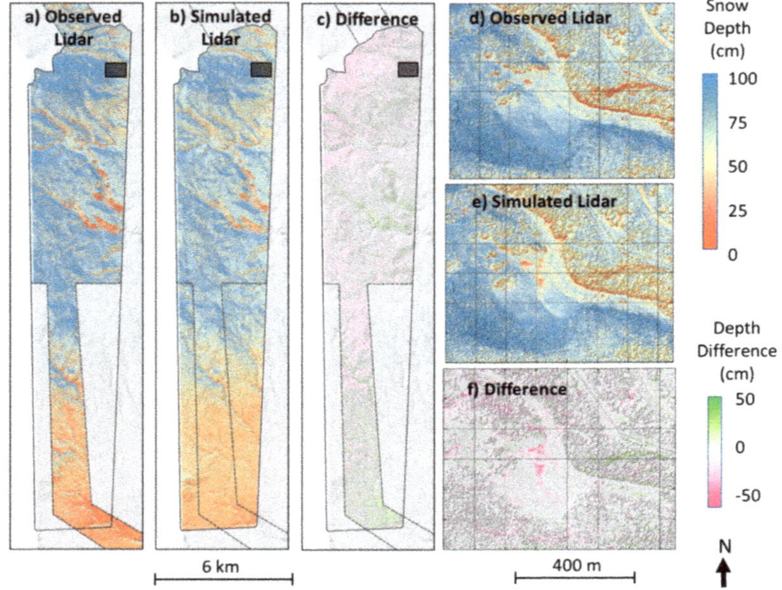

Figure 7. Observed lidar-based snow depths for 4 March 2019, simulated data (using the regression between 4 March 2019 SfM data and 1 February 2017 lidar data shown in Figure 5d), and the difference (simulated-observed) for the high-elevation lidar coverage (**a–c**) and the area around the montane plots (**d–f**). Note that the simulated depths are for the area covered by the 1 February 2017 lidar, while the March 2019 lidar has a different footprint. The small dark boxes in (**a–c**) show the area covered by the inset maps (**d–f**) corresponding to the plots in Figure 1.

In general, spatial statistics from these simulated snow depth maps are also fairly similar to those from the observed lidar data. Table 5 shows statistics related to the distribution of snow depths (mean, cv, skewness) as well as measures of autocorrelation (measured by the fractal dimension of snow depths for distances that are smaller or larger than a well-known scale break that occurs at ~25–30 m). Note that there was no SfM survey for the high-elevation coverage on 7 March 2017, so statistics for this date are given assuming a simulated map that would be obtained if the relationships between the lidar data (red lines in Figure 5) were captured. For the high-elevation coverage, the distribution parameters for the simulated maps are very close to those for the observed maps: The mean snow depth is within 5%, the CVs are ~10% lower, and the skewness is similar. Likewise, both the short- and long-range fractal dimensions are similar, with the latter above 2.9, suggesting spatially random variability for longer correlation distances, and the former ~2.7–2.8, indicating more spatial complexity at shorter correlation distances. The agreement between these statistics for the simulated and observed lidar maps is a little worse for the mid-elevation coverage, as the simulated maps had slightly lower (by ~10%) mean snow depths, lower (by ~25%) CV, but similar skewness. Compared with the high-elevation coverage, there were also slightly larger differences between the short-range fractal dimensions for the observed and simulated maps, suggesting that the simulated maps for the mid-elevation coverage were not as good at capturing small scale snow depth variability as for the high-elevation coverage.

Because the fine scale (1 m) spatial variability depicted in these simulated maps can be somewhat different than in the observed maps, we compared the observed and simulated maps at a variety of spatial scales, by averaging the 1 m pixels to larger pixel sizes (Figure 8). As expected, the simulated and observed maps have a closer fit for larger pixel sizes, with the largest increases occurring between 1 and 100 m, and smaller differences occurring after that. For the mid-elevation lidar coverage, the R^2

values increase from ~0.7–0.8 for 1 m pixels to > 0.9 for 100 m pixels, while RMSEs decrease from 10–15 cm for 1 m pixels to ~7–8 cm for 100 m pixels. For the high-elevation lidar coverage, the R^2 values increase from ~0.9 for 1 m pixels to > 0.95 for 100 m pixels, while RMSEs again decrease from 10–12 cm for 1 m pixels to ~6–7 cm for 100 m pixels. Figure 8 also shows the performance of simulated maps that would be obtained if the relationships between the lidar data (red lines in Figure 5) were captured perfectly. For these scenarios, there is some reduction in RMSE (by ~20–30%), suggesting that some of the error between the simulated (using SfM data) and observed snow depth data is the result of this relationship not being fully captured by the SfM data. Similar results can be obtained for predictions of SWE values (compare Figures S5—S8 with Figures 5–8).

Table 5. Mean, coefficient of variation (CV), and skewness, and short- and long-range fractal dimensions (D) for observed and simulated snow depth maps for the mid- and high-elevation lidar coverages for 7 March 2017 and 4 March 2019.

		Mean (cm)		CV		Skewness		Short-Range D [1]		Long-Range D [1]	
		Obs	Sim	Obs	Sim	Obs	Sim	Obs	Sim	Obs	Sim
Mid-Elevation	7 March 2017	19.3	17.3	1.32	1.00	2.8	2.4	2.60	2.77	2.93	2.93
	4 March 2019	21.5	19.0	0.92	0.64	1.4	1.8	2.78	2.73	2.96	2.96
High-Elevation	7 March 2017	41.4	42.4	0.88	0.78	0.6	0.9	2.68	2.71	2.90	2.92
	4 March 2019	67.5	64.6	0.45	0.40	−0.3	−0.2	2.78	2.81	2.91	2.90

[1] The fractal dimensions (D) are estimated from the slope of log-log unidirectional semi-variogram plots as in Reference [63], where D = 3-b/2, and b is the slope of the log-log semi-variogram over a particular interval. The scale break that separates the short- and long-range segments (30 m for the mid-elevation coverage and 25 m for the high-elevation coverage) corresponds to a well-known scale break (e.g., References [64,65]) that separates rich spatial-complexity of snow depths at smaller scales, and more spatially random variability at larger-scales.

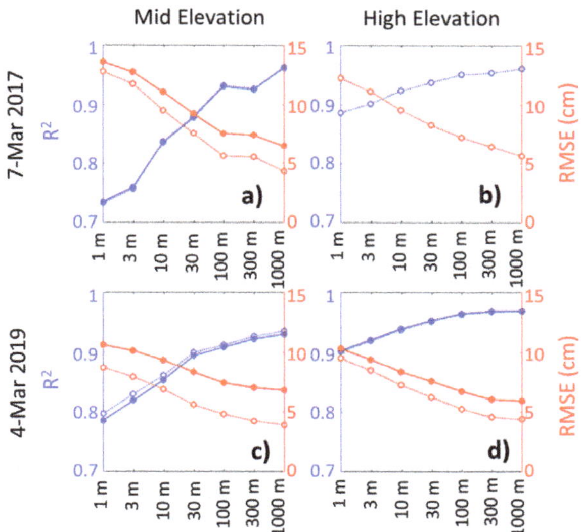

Figure 8. Performance of the simulated snow depth maps for the mid- and high-elevation lidar coverages on 7 March 2017 (**a–b**) and 4 March 2019 (**c–d**) for different pixel scales. The original 1 m maps are aggregated (by averaging) to the larger pixel scales. The solid lines depict the performance of the simulated maps, created using the relationships between the 1 February 2017 and subsequent SFM data (to assess the actual ability of the SfM data to predict snow depth distributions over the larger lidar domains). The dashed lines with open circle markers depict the performance of the simulated maps, created using the relationships between the 1 February 2017 and subsequent SFM data (to assess the potential ability of the SfM data to predict snow depth distributions over the larger lidar domains).

4. Discussion

In this study, we show that RPAS-SfM monitoring of snowpack can be a useful tool to provide temporal snowpack monitoring. This monitoring provides accurate snow depth maps over small study plots that do not have too much forest cover, and also offers the ability to capture relationships between current and past snow depths that can be used to estimate snow depths for larger areas. In particular, we show that for our study sites, limited spatial extent RPAS-SfM snow depth observations (when combined with past lidar snow depth observations) can be used to generate fairly accurate estimates of snow depth across the lidar domains (having R^2 and RMSE values of ~0.75–0.9 and 10–15 cm, respectively, as compared with observed lidar snow depth distributions). These estimates improve further when aggregating to larger pixel sizes (Figure 8). Note that even at the 1 m pixel scale, these values are comparable with those of the SfM snow depth maps themselves and a little better (higher R2, lower RMSE) than those for the first guess snow depth maps (compare Figure 8 with Table 1 and Table S1). This implies that SfM should be useful for multitemporal snow monitoring that is applicable at basin-scales. This is important given the high costs of repeated airborne lidar acquisitions. With the exception of a very few high-budget operations (e.g., NASA's Airborne Snow Observatory [29]) this high cost makes it infeasible to collect many repeated snow-on lidar data for most sites. In addition, RPAS-SfM may be especially suitable for operational snowpack monitoring because the SfM data can be processed relatively quickly [39].

The success of the extrapolations shown here depends on two critical factors. First, they rely on the fact that there is a strong relationship between snow depth distributions at different times. We tested this by comparing the snow depth data across the mid- and high- elevation lidar coverages at three different times, representing a range of snowpack conditions, from mid-accumulation to mid-ablation (Figure 5). Even though the shape of these relationships changed considerably at different times (i.e., it was more linear when comparing snow depths within the accumulation season and nonlinear when comparing accumulation season to ablation season snow depths), their strengths were consistent at different times (R^2 ~0.77 and ~0.90 for the mid- and high-elevation lidar coverage, respectively). This suggests that they should be useful for predicting snow depth distributions for other times as well.

The second critical factor affecting the success of these extrapolations is the ability of the distributed SfM data to capture these relationships. Our SfM data were generally able to capture these relationships, however, they did a better job for the montane plots than for the dense forest plot. This can be seen by the higher agreement statistics (Figure 8) and smaller differences between statistics describing the spatial distribution and spatial autocorrelation (Table 5) of the observed and simulated maps for the high-elevation lidar coverage. This ability to reproduce these spatial statistics is particularly important for applications that are based on statistics from lidar snow data (such as snow-cover depletion models [66]). The lower agreement between the observed and simulated maps for the mid-elevation lidar coverage could be because the dense forest plot (from which the SfM data is extrapolated) has a flat topography and mostly uniform forest cover, and so contains a relatively small range of snow conditions. However, it could also be because SfM data quality is much poorer for densely forested conditions [20,67]. We tried to mitigate this by only considering pixels without an overhead canopy. While this improved the ability of the SfM data to capture these relationships, it also ignores under-canopy areas such as tree wells in their derivation. Nevertheless, this does not seem to have too much of an effect on the relationships themselves as they are only sensitive to snow depth changes between the survey dates, which occur similarly for both canopy-covered and open areas (note the similarity between the thick and thin red lines in Figure 8).

Although this method works well at our research sites, more research is needed to understand how well it works at other sites. It is well known that there is substantial similarity between snow distributions at similar points in the snow season (e.g., peak SWE) [68–73]. However, there are also substantial spatial differences between snow distributions during the accumulation and melt seasons [46,74–76]. This does not necessarily mean, though, that there are not strong relationships relating melt-season snow depth to accumulation-season snow depth (for our study sites; there *were* strong relationships; Figure 5), especially since spatial patterns of snow disappearance are related to the distribution of snowpack at the start of the melt season [77].

One thing that is likely to lead to greater difficulties when using the pre-existing lidar observations of snow distributions is land cover change (such as forest thinning, wildfire, or normal forest growth) because these changes are known to alter the snow depth patterns [4,6,9,78]. We showed in Figure 6 that the observed and predicted snow depth distributions were different following forest thinning treatments at the forest thinning comparison site. Climate variability and climate change may also cause issues, since snow distributions for a particular year are influenced by differences in the sequence, timing and magnitudes of snowfall/accumulation events, redistribution (due to wind), interactions with canopy, and melt [50,79,80]. Nevertheless, the high correlations for the snow depth relationships, shown in Figure 5, and for the simulated snow depth maps resulting from these relationships (in Figures 6 and 7) suggest that these differences do not necessarily make it harder to simulate snow depth patterns for different years. That is, even though the shape of these relationships can change under changing climate conditions, it is unclear how their strength would change.

In this study, the SfM snow depth data, themselves are reasonably accurate, and under ideal conditions, they have comparable quality to airborne lidar. For example, at our montane plots, which have relatively gentle topography and sparse tree cover (average slope ~10% and average canopy cover~15%), the SfM-generated snow depth maps generally have high correspondence with the airborne lidar-generated snow depth maps (with R^2 ~0.75 to 0.85 between the two) and comparable agreement with ground observations (RMSE ~10 cm for both technologies). For comparison, most studies that evaluate the performance of SfM for snow depth mapping generally find RMSEs of ~5–15 cm in open settings [36,37,40,48,67,81]. SfM can even produce reasonable snowpack maps under very low snow conditions [37]. For example, at the thinning comparison plot in March 2019 (average snow depth ~10 cm), there is good correspondence between the RPAS-SfM, airborne lidar, and ground snowpack measurements (Figure 3 and Tables 2 and 3). The performance of SfM (and, to a lesser extent, lidar) is markedly lower under dense forest conditions [20,67]). At our dense forest plot (average canopy cover ~45%), the correspondence between the SfM data and the ground measurements is lower than between the lidar data and the ground measurements (RMSE ~10 to 20 cm for the SfM maps vs ~10 cm for the lidar maps). This performance difference is likely because SfM requires line of sight to collect visual data in order to reconstruct the ground surface, while lidar pulses can split and provide multiple returns, and thus, have a greater ability to penetrate dense tree canopies [27].

The quality of these SfM surveys depends on a variety of factors, including the conditions and parameters used during drone image acquisition, the type of environment being surveyed, as well as the success of each of the processing steps in Section 2.3. In terms of image acquisition, we found that our acquisition parameters (RPAS flight ~75 m above ground at ~5–7 m per second with ~75–80% overlap for the images, and using the RPAS's automatic camera exposure adjustments) were sufficient to capture imagery sufficient to produce good quality point clouds for our study plots. The most significant issues that we experienced were due to deep tree shadows. Although these were not a problem under sparse canopy conditions (in the montane plots), they caused problems for point reconstruction under dense canopy conditions (in the dense vegetation plot). In heavily shaded areas, the dark and poorly resolved ground surface caused spurious noise in the point clouds, which in turn led to problems with the ground filtering (already a substantial source of uncertainty [82–84]). Ultimately, this led to spurious snow depth variability for the dense forest plot (such as the artifacts seen in Figure 3).

The automated post-processing steps described in Section 2.3 were also quite important for producing high-quality snow depth maps. Since many of our point clouds were directly georeferenced with onboard GPS data, we relied upon these steps to ensure accurate georegistration of the SfM point clouds, relative to the existing snow-off data. While this approach is capable of producing accurately georeferenced point clouds [85], it increases the reliance on these post-processing steps to produce accurate snow depth maps. In particular, it relies on sufficiently accurate first guess snow depth maps to add to the snow-off data so that they could be used to remove tilting and warping issues that sometimes affect SfM point clouds [86]. For our study sites, these maps (Figure S2) were reasonably accurate (typically having RMSEs of 10–15 cm, a little higher than the final SfM maps; compare Table 2 and Table S1), leading to the ability to use them reliably in our SfM post-processing steps.

Although the high-resolution snow depth mapping afforded by lidar and SfM are already some of the best ways to monitor distributed snowpack properties in mountain forests, taking the extra step of estimating SWE is enormously valuable because SWE is the snowpack variable that is of most interest in hydrological applications [28,61]. Broxton et al. [54] showed that it is important to consider snow density differences across space at our sites, because even though snow depth variability accounts for a majority of SWE variability, snow density can still show substantial spatial variability on a given date (varying by more than 75% across these research sites). This is in line with other studies finding substantial spatial variability of snow density [62,87–90]. Following [54], we use machine learning of how field snow density measurements relate to various lidar-derived physiographic variables to produce maps of snow density, which are then combined with the snow depth maps to produce maps of SWE (see Section 2.3). These maps validate fairly well with snow density / SWE observations at our research sites. The RMSEs of these snow density maps range from 0.024 to 0.033 g/cm^3, and the RMSEs of the resulting SWE estimates range from ~2 to 4 cm [54]. For comparison, the RMSEs of the extrapolated SWE maps range from ~4-5 cm at the 1 m scale to ~2-3 cm at the 1000 m scale (Figure 8).

Overall, for multitemporal monitoring of snowpack where SfM is used as a tool to compliment airborne lidar, we would recommend:

(1) Conducting an airborne lidar survey representing mid-winter to peak snowpack conditions, to provide a baseline coverage before too much melt so that there are no major snow-free areas over the study area (which would require an alternate technique to extrapolate snowpack conditions in these areas)
(2) Planning SfM surveys in areas where SfM has a high chance of success (e.g., sparse canopy, gentle topography, not a whole lot of brush / low vegetation) and where a large range of snow depth variability can be observed (which can include multiple SfM plots), giving a greater ability for successful extrapolation to the larger airborne lidar domain
(3) Designing complimentary ground snow surveys that capture snow conditions across a range of physiographic conditions to be able to generate a reliable first-guess snow depth map to aid in the creation of the SfM data (as well as allowing for good characterization of snow densities). These surveys should capture snow variability related to the important physiographic characteristics and snow processes for a given field site. For our area, snow depth variability was primarily influenced by the tradeoff between shading (from both terrain and vegetation and terrain influences) and interception [6]. Therefore, it was ideal to have transects spanning gradients of vegetation cover and tree shading (e.g., across forest gaps and into forest edges) and on opposing aspects.

The existence of snow-on lidar data makes these planning tasks easier as objective criteria (based on these data) can be used for site selection. The RPAS flight acquisition parameters (see Section 2.1) and processing steps (see Section 2.2) used in this study seemed to be adequate to produce good quality SfM snow maps, but it is also likely that additional improvements could be made. For example, using more exact positioning (e.g., using RTK GPS), which can provide cm level precision [91–93]), could help with model positioning and reducing the warping and tilting of the SfM models, thus reducing the importance of post-processing to correct the point cloud data. Also, the use of drone-based lidar would perform better than SfM in areas with thick canopy [20] (with the obvious drawback that it is much more expensive). Finally, using a fixed-wing aircraft that can cover larger areas may improve the ability to capture the snow spatial variability needed for basin-scale snow estimation, though this option may be limited in some denser forests, due to line-of-sight requirements in places such as the US.

5. Conclusions

Although airborne lidar has revolutionized our ability to make distributed measurements of snowpack in mountain forests, it is expensive and time-consuming to process, making it hard to use for real-time or multi-temporal snow monitoring. This study demonstrates that SfM based on multi-angle imagery from a multi-rotor RPAS and airborne lidar can provide complementary information for high-quality snowpack monitoring, as the use of RPAS allows for relatively low cost, on-demand snow monitoring over small plots, and airborne lidar provides monitoring over a much larger area and at a higher cost. The existence of snow-on lidar data may allow subsequent SfM data to be extrapolated to much larger areas by comparing the spatially distributed snow thicknesses from the SfM data with those of already-measured lidar snow depths. This would make SfM data much more useful for basin-scale hydrological applications such as Land Surface Model evaluation and water supply monitoring. At our study sites, using the relationships developed between the small plot SfM data and previously flown airborne lidar resulted in fairly accurate snowpack estimates for the larger domains (R^2 values between the extrapolated maps and observed lidar data that is used for verification are ~0.7–0.9, and RMSE's between the two are ~10–15 cm for a 1 m pixel scale, with even higher performance when aggregating the data to larger pixels). However, the methodology proposed here may not always be appropriate (e.g., after extensive land cover changes), and more research is needed to understand its effectiveness at other research sites. Ultimately, successful incorporation of SfM and airborne lidar will be important for cost-effective multitemporal monitoring of snowpack.

Supplementary Materials: Supplemental figures and tables are available online at http://www.mdpi.com/2072-4292/12/14/2311/s1, Codes used to process the SfM data are available at https://github.com/broxtopd/SFM-Processing and snow depth maps are available at https://climate.arizona.edu/data/SfM/.

Author Contributions: Conceptualization, P.D.B. and W.J.D.v.L.; Data curation, P.D.B. and W.J.D.v.L.; Formal analysis, P.D.B.; Funding acquisition, W.J.D.v.L.; Investigation, P.D.B. and W.J.D.v.L.; Methodology, P.D.B.; Project administration, W.J.D.v.L.; Resources, W.J.D.v.L.; Software, P.D.B.; Supervision, W.J.D.v.L; Visualization, P.D.B.; Writing—original draft, P.D.B.; Writing—review & editing, W.J.D.v.L. All authors have read and agreed to the published version of the manuscript.

Funding: This research is supported by the Salt River Project (SRP) Agricultural Improvement and Power District in Tempe, Arizona.

Acknowledgments: SfM data were generated from imagery acquired by the authors using a DJI Phantom 3 and 4 Pro RPAS. Airborne lidar data are derived from a snow-free lidar acquisition for the Four Forests Restoration Initiative (4Fri), and snow-on lidar acquisitions and field snow surveys acquired through funding provided by SRP. Computing resources are provided by the Arizona Remote Sensing Center. We would like to thank the many volunteers from the NRCS, USFS, USDA-ARS, University of Arizona, Northern Arizona University, and SRP who helped with our field data collections. We also thank Jeff Gillan for providing advice for acquiring and processing the SfM data.

Conflicts of Interest: The authors declare no conflict of interest. The funder had no role in the design of the study; in the collection, analyses, or interpretation of data; in the writing of the manuscript, or in the decision to publish the results.

References

1. Bales, R.C.; Molotch, N.P.; Painter, T.H.; Dettinger, M.D.; Rice, R.; Dozier, J. Mountain hydrology of the western United States. *Water Resour. Res.* **2006**, *42*, W08432. [CrossRef]
2. Bolin, B.; Seetharam, M.; Pompeii, B. Water resources, climate change, and urban vulnerability: A case study of Phoenix, Arizona. *Local Environ.* **2010**, *15*, 261–279. [CrossRef]
3. Clark, S.S.; Chester, M.V.; Seager, T.P.; Eisenberg, D.A. The vulnerability of interdependent urban infrastructure systems to climate change: Could phoenix experience a katrina of extreme heat? *Sustain. Resilient Infrast.* **2019**, *4*, 21–35. [CrossRef]
4. Biederman, J.; Harpold, A.; Gochis, D.; Ewers, B.; Reed, D.; Papuga, S.; Brooks, P. Increased evaporation following widespread tree mortality limits streamflow response. *Water Resour. Res.* **2014**, *50*, 5395–5409. [CrossRef]

5. Brooks, P.; Vivoni, E.R. Mountain ecohydrology: Quantifying the role of vegetation in the water balance of montane catchments. *Ecohydrol. Ecosyst. Land Water Process. Interact. Ecohydrogeomorphology* **2008**, *1*, 187–192. [CrossRef]
6. Broxton, P.D.; van Leeuwen, W.J.; Biederman, J.A. Forest cover and topography regulate the thin, ephemeral snowpacks of the semiarid southwest United States. *Ecohydrology* **2020**, *13*, e2202. [CrossRef]
7. Ffolliott, P.F.; Gottfried, G.J.; Baker, M.B., Jr. Water yield from forest snowpack management: Research findings in Arizona and New Mexico. *Water Resour. Res.* **1989**, *25*, 1999–2007. [CrossRef]
8. Gleason, K.E.; Nolin, A.W. Charred forests accelerate snow albedo decay: Parameterizing the post-fire radiative forcing on snow for three years following fire. *Hydrol. Process.* **2016**, *30*, 3855–3870. [CrossRef]
9. Harpold, A.A.; Biederman, J.A.; Condon, K.; Merino, M.; Korgaonkar, Y.; Nan, T.; Sloat, L.L.; Ross, M.; Brooks, P.D. Changes in snow accumulation and ablation following the las conchas forest fire, New Mexico, USA. *Ecohydrology* **2014**, *7*, 440–452. [CrossRef]
10. Jenicek, M.; Pevna, H.; Matejka, O. Canopy structure and topography effects on snow distribution at a catchment scale: Application of multivariate approaches. *J. Hydrol. Hydromech.* **2018**, *66*, 43–54. [CrossRef]
11. Molotch, N.P.; Brooks, P.D.; Burns, S.P.; Litvak, M.; Monson, R.K.; McConnell, J.R.; Musselman, K. Ecohydrological controls on snowmelt partitioning in mixed-conifer sub-alpine forests. *Ecohydrology* **2009**, *2*, 129–142. [CrossRef]
12. Robles, M.D.; Marshall, R.M.; O'Donnell, F.; Smith, E.B.; Haney, J.A.; Gori, D.F. Effects of climate variability and accelerated forest thinning on watershed-scale runoff in southwestern USA ponderosa pine forests. *PLoS ONE* **2014**, *9*, e111092. [CrossRef]
13. Svoma, B.M. Canopy effects on snow sublimation from a central Arizona Basin. *J. Geophys. Res. Atmos.* **2017**, *122*, 20–46. [CrossRef]
14. Broxton, P.; Harpold, A.; Biederman, J.; Troch, P.A.; Molotch, N.; Brooks, P. Quantifying the effects of vegetation structure on snow accumulation and ablation in mixed-conifer forests. *Ecohydrology* **2015**, *8*, 1073–1094. [CrossRef]
15. Gustafson, J.R.; Brooks, P.; Molotch, N.; Veatch, W. Estimating snow sublimation using natural chemical and isotopic tracers across a gradient of solar radiation. *Water Resour. Res.* **2010**, *46*, W12511. [CrossRef]
16. Musselman, K.; Molotch, N.P.; Brooks, P.D. Effects of vegetation on snow accumulation and ablation in a mid-latitude sub-alpine forest. *Hydrol. Process. An. Int. J.* **2008**, *22*, 2767–2776. [CrossRef]
17. Rinehart, A.J.; Vivoni, E.R.; Brooks, P.D. Effects of vegetation, albedo, and solar radiation sheltering on the distribution of snow in the Valles Caldera, New Mexico. *Ecohydrol. Ecosyst. Land Water Process. Interact. Ecohydrogeomorphology* **2008**, *1*, 253–270. [CrossRef]
18. Veatch, W.; Brooks, P.; Gustafson, J.; Molotch, N. Quantifying the effects of forest canopy cover on net snow accumulation at a continental, mid-latitude site. *Ecohydrology* **2009**, *2*, 115–128. [CrossRef]
19. Molotch, N.; Colee, M.; Bales, R.; Dozier, J. Estimating the spatial distribution of snow water equivalent in an alpine basin using binary regression tree models: The impact of digital elevation data and independent variable selection. *Hydrol. Process. An. Int. J.* **2005**, *19*, 1459–1479. [CrossRef]
20. Harder, P.; Pomeroy, J.W.; Helgason, W.D. Advances in mapping sub-canopy snow depth with unmanned aerial vehicles using structure from motion and lidar techniques. *Cryosphere Discuss.* **2019**, *2019*, 1–31.
21. Nolin, A.W. Recent advances in remote sensing of seasonal snow. *J Glaciol.* **2010**, *56*, 1141–1150. [CrossRef]
22. Vikhamar, D.; Solberg, R. Subpixel mapping of snow cover in forests by optical remote sensing. *Remote Sens. Environ.* **2003**, *84*, 69–82. [CrossRef]
23. Dong, C.; Menzel, L. Snow process monitoring in montane forests with time-lapse photography. *Hydrol. Process.* **2017**, *31*, 2872–2886. [CrossRef]
24. McCabe, M.F.; Rodell, M.; Alsdorf, D.E.; Miralles, D.G.; Uijlenhoet, R.; Wagner, W.; Lucieer, A.; Houborg, R.; Verhoest, N.E.C.; Franz, T.E.; et al. The future of Earth observation in hydrology. *Hydrol. Earth Syst. Sci.* **2017**, *21*, 3879–3914. [CrossRef] [PubMed]
25. Appel, I. Uncertainty in satellite remote sensing of snow fraction for water resources management. *Front. Earth Sci.* **2018**, *12*, 711–727.
26. Piazzi, G.; Tanis, C.M.; Kuter, S.; Simsek, B.; Puca, S.; Toniazzo, A.; Takala, M.; Akyürek, Z.; Nadir Arslan, A.; Gabellani, S.; et al. Cross-country assessment of H-SAF snow products by Sentinel-2 imagery validated against in-situ observations and webcam photography. *Geosciences* **2019**, *9*, 129. [CrossRef]

27. Deems, J.S.; Painter, T.H.; Finnegan, D.C. Lidar measurement of snow depth: A review. *J. Glaciol.* **2013**, *59*, 467–479. [CrossRef]
28. Hedrick, A.R.; Marks, D.; Havens, S.; Robertson, M.; Johnson, M.; Sandusky, M.; Marshall, H.P.; Kormos, P.R.; Bormann, K.J.; Painter, T.H. Direct insertion of NASA airborne snow observatory-derived snow depth time series into the iSnobal energy balance snow model. *Water Resour. Res.* **2018**, *54*, 8045–8063. [CrossRef]
29. Painter, T.H.; Berisford, D.F.; Boardman, J.W.; Bormann, K.J.; Deems, J.S.; Gehrke, F.; Hedrick, A.; Joyce, M.; Laidlaw, R.; Marks, D.; et al. The airborne snow observatory: Fusion of scanning lidar, imaging spectrometer, and physically-based modeling for mapping snow water equivalent and snow albedo. *Remote Sens. Environ.* **2016**, *184*, 139–152. [CrossRef]
30. Harpold, A.; Guo, Q.; Molotch, N.; Brooks, P.; Bales, R.; Fernandez-Diaz, J.; Musselman, K.; Swetnam, T.; Kirchner, P.; Meadows, M.W.; et al. LiDAR-derived snowpack data sets from mixed conifer forests across the western United States. *Water Resour. Res.* **2014**, *50*, 2749–2755. [CrossRef]
31. McClelland, M.P., II; Hale, D.S.; van Aardt, J. A comparison of manned and unmanned aerial Lidar systems in the context of sustainable forest management. In Proceedings of the Autonomous Air and Ground Sensing Systems for Agricultural Optimization and Phenotyping III 106640S, Orlando, FL, USA, 18–19 April 2018.
32. Hummel, S.; Hudak, A.; Uebler, E.; Falkowski, M.; Megown, K. A comparison of accuracy and cost of LiDAR versus stand exam data for landscape management on the Malheur National Forest. *J. For.* **2011**, *109*, 267–273.
33. Lendzioch, T.; Langhammer, J.; Jenicek, M. Estimating snow depth and leaf area index based on UAV digital photogrammetry. *Sensors* **2019**, *19*, 1027. [CrossRef]
34. Lendzioch, T.; Langhammer, J.; Jenicek, M. Tracking forest and open area effects on snow accumulation by unmanned aerial vehicle photogrammetry. *Int. Arch. Photogramm. Remote Sens. Spat. Inf. Sci.* **2016**, *41*, 917–923. [CrossRef]
35. Westoby, M.J.; Brasington, J.; Glasser, N.F.; Hambrey, M.J.; Reynolds, J.M. 'Structure-from-motion' photogrammetry: A low-cost, effective tool for geoscience applications. *Geomorphology* **2012**, *179*, 300–314. [CrossRef]
36. Cimoli, E.; Marcer, M.; Vandecrux, B.; Bøggild, C.E.; Williams, G.; Simonsen, S.B. Application of low-cost UASs and digital photogrammetry for high-resolution snow depth mapping in the Arctic. *Remote Sen.* **2017**, *9*, 1144. [CrossRef]
37. Goetz, J.; Brenning, A. Quantifying uncertainties in snow depth mapping from structure from motion photogrammetry in an alpine area. *Water Resour. Res.* **2019**, *55*, 7772–7783. [CrossRef]
38. Meyer, J.; Skiles, S.M. Assessing the ability of structure from motion to map high-resolution snow surface elevations in complex terrain: A case study from senator beck basin, CO. *Water Resour. Res.* **2019**, *55*, 6596–6605. [CrossRef]
39. Nolan, M.; Larsen, C.; Sturm, M. Mapping snow-depth from manned-aircraft on landscape scales at centimeter resolution using structure-from-motion photogrammetry. *Cryosphere Discuss.* **2015**, *9*, 333–381. [CrossRef]
40. Bühler, Y.; Adams, M.S.; Bösch, R.; Stoffel, A. Mapping snow depth in alpine terrain with unmanned aerial systems (UASs): Potential and limitations. *Cryosphere* **2016**, *10*, 1075–1088. [CrossRef]
41. Federman, A.; Quintero, M.S.; Kretz, S.; Gregg, J.; Lengies, M.; Ouimet, C.; Laliberte, J. UAV photogrammetric workflows: A best practice guideline. *Int. Arch. Photogramm. Remote Sens. Spat. Inf. Sci.* **2017**, *42*, 237–244. [CrossRef]
42. Adams, M.S.; Bühler, Y.; Fromm, R. Multitemporal accuracy and precision assessment of unmanned aerial system photogrammetry for slope-scale snow depth maps in alpine terrain. *Pure Appl. Geophys.* **2018**, *175*, 3303–3324. [CrossRef]
43. Avanzi, F.; Bianchi, A.; Cina, A.; De Michele, C.; Maschio, P.; Pagliari, D.; Passoni, D.; Pinto, L.; Piras, M.; Rossi, L. Centimetric accuracy in snow depth using unmanned aerial system photogrammetry and a multistation. *Remote Sens.* **2018**, *10*, 765. [CrossRef]
44. Baños, I.M.; García, A.R.; i Alavedra, J.M.; i Figueras, P.O.; Iglesias, J.P.; i Figueras, P.M.; López, J.T. Assessment of airborne LIDAR for snowpack depth modeling. *Boletín de la Soc. Geológica Mex.* **2011**, *63*, 95–107. [CrossRef]
45. Deems, J.S.; Painter, T.H. Lidar measurement of snow depth: Accuracy and error sources. In Proceedings of the 2006 International Snow Science Workshop, Telluride, CO, USA, 1–6 October 2006; pp. 330–338.
46. Grünewald, T.; Schirmer, M.; Mott, R.; Lehning, M. Spatial and temporal variability of snow depth and SWE in a small mountain catchment. *Cryosphere* **2010**, *4*, 215–225. [CrossRef]

47. López-Moreno, J.I.; Fassnacht, S.; Beguería, S.; Latron, J. Variability of snow depth at the plot scale: Implications for mean depth estimation and sampling strategies. *Cryosphere* **2011**, *5*, 617–629. [CrossRef]
48. Redpath, T.A.; Sirguey, P.; Cullen, N.J. Repeat mapping of snow depth across an alpine catchment with RPAS photogrammetry. *Cryosphere* **2018**, *12*, 3477–3497. [CrossRef]
49. Tinkham, W.T.; Smith, A.M.; Marshall, H.-P.; Link, T.E.; Falkowski, M.J.; Winstral, A.H. Quantifying spatial distribution of snow depth errors from LiDAR using random forest. *Remote Sens. Environ.* **2014**, *141*, 105–115. [CrossRef]
50. Trujillo, E.; Ramírez, J.A.; Elder, K.J. Topographic, meteorologic, and canopy controls on the scaling characteristics of the spatial distribution of snow depth fields. *Water Resour. Res.* **2007**, *43*, W07409. [CrossRef]
51. Kennedy, M.C.; Johnson, M.C. Fuel treatment prescriptions alter spatial patterns of fire severity around the wildland–urban interface during the wallow fire, Arizona, USA. *For. Ecol.Manag.* **2014**, *318*, 122–132. [CrossRef]
52. QSI. *4FRI Snow Analysis 2019, Arizona LIDAR*; Technical Data Report; QSI: Corvallis, OR, USA, 2019.
53. Biederman, J.A.; Brooks, P.D.; Harpold, A.A.; Gochis, D.J.; Gutmann, E.; Reed, D.E.; Pendall, E.; Ewers, B.E. Multiscale observations of snow accumulation and peak snowpack following widespread, insect-induced lodgepole pine mortality. *Ecohydrology* **2014**, *7*, 150–162. [CrossRef]
54. Broxton, P.D.; van Leeuwen, W.J.; Biederman, J.A. Improving snow water equivalent maps with machine learning of snow survey and lidar measurements. *Water Resour. Res.* **2019**, *55*, 3739–3757. [CrossRef]
55. QSI. *4FRI LiDAR: Four Forests Restoration Initiative*; Technical Data Report; QSI: Corvallis, OR, USA, 2013.
56. QSI. *4FRI LiDAR: Four Forests Restoration Initiative*; Technical Data Report; QSI: Corvallis, OR, USA, 2014.
57. QSI. *4FRI Snow Analysis, Arizona LIDAR*; Technical Data Report; QSI: Corvallis, OR, USA, 2017.
58. QSI. *4FRI Snow Analysis II, Arizona LIDAR*; Technical Data Report; QSI: Corvallis, OR, USA, 2017.
59. Zhang, W.; Qi, J.; Wan, P.; Wang, H.; Xie, D.; Wang, X.; Yan, G. An easy-to-use airborne LiDAR data filtering method based on cloth simulation. *Remote Sens.* **2016**, *8*, 501. [CrossRef]
60. McGaughey, R. *FUSION/LDV: Software for LiDAR Data Analysis and Visualization*; Version 3.01; US Department of Agriculture, Forest Service, Pacific Northwest. Research Station; University of Washington: Seattle, WA, USA, 2012.
61. Dozier, J. Mountain hydrology, snow color, and the fourth paradigm. *Eostransactions Am. Geophys. Union* **2011**, *92*, 373–374. [CrossRef]
62. Wetlaufer, K.; Hendrikx, J.; Marshall, L. Spatial heterogeneity of snow density and its influence on snow water equivalence estimates in a large mountainous basin. *Hydrology* **2016**, *3*, 3. [CrossRef]
63. Gao, J.; Xia, Z.-G. Fractals in physical geography. *Prog. Phys. Geogr.* **1996**, *20*, 178–191. [CrossRef]
64. Deems, J.S.; Fassnacht, S.R.; Elder, K.J. Fractal distribution of snow depth from LiDAR data. *J. Hydrometeorol.* **2006**, *7*, 285–297. [CrossRef]
65. Mott, R.; Schirmer, M.; Lehning, M. Scaling properties of wind and snow depth distribution in an alpine catchment. *J. Geophys. Res. Atmos.* **2011**, 116. [CrossRef]
66. Clark, M.P.; Hendrikx, J.; Slater, A.G.; Kavetski, D.; Anderson, B.; Cullen, N.J.; Kerr, T.; Örn Hreinsson, E.; Woods, R.A. Representing spatial variability of snow water equivalent in hydrologic and land-surface models: A review. *Water Resour. Res.* **2011**, 47. [CrossRef]
67. Harder, P.; Schirmer, M.; Pomeroy, J.; Helgason, W. Accuracy of snow depth estimation in mountain and prairie environments by an unmanned aerial vehicle. *Cryosphere* **2016**, *10*, 2559–2571. [CrossRef]
68. Deems, J.S.; Fassnacht, S.R.; Elder, K.J. Interannual consistency in fractal snow depth patterns at two Colorado mountain sites. *J. Hydrometeorol.* **2008**, *9*, 977–988. [CrossRef]
69. Schirmer, M.; Wirz, V.; Clifton, A.; Lehning, M. Persistence in intra-annual snow depth distribution: 1. Measurements and topographic control. *Water Resour. Res.* **2011**, 47. [CrossRef]
70. Erickson, T.A.; Williams, M.W.; Winstral, A. Persistence of topographic controls on the spatial distribution of snow in rugged mountain terrain, Colorado, United States. *Water Resour. Res.* **2005**, 41. [CrossRef]
71. Kuchment, L.; Gelfan, A. Statistical self-similarity of spatial variations of snow cover: Verification of the hypothesis and application in the snowmelt runoff generation models. *Hydrol. Process.* **2001**, *15*, 3343–3355. [CrossRef]
72. Sturm, M.; Wagner, A.M. Using repeated patterns in snow distribution modeling: An Arctic example. *Water Resour. Res.* **2010**, 46. [CrossRef]

73. Revuelto, J.; López-Moreno, J.I.; Azorin-Molina, C.; Vicente-Serrano, S.M. Topographic control of snowpack distribution in a small catchment in the central Spanish Pyrenees: Intra-and inter-annual persistence. *Cryosphere* **2014**, *8*, 1989–2006. [CrossRef]
74. Elder, K.; Dozier, J.; Michaelsen, J. Snow accumulation and distribution in an alpine watershed. *Water Resour. Res.* **1991**, *27*, 1541–1552. [CrossRef]
75. Egli, L.; Jonas, T. Hysteretic dynamics of seasonal snow depth distribution in the Swiss Alps. *Geophys. Res. Lett.* **2009**, 36. [CrossRef]
76. Magand, C.; Ducharne, A.; Le Moine, N.; Gascoin, S. Introducing hysteresis in snow depletion curves to improve the water budget of a land surface model in an Alpine catchment. *J. hydrometeorol.* **2014**, *15*, 631–649. [CrossRef]
77. Anderton, S.; White, S.; Alvera, B. Evaluation of spatial variability in snow water equivalent for a high mountain catchment. *Hydrol. Process.* **2004**, *18*, 435–453. [CrossRef]
78. Jost, G.; Weiler, M.; Gluns, D.R.; Alila, Y. The influence of forest and topography on snow accumulation and melt at the watershed-scale. *J. Hydrol.* **2007**, *347*, 101–115. [CrossRef]
79. Winstral, A.; Marks, D. Simulating wind fields and snow redistribution using terrain-based parameters to model snow accumulation and melt over a semi-arid mountain catchment. *Hydrol. Process.* **2002**, *16*, 3585–3603. [CrossRef]
80. López-Moreno, J.; Pomeroy, J.; Revuelto, J.; Vicente-Serrano, S. Response of snow processes to climate change: Spatial variability in a small basin in the Spanish Pyrenees. *Hydrol. Process.* **2013**, *27*, 2637–2650. [CrossRef]
81. De Michele, C.; Avanzi, F.; Passoni, D.; Barzaghi, R.; Pinto, L.; Dosso, P.; Ghezzi, A.; Gianatti, R.; Vedova, G.D. Using a fixed-wing UAS to map snow depth distribution: An evaluation at peak accumulation. *Cryosphere* **2016**, *10*, 511–522. [CrossRef]
82. Liu, X. Airborne LiDAR for DEM generation: Some critical issues. *Prog. Phys. Geogr.* **2008**, *32*, 31–49.
83. Gould, S.B.; Glenn, N.F.; Sankey, T.T.; McNamara, J.P. Influence of a dense, low-height shrub species on the accuracy of a LiDAR-derived DEM. *Photogramm. Eng. Remote Sens.* **2013**, *79*, 421–431. [CrossRef]
84. Meng, X.; Currit, N.; Zhao, K. Ground filtering algorithms for airborne LiDAR data: A review of critical issues. *Remote Sens.* **2010**, *2*, 833–860. [CrossRef]
85. Miziński, B.; Niedzielski, T. Fully-automated estimation of snow depth in near real time with the use of unmanned aerial vehicles without utilizing ground control points. *Cold Reg. Sci. Technol.* **2017**, *138*, 63–72. [CrossRef]
86. Eltner, A.; Schneider, D. Analysis of different methods for 3D reconstruction of natural surfaces from parallel-axes UAV images. *Photogramm. Rec.* **2015**, *30*, 279–299. [CrossRef]
87. Bormann, K.J.; Westra, S.; Evans, J.P.; McCabe, M.F. Spatial and temporal variability in seasonal snow density. *J. Hydrol.* **2013**, *484*, 63–73. [CrossRef]
88. Dawson, N.; Broxton, P.; Zeng, X. A new snow density parameterization for land data initialization. *J. Hydrometeorol.* **2017**, *18*, 197–207. [CrossRef]
89. López-Moreno, J.I.; Fassnacht, S.; Heath, J.; Musselman, K.; Revuelto, J.; Latron, J.; Morán-Tejeda, E.; Jonas, T. Small scale spatial variability of snow density and depth over complex alpine terrain: Implications for estimating snow water equivalent. *Adv. Water Resour.* **2013**, *55*, 40–52. [CrossRef]
90. Raleigh, M.S.; Small, E.E. Snowpack density modeling is the primary source of uncertainty when mapping basin-wide SWE with lidar. *Geophys. Res. Lett.* **2017**, *44*, 3700–3709. [CrossRef]
91. Fazeli, H.; Samadzadegan, F.; Dadrasjavan, F. Evaluating the potential of RTK-UAV for automatic point cloud generation in 3D rapid mapping. *Int. Arch. Photogramm. Remote Sens. Spat. Inf. Sci.* **2016**, *41*, 221. [CrossRef]
92. Zhang, H.; Aldana-Jague, E.; Clapuyt, F.; Wilken, F.; Vanacker, V.; Van Oost, K. Evaluating the potential of post-processing kinematic (PPK) georeferencing for UAV-based structure-from-motion (SfM) photogrammetry and surface change detection. *Earth Surf. Dyn.* **2019**, *7*, 807–827. [CrossRef]
93. Forlani, G.; Dall'Asta, E.; Diotri, F.; di Cella, U.M.; Roncella, R.; Santise, M. Quality assessment of DSMs produced from UAV flights georeferenced with on-board RTK positioning. *Remote Sens.* **2018**, *10*, 311. [CrossRef]

Article

Use of UAV-Photogrammetry for Quasi-Vertical Wall Surveying

Patricio Martínez-Carricondo [1,2,*], Francisco Agüera-Vega [1,2] and Fernando Carvajal-Ramírez [1,2]

1 Department of Engineering, University of Almería (Agrifood Campus of International Excellence, ceiA3), La Cañada de San Urbano, s/n, 04120 Almería, Spain; faguera@ual.es (F.A.-V.); carvajal@ual.es (F.C.-R.)

2 Peripheral Service of Research and Development Based on Drones, University of Almeria, La Cañada de San Urbano, s/n, 04120 Almería, Spain

* Correspondence: pmc824@ual.es

Received: 8 June 2020; Accepted: 9 July 2020; Published: 10 July 2020

Abstract: In this study, an analysis of the capabilities of unmanned aerial vehicle (UAV) photogrammetry to obtain point clouds from areas with a near-vertical inclination was carried out. For this purpose, 18 different combinations were proposed, varying the number of ground control points (GCPs), the adequacy (or not) of the distribution of GCPs, and the orientation of the photographs (nadir and oblique). The results have shown that under certain conditions, the accuracy achieved was similar to those obtained by a terrestrial laser scanner (TLS). For this reason, it is necessary to increase the number of GCPs as much as possible in order to cover a whole study area. In the event that this is not possible, the inclusion of oblique photography ostensibly improves results; therefore, it is always advisable since they also improve the geometric descriptions of break lines or sudden changes in slope. In this sense, UAVs seem to be a more economic substitute compared to TLS for vertical wall surveying.

Keywords: UAV; photogrammetry; 3D-model; surveying; vertical wall

1. Introduction

Topographical surveys of surfaces with high angles of inclination such as nearly vertical slopes on roads or motorways, as well as the survey on architectural facades, are a technical challenge, the main concern of which is ensuring the safety of the operators responsible for carrying out the survey. Technical equipment has traditionally been used to guarantee this safety by preventing the operator from having to access the survey area. Thus, for road slopes, for example, total stations without prisms have been used, or architectural facades have been counted, in addition, with the resource of the rectification of photographs. However, since terrestrial laser scanners (TLS) came on the market, their use has contributed to improving the effectiveness in these kinds of studies, as well as those in other fields (e.g., cartography, geographic information systems, spatial planning, industry, forestry). TLS measures the time of flight of an emitted laser pulse that is reflected off of an intervening feature and returned to the sensor, thus resulting in a range measurement [1]. Because lasers arrive directly at the surface of the object and are reflected from it, this technology can precisely acquire spatial coordinates with an error that depends on the range, which usually varies between 1 and 10 mm. However, the TLS is expensive, and there are times when its use is limited due to certain circumstances that can distort and introduce error in measurements, including penetration and diffused reflection of the beam [2], or shadows that produce gaps in the point cloud. At present, there is a trend to complement this technology with unmanned aerial vehicles (UAVs) carrying digital cameras.

In recent years there has been a growing interest in UAVs from the scientific community, as well as geomatics professionals and software developers, which has led to their use in increasing applications related to architecture and engineering [3,4]. In fact, UAVs were first used for military applications [5] and then for civilian purposes [6] such as precision agriculture [7,8], forestry studies [9,10], fire monitoring [11,12], cultural heritage and archaeology [13–15], traffic monitoring [16,17], environmental surveying [18,19], and 3D reconstruction [20–22]. UAV photogrammetry is gaining ground in the gap between traditional surveying methods and photogrammetric flights performed with conventional aircraft. Depending on the extent of the area, UAVs are more competitive because they offer greater flexibility while requiring less time to acquire data, and, additionally, they represent a significant cost reduction compared to the use of traditional aircraft [23].

The combination of computer vision and photogrammetry [24] has allowed great advances in the automation process by highlighting the use of images with different tilt angles and at different heights [25]. There are several software packages that allow photographs taken with conventional cameras to obtain 3D reconstruction by means of point clouds. The majority of these software packages are based on special algorithms, such as Structure-from-Motion (SfM) [26–28]. SfM is an algorithm that automatically reconstructs the geometry of the scene, the positions and the orientation from which the photographs were taken, without the need to establish a point network with known 3D coordinates [29,30]. SfM incorporates multi-view stereopsis (MVS) techniques [31], which derive a 3D structure from overlapping photography, acquired from multiple locations and angles [32], and applies them to a scale-invariant feature transform (SIFT) operator for key-point detection. This generates 3D point clouds from photographs. Contrary to classical aerial photogrammetry, which requires sophisticated flight planning and pre-calibration of cameras [33], SfM simplifies the process, thus eliminating the need for exhaustive planning or camera calibration, and allowing for the use of images from different cameras. The result of processing this algorithm is a point cloud without scale or orientation, whose georeferencing can be obtained by direct methods through the use of photographs with EXIF data or by indirect methods using ground control points (GCPs) [34]. There are numerous studies that analyze the effect of different parameters on the accuracy of products obtained by UAV photogrammetry [35–40]. Of all of them, the number of GCPs is of special importance [41], as is their distribution and the use of photographs with different inclination angles [42]. In summary, UAV photogrammetry has shown a great deal of development in recent years and is increasingly used in situations where other techniques are less efficient or simply not feasible. Therefore, it is necessary to continue developing specific methodologies to obtain accurate results using UAV photogrammetry in extreme topographic situations, such as dealing with quasi-vertical walls.

The goal of this study is to validate the specific use of point clouds obtained by UAV photogrammetry for the topographic survey of walls or facades that have an inclination close to vertical. For this purpose, 18 scenarios have been proposed that use different combinations of GCPs and photograph orientations and adopt adequate (or inadequate) GCP distributions. All of these scenarios have been compared with the one obtained through a TLS, resulting in certain parameters for UAV photogrammetry, with accuracy that is comparable to that obtained through TLS.

2. Materials and Methods

The flow followed in this work is shown in Figure 1. In summary, work begins with a global navigation satellite system (GNSS) survey of the dam edges and georeferencing targets. The points obtained on the dam edges are then used to interpolate seven profiles along the dam cross section that are then used to validate the point cloud obtained by the TLS. The targets are used indistinctly for the georeferencing of the 18 UAV photogrammetry projects and for quality control. Once the point clouds are obtained by UAV photogrammetry, they are compared with that obtained by TLS.

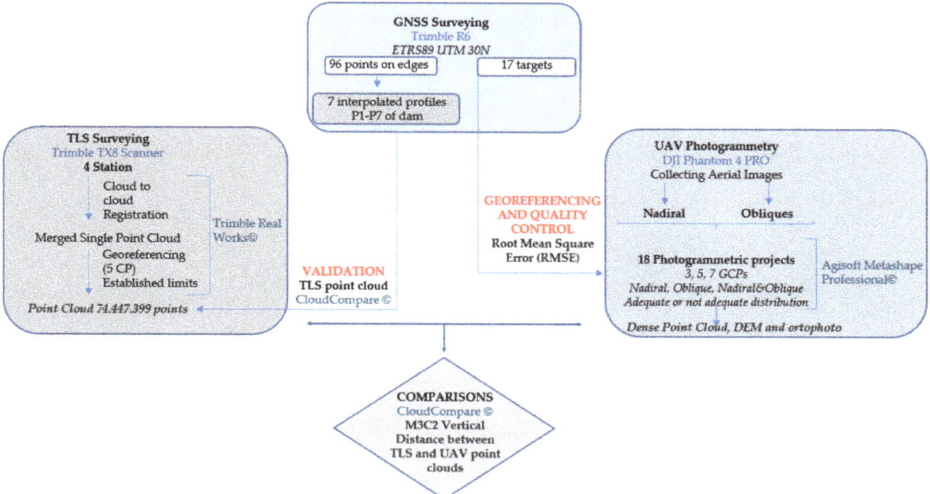

Figure 1. Workflow followed in this work.

2.1. Study Site

The study area was centered on the Isabel II Dam (1841–1857), located in the province of Almería (Spain) about 7 km from the village of Níjar (Figure 2). Construction on the dam began with the foundation works in 1841, although it was not inaugurated until 8 May 1850 without the completion of the canals or offices, which would last until 1857. The reservoir was never completely filled, and the project failed due to several faults in the calculations made in the hydrological and pluviometric studies of the area.

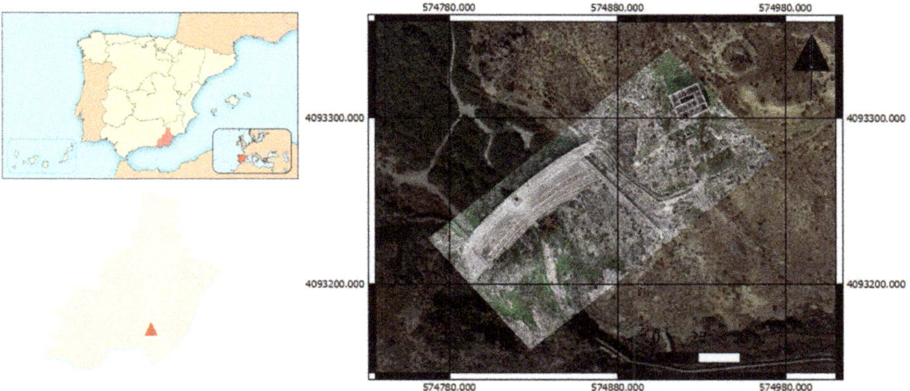

Figure 2. Location of the study site.

However, this dam is one of the most spectacular, and simultaneously least known, elements of Spain's hydraulic heritage. It is one of the few examples of the great hydraulic work undertaken in the 19th century, as well a key world reference for stone arch-gravity dams [43].

It was selected because of its downstream profile with walls that are very close to the vertical; its total height at its central point 31 m above the riverbed, as is shown in Figure 3.

Figure 3. View perspective of the study area.

2.2. GNSS Surveying of Dam Edges and Ground Control Points

Before performing the photogrammetric flights and TLS scanning, a traditional GNSS survey was carried out, materializing a total of 113 points distributed along the dam's downstream face, mostly on the edges of the steps, which allowed for their subsequent complete interpolation by means of cabinet work, as shown in Figure 4a. Of all these points, 17 were materialized by means of targets that allowed for their later viewing in photographs taken by the UAV. The targets used consisted of a red paper of size A3 (420 × 297 mm) with four quadrants, two of them black. Figure 4b shows an example of one of these targets. The three-dimensional (3D) coordinates of these targets were measured with a GNSS receiver operating in post-processing kinematic mode (PPK), with the base emitting corrections at a point near the dam, as shown in Figure 4c. Both rover and base GNSS receivers were Trimble R6 systems. The 3D coordinates of the base, corrected via the Trimble Centerpoint RTX Post-Processing Service, were 574909.418, 4093250.721, and 372.012 m (European Terrestrial Reference System 1989, ETRS89 and EGM08 geoid model).

From the data obtained by GNSS, seven theoretical profiles were made along the entire downstream face of the dam, numbered P1 to P7. These theoretical profiles obtained by GNSS are shown in Figure 5.

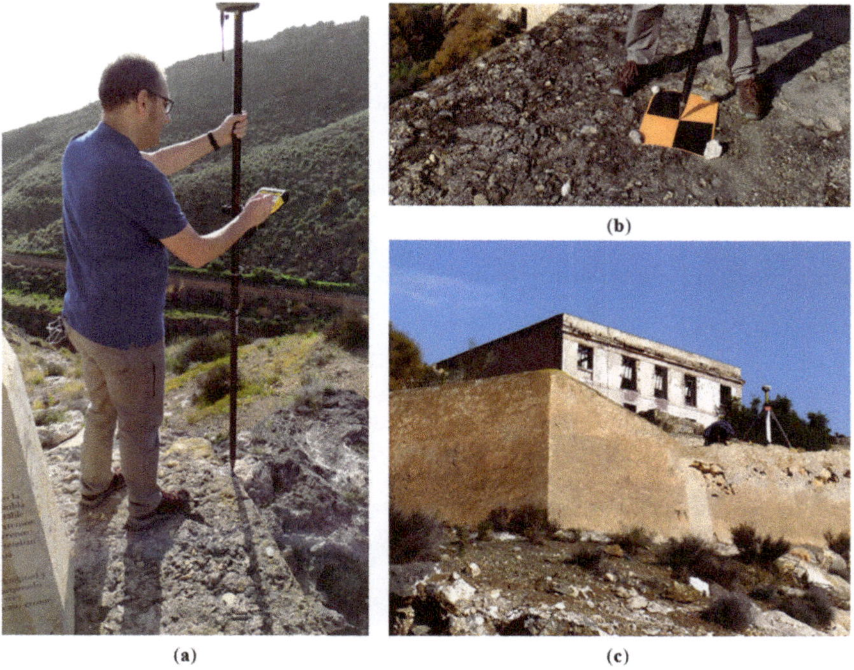

Figure 4. (a) Surveying the edges of the dam using global navigation satellite system (GNSS); (b) example of target; (c) GNSS receiver base applied to this work.

Figure 5. (a) Plant view of theoretical profiles acquired by GNSS; (b) south view of seven profiles P1—P7.

2.3. Topographic Surveying Using TLS

In order to meet the objective of this research, it was necessary to obtain a complete point cloud of the study area that would allow its later comparison with the products obtained by UAV photogrammetry. To this end, a complete scan of the dam's downstream face was carried out using TLS.

The TLS provides a three-dimensional quasi-constant point cloud of the observed objects. The accuracy of this point cloud depends mainly on the distance between the scanner and the observed object. In recent years, the application of this equipment is increasing exponentially [44–47]. TLS measurements are based on the time elapsed by a laser ray emitted by the scanner and reflected by the object. From the time measurement, the distance is obtained and transformed into real-time coordinates. The working guidelines are described in depth in [48].

2.3.1. Data Acquisition

In this paper, a Trimble TX8 Scanner (Figure 6) was utilized for the topographical survey of the dam. This scanner measures almost 1 million points per second and its maximum scanning range is 120 m, but can extend up to 340 m under favorable conditions. The wavelength of the laser is 1.5 µm, and the scanning frequency is 1 MHz. Advertised measurement accuracy is ±2 mm, and the angular resolution is 0.07 mrad. It also includes an integrated HDR camera with 10 MP resolution. This scanner was used by [49] to measure the deflections of a technological suspension bridge above the Odra River (Southern Poland).

Figure 6. Trimble TX-8 Scanner.

Due to the geometry of the study area, it was not possible to capture the entire wall from a single location. Therefore, the measurements for the TLS point cloud acquisition were taken from four different station points, as shown in Figure 7.

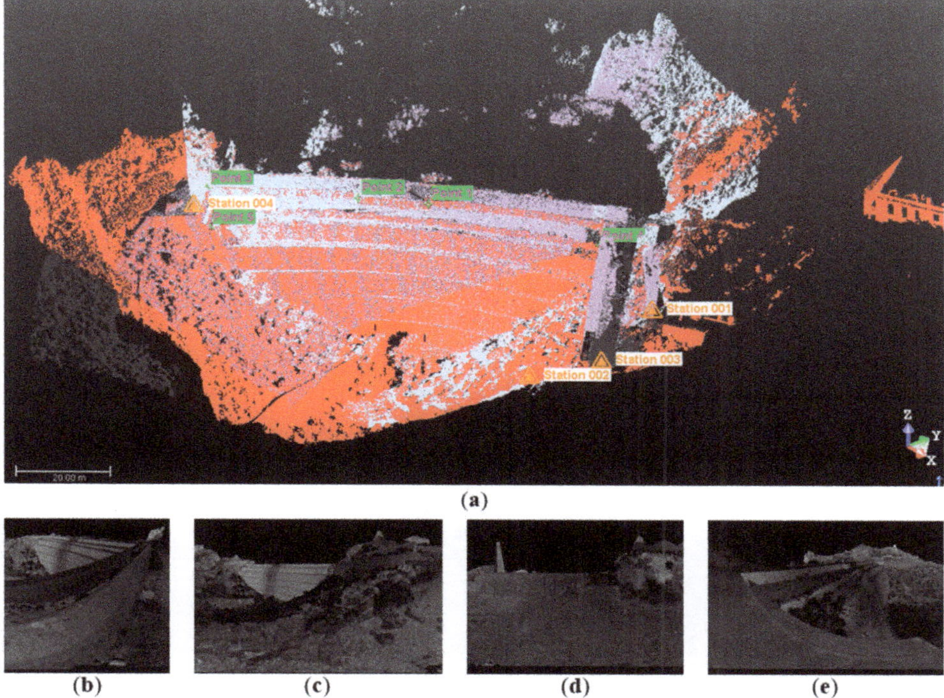

Figure 7. (**a**) View of the 4 stations from which the complete scanning was carried out and 5 points for georeferencing; (**b**) view from Station 1; (**c**) view from Station 2; (**d**) view from Station 3; (**e**) view from Station 4.

The point cloud for Station 1 had a total of 57,348,448 points. Station 2 had 119,268,180 points, Station 3 had 96,323,508 points, and Station 4 had 274,417,074 points.

2.3.2. Data Processing

Processing of the point clouds obtained by TLS was carried out using Trimble RealWorks software. The first step carried out once the clouds were loaded was the relative registration between them. To obtain a fine registration of all TLS datasets, cloud-to-cloud registration was used. This required searching for common tie points between each pair of clouds. Then, the four point clouds were merged into a single point cloud. This processing methodology was used by [50] who reported errors between 11–19 mm, [51] with an error of 13 mm, or [52] where a maximum error of 42 mm was found between scans from two different stations. The results of the registration of this work are shown in Table 1.

Table 1. Result of the registration of the 4 scanning stations.

Station	Cloud to Cloud Error (cm)	Common Points (%)	Confidence (%)
SS-1			
SS-2	0.237	35%	97%
SS-3	0.278	18%	80%
SS-4	0.469	48%	99%
SS-2			
SS-1	0.237	35%	97%
SS-4	0.381	57%	100%
SS-3			
SS-1	0.278	18%	80%
SS-4			
SS-1	0.469	48%	99%
SS-2	0.381	57%	100%

Once registration of the four scanning stations was complete, the indirect georeferencing of the merged point cloud was carried out in absolute coordinates using the ETRS89 UTM zone 30 N reference system. For this purpose, five control points were used, measured by GNSS, the position of which is shown in Figure 7a. The results of the indirect georeferencing adjustment are shown in Table 2.

Table 2. Result of the indirect georeferencing of the project.

Point	X-Error (cm)	Y-Error (cm)	Z-Error (cm)	Total Error (cm)
Point-1				
SS-1	−1.062	0.491	0.404	1.238
Point-2				
SS-2	−1.841	−2.341	−1.459	3.316
Point-3				
SS-2	0.741	2.998	1.619	3.487
Point-4				
SS-4	1.741	0.973	0.354	2.026
Point-5				
SS-2	0.420	−2.120	−0.917	2.348

The final TLS point cloud resulting from combining the four individual point clouds can also be seen in Figure 7a. To reduce the size of the resulting point cloud, limits were established according to the area of interest, resulting in a point cloud of 74,447,399 points.

2.4. Topographic Surveying Using UAV Photogrammetry

2.4.1. Image Capture

The images used in this study were captured by a rotary wing with four rotors, DJI Phantom 4 Pro UAV. This equipment has a navigation system that uses GPS and GLONASS. In addition, it is equipped with front, rear, and lower vision systems that allow it to detect surfaces with defined patterns and adequate lighting and avoid obstacles with a range between 0.2 and 7 m. The Phantom 4 RGB camera is equipped with a one-inch, 20-megapixel (5472 × 3648) sensor and has a manually adjustable aperture (from F2.8 to F11). The lens has a fixed focal length of 8.8 mm and a horizontal field of view (FOV) of 84°. The acquisition of photographs from a UAV takes place via airborne photogrammetry from the aircraft, whereby a block of photographs is taken from parallel flight lines that are flown in a snake pattern at a stable altitude with constant overlap and a vertical camera angle (90°) [53]. However, the integration of oblique photographs can reduce the systematic deformation resulting from inaccurate calculations to determine the internal geometry of the camera in modern SfM–MVS photogrammetry [54,55]. There is evidence that oblique photographs contribute to the integrity of the point cloud reconstruction [51,56]. For example, [51] studied the influence of the angle of the oblique images, ranging from 0–35°, and compared it with the TLS. They concluded that oblique images are indispensable to the improvement of accuracy and the decreasing of systematic errors in the endpoint cloud. The results also suggested that oblique camera angles of between 20 and 35° increased accuracy by almost 50%, in relation to blocks with only nadir images.

The images were obtained from two independent flights. The first was carried out in automatic pilot mode through the DJI GS Pro application, and a total of 207 nadir photographs were obtained in 13 passes. The flight height was set at 36 m above the dam crest, which is equivalent to a ground sample distance (GSD) of 1.3 cm. To obtain side and forward overlaps of 65% and 80%, respectively, the camera took a shot every two seconds. The second flight was made in manual mode to obtain oblique photographs, in order to provide photographic capture of all the details of the dam´s geometry. This flight was carried out at an approximate distance of 30 m from the downstream face of the dam and was executed in seven different passes, parallel to the dam and at varying altitudes. A total of 372 photographs (including those of the control building and the guard, which are not relevant to this study) were taken. In addition, due to the camera's wide FOV, it was possible to cover the entire dam without going too far away from it. To avoid the appearance of the horizon in the photographs, an inclination angle of about 45° was adopted. No frontal photographs were taken, as much of the study area had horizontal surfaces. A total of 579 photographs with different points of view and scales were used to process the photogrammetric projects. Figure 8 shows the image overlap and camera locations for both flights.

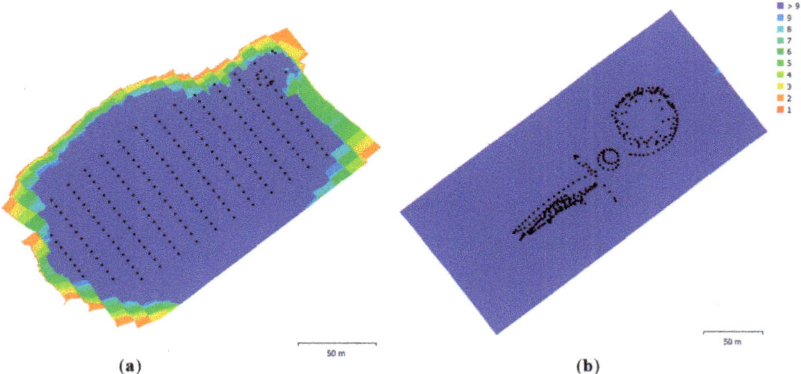

Figure 8. (**a**) Image overlaps and camera locations from nadiral flight; (**b**) image overlaps and camera locations from oblique flight.

2.4.2. Image Processing

The photogrammetric projects were executed using Agisoft Metashape Professional software, version 1.6.1.10009. This software is based on the SfM algorithm and runs in three independent steps. In the first step, all images are aligned by identifying and tying common points. During this process, the software estimates the camera's internal and external orientation parameters, including non-linear radial distortion. The software only needs the focal length value, which it obtains directly from the EXIF data of the photographs. This was carried out with the PhotoScan accuracy set to "medium" in order to reduce the processing time. The result of this step is the camera position and orientation, as well as the internal calibration parameters and the 3D relative coordinates of a sparse point cloud in the area of interest. In the second step, the sparse point cloud is referred to an absolute coordinate system, in our case ETRS89 UTM 30N, and the point cloud is densified, once the optimization and adjustment of the camera model have been completed. This was also carried out with the PhotoScan accuracy set to "medium". This point cloud needs to be cleaned up to eliminate all of the wild points not belonging in the model, which is done manually. The result is a highly detailed point cloud. From this point cloud, a mesh can be obtained; in this study, this was done using the height field method. In the third step, texture is applied to the mesh, and finally the orthophoto, digital surface model (DSM) and point cloud can be exported, in *.las formats. The bundle adjustment can be carried out using at least three ground control points (GCPs), but more accurate results can be obtained if more GCPs are used; thus, it is recommended that more GCPs be used to obtain optimal accuracy [57,58]. In this study, 17 targets placed on the dam were used as GCPs to georeference the project. The remaining targets not used in the bundle adjustment were used as control points (CPs) to evaluate the photogrammetric project accuracy, according to the root mean square error (RMSE) formula, described in [35]. There are numerous studies in which it has been corroborated that increasing the number of GCPs improves the accuracy of results of UAV photogrammetric projects. For example, [59] needed approximately six to seven GCPs to obtain accuracy of about 15.6 cm, and specified that it was necessary to increase the number of GCPs to reduce error. In [60], a study was conducted to analyze the optimal number of GCPs for volumetric measurements in open pit mines; for accuracies of about 5.0 cm, they concluded that about 15 GCP/km^2 were needed. In [61], an extensive study was carried out with over 3000 combinations, concluding that an increase in GCPs improves accuracy, with a limited project ground sample distance (GSD). In addition, [62] studied the influence of GCP distribution on the accuracy obtained in photogrammetric projects and concluded that the vertical error is proportional to distance, to the nearest GCP. In their case, they obtained vertical RMSEs that ranged from 15.6 cm for three GCPs to 5.9 cm for 101 GCPs.

In total, 18 photogrammetric projects were carried out, differentiating the types of orientation used in the photographs (nadiral, oblique, or both), the number of GCPs used for georeferencing (3, 5, and 7), and an adequate or inadequate distribution of GCPs, according to [41], who established that an adequate distribution of GCPs is one in which GCPs are arranged at the edge of the study area, while the interior area is covered homogeneously with GCPs; distributions of GCPs that do not meet these criteria are considered inadequate. Table 3 shows a summary of the executed photogrammetric projects, and Figure 9 shows the different combinations of numbers of GCPs and type of distribution of the GCPs.

Table 3. Summary of photogrammetric projects carried out.

Id Photogrammetric Project	Photographic Orientation	Number of GCPs Used	Adequate Distribution
1	Nadiral	3	Yes
2	Oblique	3	Yes
3	Nadiral & Oblique	3	Yes
4	Nadiral	3	Not
5	Oblique	3	Not
6	Nadiral & Oblique	3	Not
7	Nadiral	5	Yes
8	Oblique	5	Yes
9	Nadiral & Oblique	5	Yes
10	Nadiral	5	Not
11	Oblique	5	Not
12	Nadiral & Oblique	5	Not
13	Nadiral	7	Yes
14	Oblique	7	Yes
15	Nadiral & Oblique	7	Yes
16	Nadiral	7	Not
17	Oblique	7	Not
18	Nadiral & Oblique	7	Not

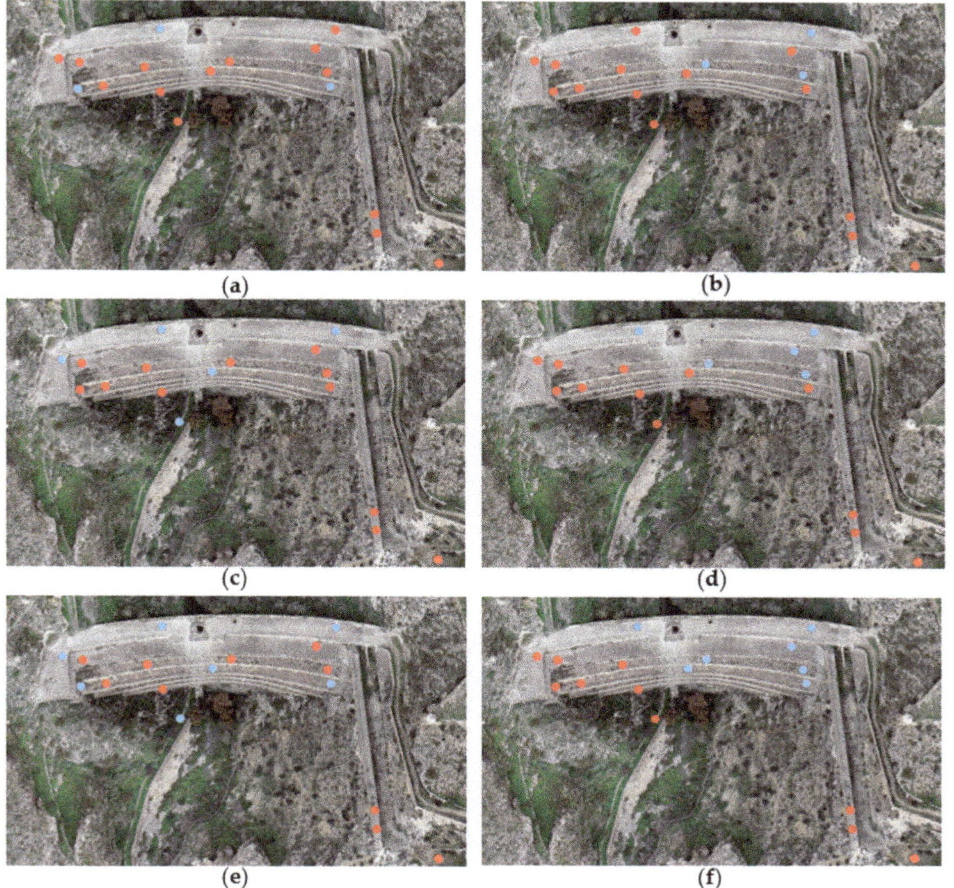

Figure 9. Number and distribution of ground control points (GCPs). Blue plots represent GCPs and red plots represent control points (CPs). (**a**) Three GCPs and adequate distribution; (**b**) three GCPs and inadequate distribution; (**c**) five GCPs and adequate distribution; (**d**) five GCPs and inadequate distribution; (**e**) seven GCPs and adequate distribution; (**f**) seven GCPs and inadequate distribution.

2.4.3. Accuracy Assessment

The evaluation of the accuracy was carried out by measuring the values of $RMSE_X$, $RMSE_Y$ and $RMSE_Z$, as well as $RMSE_T$ (total error) measured on each CP.

2.5. Point Cloud Management

Prior to the above, and in order to compare the profiles obtained from GNSS with the point cloud obtained from TLS, the cloud-to-mesh (C2M) tool, offered by CloudCompare v2.8 [63], was used. The mesh was first obtained from the point cloud of the TLS, using the Dalaunay 2.5 tool, and once obtained, the C2M algorithm was applied to corroborate the adequate georeferencing of the TLS cloud and its validity for use as a reference for comparison against photogrammetric projects.

A multiscale model-to-model cloud comparison (M3C2) tool was used to compare the point clouds generated by the photogrammetric projects with the point cloud generated by the TLS. This tool calculates in a robust way, the distance, positive or negative, between two point clouds [64].

The principles of operation of this tool are described in [41]. M3C2 output consists of, amongst other data, a text file with the *x*-, *y*-, and *z*-coordinates for each point of the reference cloud and the 3D distance associated with the comparison. All data can be displayed using a color scale to highlight the resulting scalar field.

3. Results

3.1. Validation of Data Derived From TLS in Comparison With Theoretical Profiles Obtained by GNSS

Table 4 shows a summary of the comparison between the profiles obtained by GNSS and the model derived from the point cloud obtained by TLS. For the seven profiles studied, the average of the mean differences is below 0.03 cm with a standard deviation of less than 3 cm, which ensures the correct georeferencing of the TLS point cloud.

Table 4. Summary of comparison between profiles obtained by GNSS and data derived from a terrestrial laser scanner (TLS).

	Errors TLS vs GNSS	
Profiles	**Mean (cm)**	**Standard Deviation (cm)**
P-1	−0.180	2.888
P-2	−0.122	2.602
P-3	−0.244	2.820
P-4	0.003	3.487
P-5	0.166	2.617
P-6	−0.139	1.877
P-7	0.302	4.463
Mean of profiles	**−0.030**	**2.965**

Figure 10a graphically represents the differences across the seven profiles obtained by GNSS with respect to the model derived from the point cloud obtained by TLS. Figure 10b shows an example of the error distribution for Profile 2, in which it can be seen that most of the differences are close to 0 and that, in some cases, larger errors appeared in most cases due to the presence of shrubs growing on the dam face, as shown in Figure 10c.

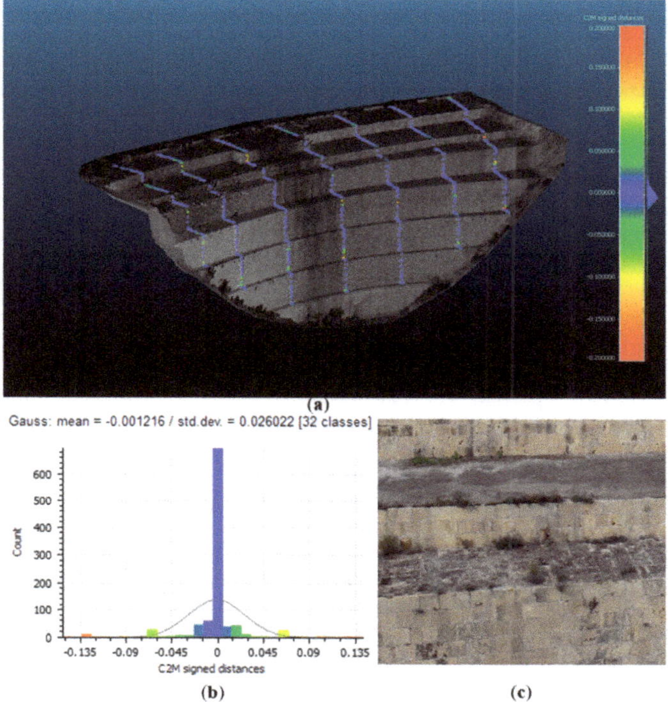

Figure 10. (**a**) Errors between GNSS and TLS along the seven profiles studied (units in meters); (**b**) errors in Profile 2 (units in meters); (**c**) example of bushes growing on the face of the dam.

3.2. RMSE UAV-PhotogrAmmetric Projects

Figure 11 shows the precision results obtained in the 18 photogrammetric projects through the evaluation of the RMSE.

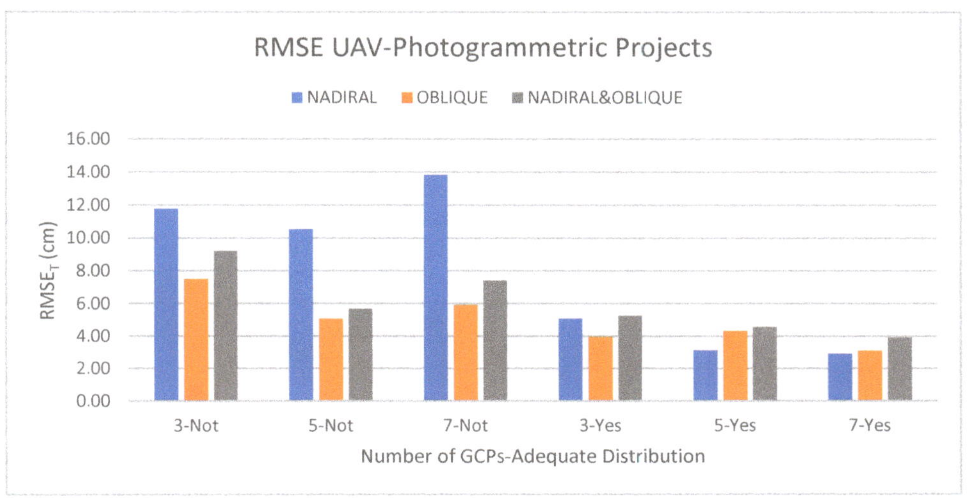

Figure 11. Root mean square error (RMSE) unmanned aerial vehicle (UAV)-photogrammetric projects.

From the analysis of the data obtained, it is clear that an adequate distribution of GCPs is important for the minimization of errors. The results obtained were in line with those obtained by [41]. Therefore, the average error in the projects with inadequate distribution was around 8.5 cm, while in the projects with adequate distribution, it was in an average of 4.0 cm, which represents an improvement of more than 50%. As indicated in [57], increasing the number of GCPs improves the accuracy of photogrammetric projects, regardless of whether the distribution of GCPs is adequate or not. In this study, a slight worsening was found for 7 GCPs against 5 GCPs. Thus, for 3 GCPs the average error was 7.1, for 5 GCPs it was 5.5, and for 7 GCPs it was 6.2 cm. This alteration was only found for the projects with inadequate distribution, while for the projects with adequate distribution, the error decreased as the number of GCPs increased. Thus, for 3 GCPs the average error was 4.8, for 5 GCPs it was 4.0, and for 7 GCPs it was 3.3 cm. Regarding the orientation of the photographs, an average error of 7.9 was obtained for projects with nadiral photographs, 5.0 for projects with oblique photographs, and 6.0 cm for projects that combined both orientations of photographs. For the projects with inadequate distribution, the best results were obtained for oblique photographs with an average of 6.2 cm. However, for an adequate distribution, the best results were obtained for nadiral photographs with an average of 3.7 as opposed to 3.8 obtained for oblique photographs, or 4.6 cm for projects that combined both orientations.

3.3. Vertical Distances between the Point Clouds Obtained by TLS and UAV Photogrammetry

For all studied scenarios, the point clouds were evaluated based on reference data acquired by the TLS. There are several studies in scientific literature concerning the evaluation of vertical distances between clouds obtained by UAV photogrammetry and reference clouds. For example, [65] evaluated the capacity of UAV photogrammetry to obtain point clouds in quarries where there were walls with near-vertical inclination. In this study, they corroborated the improvement in the description of the break lines and in the accuracy of cross profiles with vertical walls, using oblique photographs and comparing the UAV photogrammetry clouds with the profiles obtained by a total station. One of their main conclusions was that the scenario with nadir photographs resulted in smoother geometry and a more gradual transition between vertical and horizontal surfaces than the scenario involving nadir and oblique photographs.

Figure 12 shows the vertical distances between the cloud obtained by TLS and the six UAV photogrammetry projects executed with 3 GCPs. For the projects with nadiral photographs and adequate distribution of GCPs, the M3C2-calculated vertical distances resulted in absolute values distributed as a Gaussian function with mean = −0.075 and standard deviation (SD) = 8.335 cm. For the project with nadiral photographs and inadequate distribution of GCPs, a mean distance of −7.806 and an SD of 13.586 cm were obtained. For the project with oblique photographs and adequate distribution of GCPs, an average distance of 0.068 and an SD of 6.406 cm were obtained. For the project with oblique photographs and inadequate distribution of GCPs, an average distance of −21.597 and an SD of 22.466 cm were obtained. For the project with nadiral and oblique photographs and adequate distribution of GCPs, a mean distance of 0.195 and an SD of 7.066 cm were obtained. For the project with nadiral and oblique photographs and inadequate distribution of GCPs, an average distance of −24.692 and an SD of 26.988 cm were obtained. For this scenario, with inadequate distribution, the use of oblique photographs increased the average absolute distance between the clouds by 47%. However, for an adequate distribution, there was a reduction of the average cloud distance by 30%.

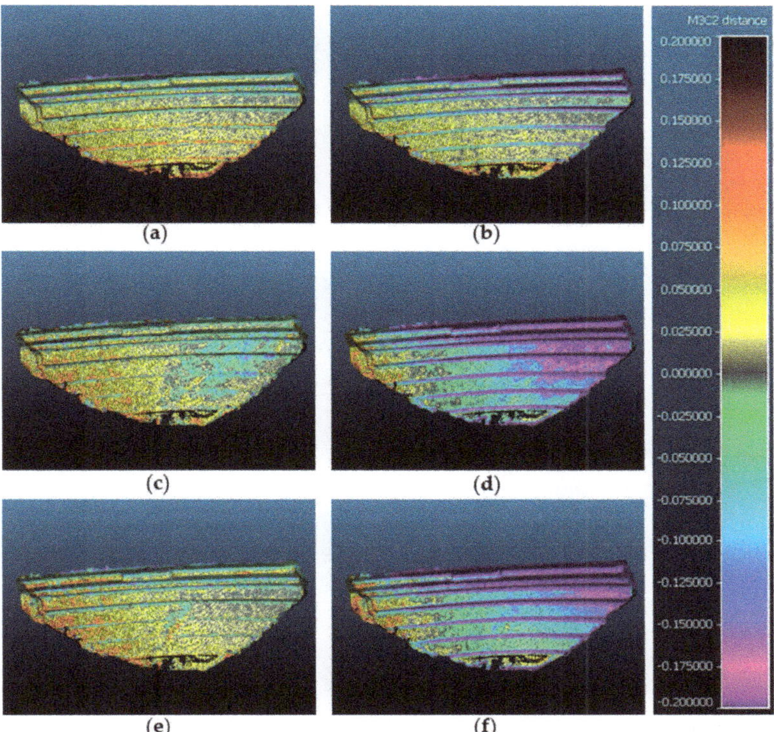

Figure 12. UAV-photogrammetric projects with 3 GCPS (units in meters). (**a**) Nadiral-3GCPs-Yes; (**b**) Nadiral-3GCPs-Not; (**c**) Oblique-3GCPs-Yes; (**d**) Oblique-3GCPs-Not; (**e**) Nadiral&Oblique-3GCPs-Yes; (**f**) Nadiral&Oblique-3GCPs-Not.

Figure 13 shows the vertical distances between the cloud obtained by TLS and the six UAV photogrammetry projects executed with 5 GCPs. For the project with nadiral photographs and adequate distribution of GCPs, an average distance of 0.492 and an SD of 7.236 cm were obtained. For the project with nadir photographs and inadequate distribution of GCPs, a mean distance of −3.015 and an SD of 7.501 cm were obtained. For the project with oblique photographs and inadequate distribution of GCPs, a mean distance of 1.109 and an SD of 6.092 cm were obtained. For the project with oblique photographs and inadequate distribution of GCPs, a mean distance of −1.418 and an SD of 6.303 cm were obtained. For the project with nadiral and oblique photographs and adequate distribution of GCPs, a mean distance of 1.361 and SD of 6.593 m were obtained. For the project with nadiral and oblique photographs and inadequate distribution of GCPs, a mean distance of −1.666 and SD of 6.661 cm were obtained. For this scenario, with inadequate distribution, the use of oblique photographs reduced the absolute average distance between the clouds by 25%, and by 9% for adequate distribution.

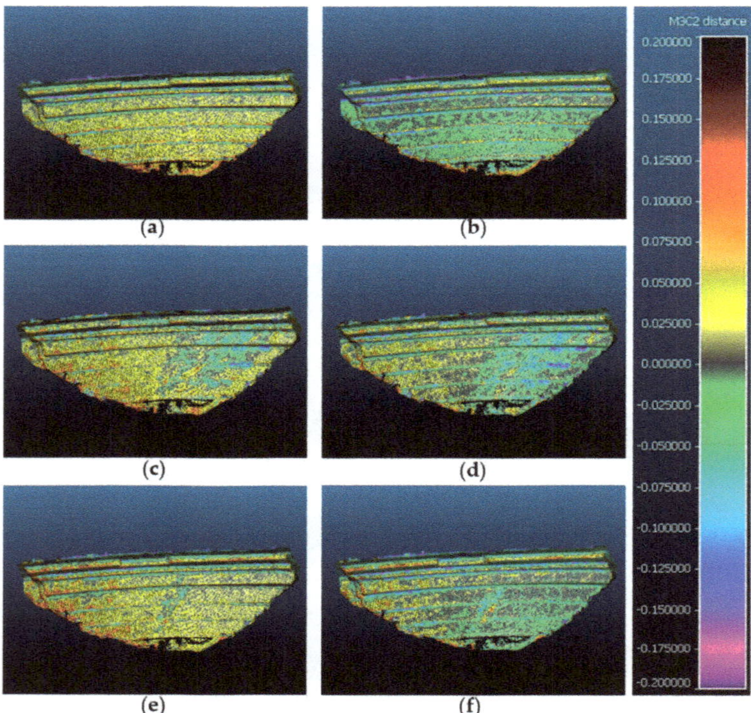

Figure 13. UAV-photogrammetric projects with 5 GCPS (units in meters). (**a**) Nadiral-5GCPs-Yes; (**b**) Nadiral-5GCPs-Not; (**c**) Oblique-5GCPs-Yes; (**d**) Oblique-5GCPs-Not; (**e**) Nadiral&Oblique-5GCPs-Yes; (**f**) Nadiral&Oblique-5GCPs-Not.

Figure 14 shows the vertical distances between the cloud obtained by TLS and the six UAV photogrammetry projects executed with 7 GCPs. For the project with nadiral photographs and adequate distribution of GCPs, an average distance of 0.325 and SD of 7.226 cm were obtained. For the project with nadiral photographs and inadequate distribution of GCPs, a mean distance of −0.637 and SD of 8.689 cm were obtained. For the project with oblique photographs and adequate distribution of GCPs, a mean distance of 0.704 and SD of 6.187 cm were obtained. For the project with oblique photographs and inadequate distribution of GCPs, a mean distance of 1.134 and SD of 6.616 cm were obtained. For the project with nadiral and oblique photographs and adequate distribution of GCPs, a mean distance of 1.130 and a SD of 6.579 cm were obtained. For the project with nadiral and oblique photographs and inadequate distribution of GCPs, a mean distance of 1.314 and SD of 7.143 cm were obtained. For this scenario, with inadequate distribution, the use of oblique photographs reduced the absolute average distance between the clouds by 18%, and by 11% for adequate distribution.

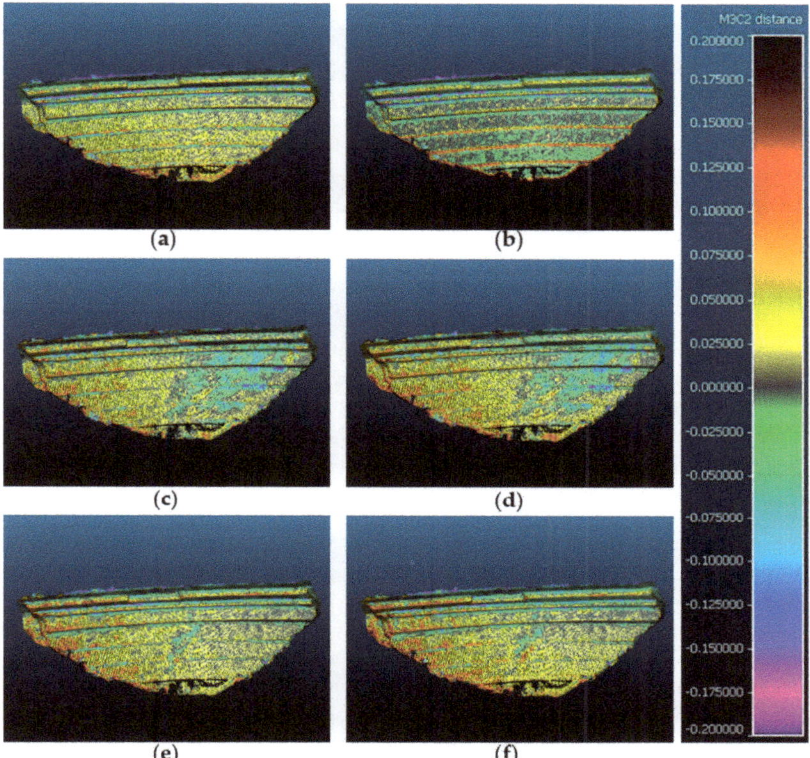

Figure 14. UAV-photogrammetric projects with 7 GCPS (units in meters). (**a**) Nadiral-7GCPs-Yes; (**b**) Nadiral-7GCPs-Not; (**c**) Oblique-7GCPs-Yes; (**d**) Oblique-7GCPs-Not; (**e**) Nadiral&Oblique-7GCPs-Yes; (**f**) Nadiral&Oblique-7GCPs-Not.

4. Discussion

Considering that the accuracy error of measurements made with GNSS was about 8 for the horizontal plane and 15 mm vertically, the adjustment obtained for the TLS point cloud is considered adequate, since the average adjustment error obtained at the control points was 25 mm. In turn, the maximum cloud-to-cloud error found in the four-station registry was below 5 mm, similar to that obtained by [50,51], and much less than that reported by [52]. Analyzing the differences with respect to the seven theoretical profiles obtained by GNSS, a mean error of −0.030 and a standard deviation (SD) of 2.965 cm were found. Similar errors were obtained by [66], who reported three different experiments to obtain indirect georeferencing of TLS point clouds by using control lines. The results they found ranged from a mean error of 0.002–0.064 and a SD between 0.928 cm and 1.729 cm. Therefore, in view of the results obtained for the adjustment and georeferencing of the point cloud by TLS, it can be adopted as a reference cloud for subsequent comparison with projects obtained by UAV photogrammetry.

Regarding the number of GCPs used for the georeferencing of photogrammetric projects, in this study, a clear trend of improved results was obtained by increasing the number of GCPs, with the only noteworthy change being for an inadequate distribution of the seven GCPs. This is approximately 50% less obvious with oblique photographs than with nadiral photographs, which demonstrates the need to incorporate oblique photographs when it is not possible to have an adequate distribution of GCPs.

According to [62], the results of our study show similar values, where, for three GCPs and an inadequate distribution, an RMSE of 9.49 cm was obtained, while with an increase of GCPs and improved distribution the RMSE decreased to 3.32 cm. This same trend is observed regardless of the orientation of the photographs.

In recent years, studies on the benefit of introducing oblique images or an adequate distribution of GCPs in UAV photogrammetry projects are common. However, there have been only few such studies applied to quasi-vertical or vertical walls. In [55], it is noted that for the network of self-calibrated images to be accurate, the spatial distribution of GCPs needs to cover the whole area of interest. In turn, the density of GCPs depends on the accuracy needed in the project, the geometry of the network, and the quality of the photographs. For example, to obtain accuracies of 5.0 cm, at a height of 100 m (GSD 23 mm), about 15 GCPs were needed, with an average minimum spacing of 50 m. They pointed out that improving the network geometry with oblique images or with a precisely pre-calibrated camera could reduce the number of GCPs. These results coincide with those obtained in this study, especially when the distribution of GCPs was inadequate. However, when the number of GCPs was increased to five or seven, and the distribution was adequate, no improvement was found with the use of oblique photographs. For projects with inadequate distribution, the use of nadir photographs dramatically worsens the results. In this case, nadir photographs do not have such a strong effect when combined with oblique photographs. In the case of adequate distribution, the opposite happens; the best results are obtained with nadir photographs, while in this case, oblique photographs slightly worsen the results.

As noted in [67], the accuracy does not only depend upon the number of GCPs, but also on their distribution pattern. Therefore, the choice of a suitable pattern and the number of GCPs for a particular mission can help obtain sufficiently accurate results with economic feasibility. In their study, the accuracy ranged from 18.2 with three GCPs to 11.3 cm with nine GCPs. They also highlighted the importance of positioning a GCP in the central region of the area. There was a substantial improvement in accuracy due to the addition of another GCP in the central area. Our study has improved those results, where, for three GCPs and an adequate distribution, the average error was 4.8 cm. In [42], the authors studied the acquisition and use of oblique images for the 3D reconstruction of a historical building, obtained by UAV for the realization of a high-level-of-detail architectural survey. For this purpose, they used several software applications, obtaining similar results, in which the differences with respect to a reference cloud ranged from 0.3–2.5 to 1.6–3.9 cm. As demonstrated in [68], the use of oblique images obtained from a low-cost UAV system and processed by SfM software was an effective method for surveying cultural heritage sites. In particular, they studied a facade, in which they obtained an average distance between the clouds (UAV vs TLS) of 4.0 cm, with an SD of 9.0 cm. Similar results were found in our study, where, for an adequate distribution of GCPs, the mean value ranged from −0.1 to 1.4 cm with an SD ranging from 6.1 to 8.3 cm.

With respect to the break lines referred to by [65], this fact can be visually corroborated with reference to Figures 11–13, wherein the major differences can be seen in the edges, at which abrupt changes of slope occur. According to [51], in our study, evaluating the vertical distances between the UAV photogrammetry cloud and the TLS cloud, the use of oblique photographs improved results by between 9% and 30% for all scenarios except three GCPs and inadequate distribution.

5. Conclusions

The SfM analysis of the UAV images is a valuable and verified tool for surveyors interested in the high-resolution reconstruction of nearly vertical walls. UAV studies are also a useful tool for risk management in accessing hazardous and inaccessible areas, such as in cases of slope cuttings on highways or roads with near vertical inclination or the analysis of architectural facades in cities. This study has addressed the influence of three factors: The number of GCPs, the distribution of GCPs, and the orientation of photographs obtained by the UAV. From the results obtained, a series of guidelines can be established to simplify the process of capturing data in the field and to improve the accuracy of photogrammetric products. In order to obtain optimal results with respect to accuracy, it is important to distribute of as many GCPs as possible throughout the study area and to make use of an adequate distribution GCPs coverage, as far as possible, over the entire area of interest. In case these two conditions are difficult to meet, the inclusion of oblique photographs in the photogrammetric project results in a substantial improvement in the results obtained. In this way, it is possible to achieve an RMSE of around 3.0 cm, which is a sufficient scale for most engineering or architectural projects. In addition, oblique photographs ostensibly improve the geometric description of break lines or sudden changes in slope. Therefore, it is always advisable to use them for projects with vertical topography. Under these circumstances, UAV photogrammetry constitutes a technique whose results are equivalent to those obtained by a TLS, but with the incentive of lower cost and greater facility for the treatment of the point cloud.

Author Contributions: The contributions of the authors were: Conceptualization, P.M.-C., F.C.-R. and F.A.-V.; methodology, P.M.-C., F.C.-R. and F.A.-V.; software P.M.-C.; validation, P.M.-C.; formal analysis, P.M.-C., F.C.-R. and F.A.-V.; investigation, P.M.-C., F.C.-R. and F.A.-V.; resources, F.C.-R. and F.A.-V.; writing—original draft preparation, P.M.-C.; writing—review and editing, P.M.-C., F.C.-R. and F.A.-V.; visualization, P.M.-C.; project administration, F.C.-R. and F.A.-V. All authors have read and agreed to the published version of the manuscript.

Funding: This research received no external funding.

Conflicts of Interest: The authors declare no conflict of interest.

References

1. Wallace, L.O.; Lucieer, A.; Malenovsky, Z.; Turner, D.; Vopěnka, P. Assessment of Forest Structure Using Two UAV Techniques: A Comparison of Airborne Laser Scanning and Structure from Motion (SfM) Point Clouds. *Forests* **2016**, *7*, 62. [CrossRef]
2. Koo, J.-B. The Study on Recording Method for Buried Cultural Property Using Photo Scanning Technique. *J. Digit. Contents Soc.* **2015**, *16*, 835–847. [CrossRef]
3. Yao, H.; Qin, R.; Chen, X. Unmanned Aerial Vehicle for Remote Sensing Applications—A Review. *Remote Sens.* **2019**, *11*, 1443. [CrossRef]
4. Liu, P.; Chen, A.Y.; Huang, Y.-N.; Han, J.-Y.; Lai, J.-S.; Kang, S.-C.; Wu, T.-H.; Wen, M.-C.; Tsai, M.-H. A review of rotorcraft Unmanned Aerial Vehicle (UAV) developments and applications in civil engineering. *Smart Struct. Syst.* **2014**, *13*, 1065–1094. [CrossRef]
5. Bento, M.D.F. Unmanned Aerial Vehicles: An Overview. *InsideGNSS* **2008**, *3*, 54–61.
6. Samad, A.M.; Kamarulzaman, N.; Hamdani, M.A.; Mastor, T.A.; Hashim, K.A. The potential of Unmanned Aerial Vehicle (UAV) for civilian and mapping application. In Proceedings of the 2013 IEEE 3rd International Conference on System Engineering and Technology, Shah Alam, Malaysia, 19–20 August 2013; Institute of Electrical and Electronics Engineers (IEEE): Piscataway, NJ, USA, 2013; pp. 313–318.
7. Mesas-Carrascosa, F.-J.; Pérez-Porras, F.; De Larriva, J.E.M.; Frau, C.M.; Agüera-Vega, F.; Carvajal-Ramirez, F.; Carricondo, P.J.M.; García-Ferrer, A. Drift Correction of Lightweight Microbolometer Thermal Sensors On-Board Unmanned Aerial Vehicles. *Remote Sens.* **2018**, *10*, 615. [CrossRef]
8. Grenzdörffer, G.; Engel, A.; Teichert, B. The photogrammetric potential of low-cost UAVs in forestry and agriculture. *Int. Arch. Photogramm. Remote Sens. Spat. Inf. Sci.* **2008**, *XXXVII*(Part B1), 1207–1214.
9. Tian, J.; Wang, L.; Li, X.; Gong, H.; Shi, C.; Zhong, R.; Liu, X. Comparison of UAV and WorldView-2 imagery for mapping leaf area index of mangrove forest. *Int. J. Appl. Earth Obs. Geoinf.* **2017**, *61*, 22–31. [CrossRef]
10. Kachamba, D.; Ørka, H.O.; Gobakken, T.; Eid, T.; Mwase, W. Biomass Estimation Using 3D Data from Unmanned Aerial Vehicle Imagery in a Tropical Woodland. *Remote Sens.* **2016**, *8*, 968. [CrossRef]
11. Carvajal-Ramirez, F.; Da Silva, J.M.; Agüera-Vega, F.; Carricondo, P.J.M.; Serrano, J.; Moral, F.J. Evaluation of Fire Severity Indices Based on Pre- and Post-Fire Multispectral Imagery Sensed from UAV. *Remote Sens.* **2019**, *11*, 993. [CrossRef]
12. Yuan, C.; Zhang, Y.; Liu, Z. A survey on technologies for automatic forest fire monitoring, detection, and fighting using unmanned aerial vehicles and remote sensing techniques. *Can. J. For. Res.* **2015**, *45*, 783–792. [CrossRef]
13. Carricondo, P.J.M.; Carvajal-Ramirez, F.; Yero-Paneque, L.; Vega, F.A. Combination of nadiral and oblique UAV photogrammetry and HBIM for the virtual reconstruction of cultural heritage. Case study of Cortijo del Fraile in Níjar, Almería (Spain). *Build. Res. Inf.* **2019**, *48*, 140–159. [CrossRef]

14. Carvajal-Ramirez, F.; Navarro-Ortega, A.D.; Agüera-Vega, F.; Martínez-Carricondo, P.; Mancini, F. Virtual reconstruction of damaged archaeological sites based on Unmanned Aerial Vehicle Photogrammetry and 3D modelling. Study case of a southeastern Iberia production area in the Bronze Age. *Measurement* **2019**, *136*, 225–236. [CrossRef]
15. Doumit, J. Structure from motion technology for historic building information modeling of Toron fortress (Lebanon). *Proc. Int. Conf. InterCarto InterGIS* **2019**, *25*. [CrossRef]
16. Salvo, G.; Caruso, L.; Scordo, A. Urban Traffic Analysis through an UAV. *Proc. Soc. Behav. Sci.* **2014**, *111*, 1083–1091. [CrossRef]
17. Kanistras, K.; Martins, G.; Rutherford, M.J.; Valavanis, K.P. Survey of unmanned aerial vehicles (uavs) for traffic monitoring. In *Handbook of Unmanned Aerial Vehicles*; Valavanis, K.P., Vachtsevanos, G.J., Eds.; Springer Reference: Dordrecht, The Netherlands, 2015; ISBN 9789048197071.
18. Eltner, A.; Mulsow, C.; Maas, H.-G. Quantitative Measurement of Soil Erosion from Tls And Uav Data. *Int. Arch. Photogramm. Remote Sens. Spat. Inf. Sci.* **2013**, 119–124. [CrossRef]
19. Piras, M.; Taddia, G.; Forno, M.G.; Gattiglio, M.; Aicardi, I.; Dabove, P.; Russo, S.L.; Lingua, A. Detailed geological mapping in mountain areas using an unmanned aerial vehicle: Application to the Rodoretto Valley, NW Italian Alps. *Geomat. Nat. Hazards Risk* **2016**, *8*, 137–149. [CrossRef]
20. Pavlidis, G.; Koutsoudis, A.; Arnaoutoglou, F.; Tsioukas, V.; Chamzas, C. Methods for 3D digitization of Cultural Heritage. *J. Cult. Heritage* **2007**, *8*, 93–98. [CrossRef]
21. Püschel, H.; Sauerbier, M.; Eisenbeiss, H. A 3D Model of Castle Landenberg (CH) from Combined Photogrametric Processing of Terrestrial and UAV Based Images. *Int. Arch. Photogramm. Remote Sens. Spat. Inf. Sci.* **2008**, *37*, 93–98.
22. Themistocleous, K.; Agapiou, A.; Hadjimitsis, D. 3D Documentation And Bim Modeling of Cultural Heritage Structures using Uavs: The Case of the Foinikaria Church. *Int. Arch. Photogramm. Remote Sens. Spat. Inf. Sci.* **2016**, *42*, 45–49. [CrossRef]
23. Aber, J.S.; Marzolff, I.; Ries, J.B. *Small-Format Aerial Photography*; Elsevier Science: Amsterdam, The Netherlands, 2010; ISBN 9780444532602.
24. Atkinson, K.B. *Close Range Photogrammetry and Machine Vision*; Whittles Publishing: Dunbeath, UK, 2001; ISBN 978-1870325738.
25. Fernández-Hernandez, J.; González-Aguilera, D.; Rodríguez-Gonzálvez, P.; Mancera-Taboada, J. Image-Based Modelling from Unmanned Aerial Vehicle (UAV) Photogrammetry: An Effective, Low-Cost Tool for Archaeological Applications. *Archaeometry* **2014**, *57*, 128–145. [CrossRef]
26. Fonstad, M.A.; Dietrich, J.T.; Courville, B.C.; Jensen, J.L.; Carbonneau, P.E. Topographic structure from motion: A new development in photogrammetric measurement. *Earth Surf. Process. Landf.* **2013**, *38*, 421–430. [CrossRef]
27. Javernick, L.; Brasington, J.; Caruso, B. Modeling the topography of shallow braided rivers using Structure-from-Motion photogrammetry. *Geomorphology* **2014**, *213*, 166–182. [CrossRef]
28. Westoby, M.J.; Brasington, J.; Glasser, N.; Hambrey, M.; Reynolds, J. 'Structure-from-Motion' photogrammetry: A low-cost, effective tool for geoscience applications. *Geomorphology* **2012**, *179*, 300–314. [CrossRef]
29. Snavely, N.; Seitz, S.M.; Szeliski, R. Modeling the World from Internet Photo Collections. *Int. J. Comput. Vis.* **2007**, *80*, 189–210. [CrossRef]
30. Vasuki, Y.; Holden, E.-J.; Kovesi, P.; Micklethwaite, S. Semi-automatic mapping of geological Structures using UAV-based photogrammetric data: An image analysis approach. *Comput. Geosci.* **2014**, *69*, 22–32. [CrossRef]
31. Furukawa, Y.; Ponce, J. Accurate, Dense, and Robust Multiview Stereopsis. *IEEE Trans. Pattern Anal. Mach. Intell.* **2009**, *32*, 1362–1376. [CrossRef]
32. Lowe, D.G. Distinctive Image Features from Scale-Invariant Keypoints. *Int. J. Comput. Vis.* **2004**, *60*, 91–110. [CrossRef]
33. Kamal, W.A.; Samar, R. A mission planning approach for UAV applications. In Proceedings of the 2008 47th IEEE Conference on Decision and Control, Cancun, Mexico, 9–11 December 2008; Institute of Electrical and Electronics Engineers (IEEE): Piscataway, NJ, USA, 2008; pp. 3101–3106.
34. Hugenholtz, C.; Brown, O.; Walker, J.; Barchyn, T.E.; Nesbit, P.; Kucharczyk, M.; Myshak, S. Spatial Accuracy of UAV-Derived Orthoimagery and Topography: Comparing Photogrammetric Models Processed with Direct Geo-Referencing and Ground Control Points. *Geomatica* **2016**, *70*, 21–30. [CrossRef]

35. Vega, F.A.; Carvajal-Ramirez, F.; Martínez-Carricondo, P. Accuracy of Digital Surface Models and Orthophotos Derived from Unmanned Aerial Vehicle Photogrammetry. *J. Surv. Eng.* **2017**, *143*, 04016025. [CrossRef]
36. Amrullah, C.; Suwardhi, D.; Meilano, I. Product Accuracy Effect of Oblique and Vertical Non-Metric Digital Camera Utilization in Uav-Photogrammetry to Determine Fault Plane. *ISPRS Ann. Photogramm. Remote Sens. Spat. Inf. Sci.* **2016**, 41–48. [CrossRef]
37. Dandois, J.P.; Olano, M.; Ellis, E. Optimal Altitude, Overlap, and Weather Conditions for Computer Vision UAV Estimates of Forest Structure. *Remote Sens.* **2015**, *7*, 13895–13920. [CrossRef]
38. Mesas-Carrascosa, F.-J.; Rumbao, I.C.; Berrocal, J.A.B.; García-Ferrer, A. Positional Quality Assessment of Orthophotos Obtained from Sensors Onboard Multi-Rotor UAV Platforms. *Sensors* **2014**, *14*, 22394–22407. [CrossRef] [PubMed]
39. Vautherin, J.; Rutishauser, S.; Schneider-Zapp, K.; Choi, H.F.; Chovancova, V.; Glass, A.; Strecha, C. Photogrammetric Accuracy and Modeling of Rolling Shutter Cameras. *ISPRS Ann. Photogramm. Remote Sens. Spat. Inf. Sci.* **2016**, 139–146. [CrossRef]
40. Murtiyoso, A.; Grussenmeyer, P. Documentation of heritage buildings using close-range UAV images: Dense matching issues, comparison and case studies. *Photogramm. Rec.* **2017**, *32*, 206–229. [CrossRef]
41. Carricondo, P.J.M.; Vega, F.A.; Carvajal-Ramirez, F.; Mesas-Carrascosa, F.-J.; García-Ferrer, A.; Pérez-Porras, F.-J. Assessment of UAV-photogrammetric mapping accuracy based on variation of ground control points. *Int. J. Appl. Earth Obs. Geoinf.* **2018**, *72*, 1–10. [CrossRef]
42. Aicardi, I.; Chiabrando, F.; Grasso, N.; Lingua, A.M.; Noardo, F.; Spanò, A. Uav Photogrammetry with Oblique Images: First Analysis on Data Acquisition and Processing. *ISPRS Int. Arch. Photogramm. Remote Sens. Spat. Inf. Sci.* **2016**, 835–842. [CrossRef]
43. García-Sánchez, J.F. El pantano de Isabel II de Níjar (Almería): Paisaje, fondo y figura. *ph Investig.* **2014**, *3*, 55–74.
44. Lovas, T.; Barsi, A.; Polgar, A.; Kibedy, Z.; Berenyi, A.; Detrekoi, A.; Dunai, L. Potential of terrestrial laserscanning in deformation measurement of structures. *Am. Soc. Photogramm. Remote Sens. ASPRS Annu. Conf. 2008 Bridg. Horiz. New Front. Geospat. Collab.* **2008**, *2*, 454–463.
45. Danson, F.M.; Hetherington, D.; Koetz, B.; Morsdorf, F.; Allgöwer, B. Forest Canopy Gap Fraction from Terrestrial Laser Scanning. *IEEE Geosci. Remote Sens. Lett.* **2007**, *4*, 157–160. [CrossRef]
46. Lumme, J.; Karjalainen, M.; Kaartinen, H.; Kukko, A.; Hyyppä, J.; Hyyppä, H.; Jaakkola, A.; Kleemola, J. Terrestrial laser scanning of agricultural crops. *Int. Arch. Photogramm. Remote Sens. Spat. Inf. Sci.* **2008**, *47*, 563–566.
47. Abellan, A.; Calvet, J.; Vilaplana, J.M.; Blanchard, J. Detection and spatial prediction of rockfalls by means of terrestrial laser scanner monitoring. *Geomorphology* **2010**, *119*, 162–171. [CrossRef]
48. Schulz, T. Calibration of a Terrestrial Laser Scanner for Engineering Geodesy. *Geod. Metrol. Eng. Geod.* **2007**. [CrossRef]
49. Kwiatkowski, J.; Anigacz, W.; Beben, D. Comparison of Non-Destructive Techniques for Technological Bridge Deflection Testing. *Materials* **2020**, *13*, 1908. [CrossRef]
50. Fryskowska, A. Accuracy Assessment of Point Clouds Geo-Referencing in Surveying and Documentation of Historical Complexes. *ISPRS Int. Arch. Photogramm. Remote Sens. Spat. Inf. Sci.* **2017**. [CrossRef]
51. Nesbit, P.; Hugenholtz, C.H. Enhancing UAV–SfM 3D Model Accuracy in High-Relief Landscapes by Incorporating Oblique Images. *Remote Sens.* **2019**, *11*, 239. [CrossRef]
52. Abbas, M.A.; Luh, L.C.; Setan, H.; Majid, Z.; Chong, A.K.; Aspuri, A.; Idris, K.M.; Ariff, M.F.M. Terrestrial Laser Scanners Pre-Processing: Registration and Georeferencing. *J. Teknol.* **2014**, *71*. [CrossRef]
53. Martin, R.; Rojas, I.; Franke, K.W.; Hedengren, J.D. Evolutionary View Planning for Optimized UAV Terrain Modeling in a Simulated Environment. *Remote Sens.* **2015**, *8*, 26. [CrossRef]
54. James, M.R.; Robson, S. Mitigating systematic error in topographic models derived from UAV and ground-based image networks. *Earth Surf. Process. Landf.* **2014**, *39*, 1413–1420. [CrossRef]
55. James, M.R.; Robson, S.; D'Oleire-Oltmanns, S.; Niethammer, U. Optimising UAV topographic surveys processed with structure-from-motion: Ground control quality, quantity and bundle adjustment. *Geomorphology* **2017**, *280*, 51–66. [CrossRef]
56. Jiang, S.; Jiang, W. Efficient structure from motion for oblique UAV images based on maximal spanning tree expansion. *ISPRS J. Photogramm. Remote Sens.* **2017**, *132*, 140–161. [CrossRef]

57. Vega, F.A.; Carvajal-Ramirez, F.; Martínez-Carricondo, P. Assessment of photogrammetric mapping accuracy based on variation ground control points number using unmanned aerial vehicle. *Measurement* **2017**, *98*, 221–227. [CrossRef]
58. Rosnell, T.; Honkavaara, E. Point Cloud Generation from Aerial Image Data Acquired by a Quadrocopter Type Micro Unmanned Aerial Vehicle and a Digital Still Camera. *Sensors* **2012**, *12*, 453–480. [CrossRef] [PubMed]
59. Assessing the Impact of the Number of GCPS on the Accuracy of Photogrammetric Mapping from UAV Imagery. *Balt. Surv.* **2019**, *10*, 43–51. [CrossRef]
60. Siqueira, H.L.; Marcato, J.; Matsubara, E.T.; Eltner, A.; Colares, R.A.; Santos, F.M.; Junior, J.M. The Impact of Ground Control Point Quantity on Area and Volume Measurements with UAV SFM Photogrammetry Applied in Open Pit Mines. In Proceedings of the IGARSS 2019—2019 IEEE International Geoscience and Remote Sensing Symposium, Yokohama, Japan, 28 July–2 August 2019. [CrossRef]
61. Sanz-Ablanedo, E.; Chandler, J.; Rodríguez-Pérez, J.R.; Ordóñez, C. Accuracy of Unmanned Aerial Vehicle (UAV) and SfM Photogrammetry Survey as a Function of the Number and Location of Ground Control Points Used. *Remote Sens.* **2018**, *10*, 1606. [CrossRef]
62. Tonkin, T.; Midgley, N. Ground-Control Networks for Image Based Surface Reconstruction: An Investigation of Optimum Survey Designs Using UAV Derived Imagery and Structure-from-Motion Photogrammetry. *Remote Sens.* **2016**, *8*, 786. [CrossRef]
63. Girardeau-Montaut, D. Cloud Compare: 3D Point Cloud and Mesh Processing Software, Open-Source Project. Available online: http://cloudcompare.org/ (accessed on 10 May 2020).
64. Lague, D.; Brodu, N.; Leroux, J. Accurate 3D comparison of complex topography with terrestrial laser scanner: Application to the Rangitikei canyon (N-Z). *ISPRS J. Photogramm. Remote Sens.* **2013**, *82*, 10–26. [CrossRef]
65. Rossi, P.; Mancini, F.; Dubbini, M.; Mazzone, F.; Capra, A. Combining nadir and oblique UAV imagery to reconstruct quarry topography: Methodology and feasibility analysis. *Eur. J. Remote Sens.* **2017**, *50*, 211–221. [CrossRef]
66. Dos Santos, D.R.; Poz, A.P.D.; Khoshelham, K. Indirect Georeferencing of Terrestrial Laser Scanning Data using Control Lines. *Photogramm. Rec.* **2013**, *28*, 276–292. [CrossRef]
67. Awasthi, B.; Karki, S.; Regmi, P.; Dhami, D.S.; Thapa, S.; Panday, U.S. Analyzing the Effect of Distribution Pattern and Number of GCPs on Overall Accuracy of UAV Photogrammetric Results. In *Lecture Notes in Civil Engineering*; di Prisco, M., Chen, S.-H., Vayas, I., Kumar Shukla, S., Sharma, A., Kumar, N., Wang, C.M., Eds.; Springer Science and Business Media LLC: Berlin, Germany, 2020; pp. 339–354.
68. Lingua, A.; Noardo, F.; Spanò, A.; Sanna, S.; Matrone, F. 3D Model Generation Using Oblique Images Acquired by Uav. *Int. Arch. Photogramm. Remote Sens. Spat. Inf. Sci.* **2017**, *42*, 107–115. [CrossRef]

Article

Deep Learning-Based Single Image Super-Resolution: An Investigation for Dense Scene Reconstruction with UAS Photogrammetry

Mohammad Pashaei [1,2], Michael J. Starek [1,2,*], Hamid Kamangir and [2] Jacob Berryhill [2]

1 Department of Computing Sciences, Texas A&M University-Corpus Christi, Corpus Christi, TX 78412, USA; mpashaei@islander.tamucc.edu

2 Conrad Blucher Institute for Surveying and Science, Texas A&M University-Corpus Christi, Corpus Christi, TX 78412, USA; hkamangir@islander.tamucc.edu (H.K.); Jacob.Berryhill@tamucc.edu (J.B.)

* Correspondence: michael.starek@tamucc.edu

Received: 24 April 2020; Accepted: 25 May 2020; Published: 29 May 2020

Abstract: The deep convolutional neural network (DCNN) has recently been applied to the highly challenging and ill-posed problem of single image super-resolution (SISR), which aims to predict high-resolution (HR) images from their corresponding low-resolution (LR) images. In many remote sensing (RS) applications, spatial resolution of the aerial or satellite imagery has a great impact on the accuracy and reliability of information extracted from the images. In this study, the potential of a DCNN-based SISR model, called enhanced super-resolution generative adversarial network (ESRGAN), to predict the spatial information degraded or lost in a hyper-spatial resolution unmanned aircraft system (UAS) RGB image set is investigated. ESRGAN model is trained over a limited number of original HR (50 out of 450 total images) and virtually-generated LR UAS images by downsampling the original HR images using a bicubic kernel with a factor $\times 4$. Quantitative and qualitative assessments of super-resolved images using standard image quality measures (IQMs) confirm that the DCNN-based SISR approach can be successfully applied on LR UAS imagery for spatial resolution enhancement. The performance of DCNN-based SISR approach for the UAS image set closely approximates performances reported on standard SISR image sets with mean peak signal-to-noise ratio (PSNR) and structural similarity (SSIM) index values of around 28 dB and 0.85 dB, respectively. Furthermore, by exploiting the rigorous Structure-from-Motion (SfM) photogrammetry procedure, an accurate task-based IQM for evaluating the quality of the super-resolved images is carried out. Results verify that the interior and exterior imaging geometry, which are extremely important for extracting highly accurate spatial information from UAS imagery in photogrammetric applications, can be accurately retrieved from a super-resolved image set. The number of corresponding keypoints and dense points generated from the SfM photogrammetry process are about 6 and 17 times more than those extracted from the corresponding LR image set, respectively.

Keywords: unmanned aircraft system (UAS); deep learning; super-resolution (SR); convolutional neural network (CNN); generative adversarial network (GAN); structure-from-motion; photogrammetry; remote sensing

---

1. Introduction

In most remote sensing (RS) applications, high-resolution (HR) images are usually more demanding in a wide range of image analysis tasks leading to more precise and accurate RS-derived products [1–3]. HR imagery is usually more desirable in all applications, including RS imagery, because improved pictorial information makes visual interpretation easier for a human and helps to purify representation for automatic machine perception [4]. In RS applications, the resolution of a digital imaging system can be classified in four different ways: spatial resolution, spectral resolution, radiometric resolution, and temporal resolution. In the context of accurate feature mapping and positioning in RS, spatial resolution is of the greatest challenge.

Spatial resolution of a digital imaging system is primarily defined by the pixel density in the image space, which is measured in pixels per unit area. Spatial resolution in the object space represents the level of spatial detail that can be discerned in an image; the higher the resolution, the more image details. Limited spatial resolution in a certain image is primarily a function of the imaging sensor or acquisition device [4]. The spatial resolution of imagery, usually referred to as ground sample distance (GSD) in RS applications, is determined by the sensor size or the dimension of the electro-optical sensor when based on the charge-coupled device (CCD) or complementary metal-oxide-semiconductor (CMOS) technologies, the number of sensor elements, the focal length of the imaging device, and its distance from the imaging target. Regardless of the other factors contributing to the spatial resolution of imagery, such as focal length and the distance from sensor to the target, GSD of an image and the quality of its high-frequency contents deteriorate mainly due to some manufacturing limitations and imperfections of an imaging sensor.

One straightforward way to improve the spatial resolution or GSD of imagery is to build a more compact sensor in which the sensor's pixel density is increased by reducing the sensor element size. However, this reduction in sensor element size may dramatically reduce the amount of light incident on each sensor element, causing the so called shot noise [5]. Furthermore, capture of high frequency image detail is also limited or degraded by the sensor optics, such as lens blur, lens aberration, and aperture diffraction, or any external sources of image degradation including image motion due to moving objects [4]. Constructing high-quality imaging sensors with perfect optical components, capturing very high spatial resolution images with high-quality image content, is restrictively expensive and not practical in most real scenarios. This is especially true when referring to the rapid rise in the use of small unmanned aircraft systems (UASs) for RS and photogrammetry applications [4]. Such small UASs are typically equipped with low-cost, consumer-grade digital RGB cameras. Besides the cost, the resolution of these typical UAS cameras is also limited by the camera speed and hardware storage. Physical constraints of the sensing platform or environment, such as with satellite imagery, can put additional constraints on the use of very high-resolution sensors. Furthermore, in some imaging systems, HR image content may not be always achievable due to inherent restrictions within the system itself including built-in downsampling procedures to handle bandwidth limitations, different types of noise related to the sensor electronics and atmosphere, compression techniques, etc. [6].

An alternative approach to hardware-based solutions for spatial resolution enhancement is to accept the image degradation and apply signal processing techniques to attempt to recover fine image details degraded or almost lost during image capture. These approaches are often referred to as Super-Resolution (SR) image reconstruction techniques. SR techniques attempt to recover HR images from LR images, and this task remains an important yet challenging topic in image processing that has a wide range of applications in computer vision and image understanding tasks [7–10]. SR techniques not only improve image perceptual quality, but also help to improve the final accuracy of many computer vision tasks [11–13]. Application of SR techniques on highly detailed and complex RS data introduces more challenges to the SR problem [14,15]. Most traditional image SR techniques use highly sophisticated signal processing algorithms with a very high computational complexity [15,16]. Considering the size and the volume of required super-resolved images for some RS applications, such as generating a precise digital surface model (DSM) using aerial or satellite photogrammetry, traditional SR techniques are highly inefficient for such applications. Furthermore, some techniques require multiple LR images from the same scene with high temporal resolution to resolve the SR problem [17,18]. However, due to costs or limitations for acquiring the necessary imagery, complexity of natural and built terrain, scarcity of multi-view sensors, and need for accurate image registration algorithms, acquiring and processing such images for SR is a difficult task [15]. In addition, complicated and versatile interaction of most RS sensors with atmosphere and objects, image displacements due to topographic anomalies, land cover characteristics, and participation of shaded areas due to the Sun-sensor-object geometry in RS images make the SR problem a highly challenging task for almost all developed techniques in this field [15].

Deep learning (DL), specifically deep convolutional neural network (DCNN), has recently been applied to a wide range of image analysis tasks [19–22] including the highly challenging and ill-posed problem of predicting HR images from LR images in an end-to-end manner. These methods have already shown their superiority over almost all traditional techniques by achieving state-of-the-art performance on various SR benchmarks [23–25]. Currently, DCNN-based single image super-resolution (SISR) techniques have been employed to increase the geometrical and interpretation quality of RS imagery [26–28]. However, few studies have focused on applying DCNN-based SISR on UAS-based imagery, typically acquired at low altitudes with high resolution, where the accuracy of the spatial information captured by the images is critical for the reliability of results drawn from subsequent analyses [29,30]. Recently, super-resolution generative adversarial network (SRGAN) [23], is considered as one of the most efficient DCNN-based SISR models for recovering very fine details in predicted HR images from corresponding LR images [23]. Offering finer image content is always one of the most important characteristics of HR images in different RS applications, which can lead to higher accuracy and reliability in almost all spatial and non-spatial RS products. SRGAN has already proved its superiority over many other DCNN-based SISR models for recovering very fine details in predicted HR images, which are highly valuable for improving human image perception. However, the quality of the recovered image details and their potential for enhancement of hyper-spatial resolution UAS imagery for photogrammetric applications, such as dense 3D reconstruction of a scene, has not yet been fully explored. With this motivation, this paper focuses on the application of DCNN to SISR for UAS image enhancement. The contributions of the paper are as follows:

1. An overview of the SR problem and DCNN approaches for SISR is provided with emphasis on generative adversarial network (GAN) architecture. GAN-based models are fully reviewed including their specific loss functions. Additionally, different learning strategies and image quality measures (IQMs) typically employed for SISR tasks are reviewed.
2. A high performance DCNN-based SISR model based on GAN architecture [31], known as enhanced SRGAN (ESRGAN) [32], is adopted and trained on a set of LR UAS images virtually generated by downsampling the original HR image set by factor $\times 4$. Additive white Gaussian noise is applied to the LR imagery to make the SISR task more challenging. Such noise can always appear in any digital imaging and image transmission systems due to the electronics, imaging sensor quality, and the interaction of the digital imaging system with the natural environment, such as the level of illumination, temperature, etc [33]. Model performance in recovering the degraded or lost image details and noise reduction in the predicted super-resolved images is then carried out using standard IQMs. In this experiment, IQMs include peak signal-to-noise ratio (PSNR), structure similarity (SSIM) index, and a qualitative analysis through visually inspecting resulting SR images.
3. A task-based IQM using Structure-from-Motion (SfM) photogrammetry is carried out on the predicted SR image set.
4. A comprehensive comparative analysis of SfM derived photogrammetric data products, resulting from processing of the LR, HR, and SR UAS image sets, is carried out. Those products include: the camera calibration and camera pose information, densified 3D point clouds, and digital surface models (DSMs).

In regard to the UAS-SfM task-based evaluation for SR described above, the primary objectives of the experiment are summarized as follows:

1. The performance of the adopted DCNN-based SISR model on retrieving both the interior and exterior geometry of the UAS imagery is investigated. In SfM photogrammetry, the accuracy and reliability of all derived parameters, within the robust bundle adjustment (BA) computations,

are closely related to the accuracy and reliability of extracted keypoint features from raw images. Any image distortions and artefacts introduced by adding noise or upsampling images can dramatically affect the reliability of derived parameters within BA computations.

2. The potential of the employed DCNN-based SISR model to downgrade the level of inherent and additional noise introduced to the original HR images is investigated. In most image-based 3D reconstruction algorithms, including SfM photogrammetry, lower level of noise in the underlying image set results in estimating the imaging and scene geometry with higher accuracy. That is due to the fact that the feature detection operators, using sophisticated image processing algorithms, extract keypoints features with higher accuracy and lower uncertainty across multiple images in an UAS image set. To do this, the naive pre-trained ESRGAN model, with upscaling factor $\times 1$, is taken as an image restoration network. The idea is to explore the effectiveness of the ESRGAN model, trained on a large number of images within several standard image sets, to downgrade the inherent noise and restore the original UAS HR images.

The remainder of this paper is organized as follows. Section 2 briefly describes image SR as an image upscaling technique to recover the degraded or lost image details in LR images. Section 3 introduces some of the pioneering DCNN-based SISR architectures. GAN-based architecture and its specific cost function for SISR task is later described in Section 3. Learning strategies in Section 4 introduce different cost functions that are usually used in DCNN-based SISR models. Different metrics developed for evaluating the quality of resulting SR images are explained in Section 5. Section 6 explains the experiment including the employed DCNN-based SISR model. Section 7 reports the qualitative and quantitative results showing the performance of ESRGAN model on virtually-generated LR UAS images based on standard IQMs and a task-based IQM using SfM photogrammetry. Section 8 discusses the results in detail. Lastly, Section 9 provides a conclusion and future perspective.

2. Image Super-Resolution

Image SR refers to techniques which aim to restore a HR image from its LR counterpart(s). Their main goal is to recover the high frequency details lost in LR images and remove the degradation caused by the imaging device and/or environment [34,35]. SR is a topic of great interest in digital image processing and many computer vision related applications including HDTV [36], medical imaging [37,38], satellite imaging [39], face recognition [40], security and surveillance [41]. The basic idea in most SR techniques is to extract the non-redundant image content in multiple LR images and combine them to generate a HR image [5]. Single image interpolation is an easy approach within many available SR techniques, which can be used to increase the image size [4]. However, several works showed that it does not provide any additional information and would dramatically decimate details of the image [4,24,42].

Generally, the SR problem assumes the LR image represents a downsampled, noisy, and blurred (by an unknown low-pass filter) version of HR data. Due to the non-invertibility of the degradation process, SR problem is inherently ill-posed [43]. In other words, it is an under-determined inverse problem, of which the solution is not unique. In the typical SR framework, as depicted in Figure 1, the LR image I_x is modeled as follows [44]:

$$I_x = \mathcal{D}(I_y; \delta) \tag{1}$$

where I_y is the corresponding HR image, $\mathcal{D}$ represents a degradation function, and δ is a set of parameters, e.g., the parameters of the unknown convolutional kernel, the scaling factor, and some noise related factors, contributing to the degradation process. Under general conditions, the degradation process from $\mathcal{D}$ is unknown and only LR image, I_x, is provided. Thus, the SR operation, the reverse path in Figure 1, is an extremely challenging task, which effectively results in a one-to-many mapping from LR to HR image space [25].

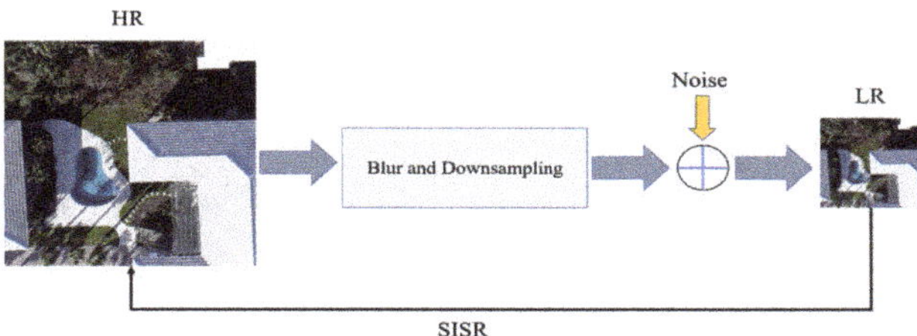

Figure 1. The overall framework for SISR.

Researchers are required to recover the corresponding HR image $\hat{I}_y$ from the LR image I_x, so that $\hat{I}_y$ is identical to the ground truth HR image I_y, as follows [44]:

$$\hat{I}_y = \mathcal{F}(I_x; \theta) \tag{2}$$

where $\mathcal{F}$ is the super-resolution model and θ represents the parameters of $\mathcal{F}$. Generally, degradation models combine several operations as follows [44]:

$$\mathcal{D}(I_y; \delta) = (I_y \otimes k) \downarrow_s + n_\zeta, \quad \{k, s, \zeta\} \subset \delta \tag{3}$$

where $(I_y \otimes k)$ represents the convolution between a blur kernel k and the HR image I_y, $\downarrow_s$ represents a downsampling process with factor s, and n_ζ is some additive white Gaussian noise with standard deviation ζ.

SR techniques typically assume that high-frequency image contents are redundant and can be reconstructed from low-frequency contents making the SR technique an inference problem [43]. Some SR techniques assume that for reconstructing a HR image of a certain scene, multiple LR instances of the same scene with different perspectives are available. These techniques are categorized as multi-image SR (MISR) approaches [16]. Such methods attempt to invert the downsampling process by exploiting the explicit redundancy and constraining the ill-posed problem with additional information. However, MISR methods are usually computationally expensive because they require complex image registration and fusion in LR image space, where the accuracy of those processes directly affects the quality of the resulting super-resolved images [43]. An alternative approach is single image super-resolution (SISR) [45]. These techniques attempt to exploit the implicit redundancy available in the LR images, in the form of local spatial correlation in an image or additional temporal correlations in a video, and recover lost or deteriorated high-frequency content from a single LR instance. In SISR techniques, prior information is usually required to constrain the solution space [46].

3. Deep Learning for SISR

Learning-based methods, also known as example-based methods [4,47–49], aim at estimating an effective mapping from LR to HR image pairs due to their fast computation and superior performance relative to many other traditional techniques [25]. These methods usually exploit machine learning (ML) algorithms to learn the statistical relationships between the HR and corresponding LR images from a substantial number of training samples [25]. Traditional methods for SISR suffer from a few drawbacks [25,43]: (1) unclear and potentially very complex definition of the mapping between the LR and HR image spaces; (2) established sub-optimal high-dimensional mapping; (3) most traditional methods rely upon handcrafted features with expert domain knowledge. Recently, deep learning-based SISR methods have achieved remarkable improvements over all traditional and ML approaches [23–25]. These methods take advantage of the huge capacity of DL models to be able to provide an extremely nonlinear mapping in a very high-dimensional space from the input space to the solution space, and efficiently explore that space to find the best solution. These methods usually take a DCNN architecture for low to high-level feature encoding and nonlinear feature mapping.

3.1. DCNN Architectures for SISR

A variety of super-resolution models based on DCNN architectures have been proposed so far. Most of those models focus on supervised super-resolution, requiring both LR images and corresponding HR images, usually as ground truth (GT). These approaches are mostly composed of a set of major components and processing strategies including the model's main framework, upsampling method, network architecture, and learning strategy.

Super-resolution convolutional neural network (SRCNN) by Dong et al. [24,50] in Figure 2 is a pioneering work in DCNN-based SISR approach. Despite its striking success, SRCNN model suffers from the following issues [25]. (1) Inputs to SRCNN are LR images upsampled to coarse HR images at a desired size using traditional methods (e.g., bicubic interpolation). Introducing interpolated images as inputs to the network have three main drawbacks: (a) severe over-smoothing and noise amplification effects introduced to interpolated inputs can result in further inaccurate estimations of the image content; (b) employing interpolated versions of images, instead of the original LR image, as input is very time-consuming and increases computational complexity almost quadratically [51]; and (c) assuming an unknown kernel in the downsampling process makes adopting a specific interpolated input, as an estimation of the output, unjustified. (2) As mentioned previously, most SR techniques undertake the assumption that the high-frequency content is redundant and can be accurately predicted from the low-frequency data [52]. Thus, exploring more contextual information within large regions of LR images to capture sufficient information for retrieving high-frequency details in predicted HR images seems inevitable. Theoretical work in DL show more contextual information can be achieved by designing very deep architectures with larger receptive fields, which can result in expanding the final solution space [19,53–56]. In some situations, effectively attaining more hierarchical representations can be achieved by increasing the DL network depth [53]. In recent years, many different CNN-based architectures have been developed, which exploit a very deep and sophisticated architecture, including residual and/or dense feature mapping [19,56], to solve complex problems more efficiently [25,44].

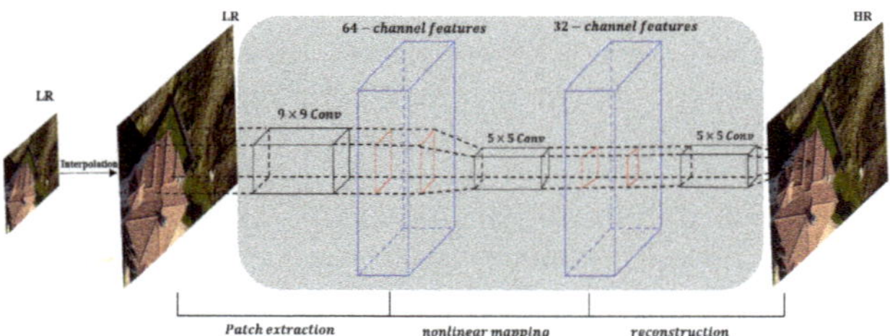

Figure 2. Sketch of the SRCNN architecture.

3.2. GAN for SISR

Introduction of recent innovative and deeper CNN-based architectures for SISR has already led to breakthroughs in accuracy and speed. Photo-realistic SISR GAN (SRGAN) [23], illustrated in Figure 3, was introduced for recovering the finer texture details when resolving at large upscaling factors. Those recovered fine details in SR images not only make predicted HR images more appealing to a human, but also have a great impact on the accuracy and reliability of imaging geometry and scene details when they are retrieved by the SfM phtotogrammetry process.

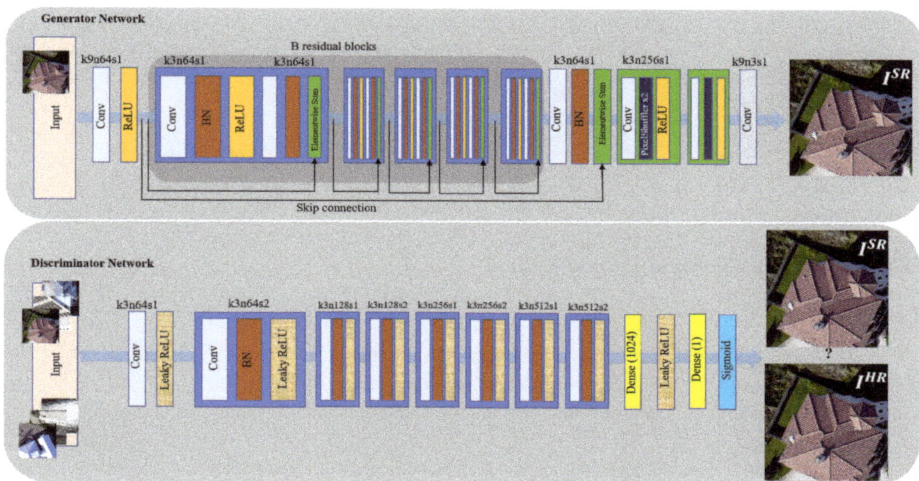

Figure 3. Architecture of Generator and Discriminator Network for SISR task with corresponding kernel size (k), number of feature maps (n), and stride (s) indicated for each convolutional layer.

The basic SRGAN model is built upon the residual blocks [19] and trained under the perceptual loss in a GAN framework, which makes it capable of predicting photo-realistic images for $\times 4$ upscaling factor [23]. The SRGAN model has shown significant improvement on overall visual quality of SR images over all previously introduced PSNR-oriented methods [23,32].

GAN [31] introduced by Goodfellow et al. tries to solve the adversarial min-max problem [23]:

$$\min_{\theta_G} \max_{\theta_D} \quad \mathbb{E}_{I^{HR} \sim p_{train}(I^{HR})} \left[\log D_{\theta_D}(I^{HR}) \right] + \mathbb{E}_{I^{LR} \sim p_G(I^{LR})} \left[\log(1 - D_{\theta_D}(G_{\theta_G}(I^{LR})) \right] \tag{4}$$

where it allows the network to train a generative model G with the purpose of fooling a discriminator D that is simultaneously trained to discriminate the SR images from the original HR images.

The formulated perceptual loss consists of a weighted sum of a content loss ($\mathcal{L}_X^{SR}$) and an adversarial loss component ($\mathcal{L}_{Gen}^{SR}$) as follows [23]:

$$\mathcal{L}^{SR} = \underbrace{\underbrace{\mathcal{L}_X^{SR}}_{\text{content loss}} + \underbrace{10^{-3}\mathcal{L}_{Gen}^{SR}}_{\text{adversarial loss}}}_{\text{perceptual loss}} \tag{5}$$

Content loss motivated by perceptual similarity chooses the solution based on the perceptual similarity from the high dimensional solution space [23]. Instead of relying on pixel-wise losses, Ledig et al. define *VGG loss* based on *ReLU* activation layers and 19 layers VGG network [53], where VGG loss is computed as the Euclidean distance between the feature representations of a reconstructed image $G_{\theta_G}(I^{LR})$ and the ground truth image I^{HR} as follows [23]:

$$\mathcal{L}_{VGG/i,j}^{SR} = \frac{1}{W_{i,j}H_{i,j}} \sum_{x=1}^{W_{i,j}} \sum_{y=1}^{H_{i,j}} (\phi_{i,j}(I^{HR})_{x,y} - \phi_{i,j}(G_{\theta_G}(I^{LR}))_{x,y})^2 \tag{6}$$

where $\phi_{i,j}$ represents the feature map obtained by the j-th convolution (after activation) before the i-th maxpooling layer within the VGG-19 network. $W_{i,j}$ and $H_{i,j}$ describe the dimensions of the respective feature maps within the VGG network.

Adversarial loss, which is the generative component of SRGAN to the perceptual loss, encourages the network to favor solutions residing on the natural image manifold [23]. The generative loss ($\mathcal{L}_{Gen}^{SR}$) is evaluated, in a probabilistic framework, based on the performance of the discriminator $D_{\theta_D}(.)$ over a training sample set as [23]:

$$\mathcal{L}_{Gen}^{SR} = \sum_{n=1}^{N} -\log D_{\theta_D}(G_{\theta_G}(I^{LR})) \tag{7}$$

where, $D_{\theta_D}(G_{\theta_G}(I^{LR}))$ represents the probability that the generated image $G_{\theta_G}(I^{LR})$ is a natural HR image. As a consequence of exploiting adversarial loss, the discriminator network is trained to push SISR solutions to the natural image manifold.

4. Learning Strategies

Learning the end-to-end mapping function $\mathcal{F}$ to map a LR image I^{LR} to the corresponding reconstructed SR image $I^{SR} = \hat{I}^{HR}$, which is an approximation of the real HR image I^{HR}, requires the estimation of network parameters θ. This is attained via minimizing the loss between the super-resolved images $I^{SR} = \mathcal{F}(I^{LR};\theta)$ and the corresponding HR images I^{HR}. In this section, different loss functions that are widely used in SISR techniques are introduced. For the sake of brevity, the subscript y is dropped from the ground truth (target) HR image I_y and the reconstructed HR image $\hat{I}_y$ in the rest of this section.

4.1. Pixel Loss

Pixel loss evaluates the pixel-wise difference between two images, mainly in the form of L_1 distance, i.e., mean absolute error (MAE), or L_2 distance, i.e., mean square error (MSE). In so doing, it attempts to capture and solve the inherent uncertainty in retrieving lost high-frequency components by minimizing related loss functions as follows [44]:

$$\mathcal{L}_{pixel-L_1}(I^{HR}, I^{SR}) = \frac{1}{hwc}\sum_{i,j,k} |I_{i,j,k}^{HR} - I_{i,j,k}^{SR}| \tag{8}$$

$$\mathcal{L}_{pixel-L_2}(I^{HR}, I^{SR}) = \frac{1}{hwc}\sum_{i,j,k} (I_{i,j,k}^{HR} - I_{i,j,k}^{SR})^2 \tag{9}$$

where h, w and c are the height, width and number of channels of the reconstructed images, respectively. Charbonnier loss [57,58], is a variant of L_1 loss, given by [44]:

$$\mathcal{L}_{pixel-Cha}(I^{HR}, I^{SR}) = \frac{1}{hwc}\sum_{i,j,k} \sqrt{(I_{i,j,k}^{HR} - I_{i,j,k}^{SR})^2 + \epsilon^2} \tag{10}$$

where ϵ is a small constant (e.g., $1e-3$) for numerical stability.

The pixel loss constraint results in a super-resolved image I^{SR}, which is close to the ground truth HR image I^{HR} in the pixel values. In comparison with L_2 loss, the L_1 loss shows higher performance and better convergence [44,59]. Using pixel loss as the loss function favors a high peak signal-to-noise ratio (PSNR). According to its definition, PSNR is heavily correlated with pixel-wise deviation, where minimizing pixel loss directly maximizes PSNR [23]. Moreover, it is partially related to the image perceptual quality. Thus, pixel loss has become the most widely used loss function in SR field.

Minimizing the pixel loss encourages finding plausible solutions, based on pixel-wise average, in the high dimensional solution space. In return, such solutions can be overly-smooth with poor perceptual quality [23,60,61]. Thus, in order to capture the reconstruction error and image quality more efficiently, a variety of other loss functions, such as content loss [61] and adversarial loss [23], were introduced to the SR field.

4.2. Perceptual/Content Loss

To evaluate image quality based on perceptual similarity, perceptual-driven approaches have also been proposed [62,63]. More convincing results from the image perceptual point of view, for both SR and artistic style-transfer tasks, are offered in this category [23,63,64]. By minimizing the error in the feature space instead of the pixel space, perceptual loss or content loss, attempts to improve the image visual quality. Denoting feature maps computed within the l-th layer of the network as $\phi^{(l)}(.)$, the content loss is evaluated using the Euclidean distance between corresponding feature maps from the original and super-resolved images as follows [44]:

$$\mathcal{L}_{content}(I^{HR}, I^{SR}; \phi, l) = \frac{1}{h_l w_l c_l} \sum_{i,j,k} \sqrt{\left(\phi^{(l)}_{i,j,k}(I^{HR}) - \phi^{(l)}_{i,j,k}(I^{SR})\right)^2} \tag{11}$$

where h_l, w_l and c_l represent the height, width and number of channels of the extracted feature maps in layer l, respectively.

Content loss encourages transferring the learned knowledge of hierarchical image features from a pre-trained classification network, usually VGG or ResNet, to the SR task [12,23,32,65].

4.3. Adversarial Loss

Adversarial learning [31] is adopted for SR task in a straightforward way, in which SR model is considered as a generator, and a discriminator network is added to the model to discriminate the generated image I^{SR} from the real image I^{HR}. Adversarial loss for SRGAN [23] is as follows [44]:

$$\mathcal{L}_{gan\_G}(I^{LR}; D_{\theta_G}) = -\log D_{\theta_D}(G_{\theta_G}(I^{LR})), \tag{12}$$

$$\mathcal{L}_{gan\_D}(I^{HR}, I^{SR}; D_{\theta_D}) = -\log D_{\theta_D}(I^{HR}) - \log D_{\theta_D}(I^{SR}) \tag{13}$$

where $\mathcal{L}_{gan\_G}$ and $\mathcal{L}_{gan\_D}$ denote the adversarial loss of the generator G_{θ_G}, which is the SR model, and the discriminator D_{θ_D}, which is a deep CNN model for binary classification, respectively. θ_G and θ_D are the parameters of the generator and discriminator, and $I^{SR} = G_{\theta_G}(I^{LR})$ is the generated image approximating the corresponding ground truth HR image.

In practice, some researchers employ a combination of multiple loss functions in their DCNN-based SISR architectures for more efficient learning and to better constrain different aspects of SR image reconstruction [12,23,57,66,67]. However, how to efficiently combine multiple loss functions with effective weights emphasizing their contribution in the learning process, remains an active area of SR research.

5. Image Quality Metrics

Image quality metrics, usually referred to as image quality measures (IQMs), are measures focusing on significant visual attributes of images where they attempt to quantify the perceptual assessments of an image when it is evaluated in a certain image quality assessment (IQA) approach [60]. IQA approaches are categorized into subjective methods, which focus on quantifying human perception, and objective methods, which are based on some computational models [60]. The subjective methods can be more accurate but they are usually inconvenient, time-consuming, and expensive to implement [60]. As a result, objective methods are currently considered the mainstream among IQMs. Since the objective methods cannot efficiently capture the human visual perception, the metrics evaluated under these methods may show some inconsistency with those from subjective methods [60].

Objective IQA methods are divided into three types [60] including: (1) full-reference methods requiring corresponding images with perfect or high quality image content; (2) reduced-reference methods, which apply IQMs on the extracted features from both images and their corresponding high quality counterparts; (3) no-reference methods, which try to evaluate image quality in a blind way without any reference images. In supervised SISR, high quality HR images are usually available for evaluating different IQMs. This section introduces some of the most commonly used IQMs, covering both subjective IQA methods and objective IQA methods.

5.1. Peak Signal-to-Noise Ratio (PSNR)

PSNR measure refers to the ratio between a signal's maximum power and the power of the signal's noise, which affects the quality of the signal's representation. Due to the very wide dynamic range (i.e., ratio of highest and lowest values) of most signals, the PSNR is usually expressed in the logarithmic decibel scale. PSNR is used to measure the reconstruction quality of lossy transformations including image compression and inpainting. For image SR task, PSNR is defined using the maximum possible pixel value in the underlying image, and the mean squared error (MSE) between two corresponding images. Given the high quality image I and the corresponding reconstructed (super-resolved) image $\hat{I}$, both of which include N pixels, the MSE and the PSNR measures are defined as follows [25]:

$$MSE = \frac{1}{N}\sum_{i=1}^{N}(I_i - \hat{I}_i)^2 \tag{14}$$

$$PSNR = 10\ \log_{10}\left(\frac{L^2}{MSE}\right) \tag{15}$$

L denotes the maximum possible pixel value in the image. For 8-bit image representations, for example, L equals to 255 and the typical values for the PSNR may vary from 20 to 40 dB, where the higher the PSNR value, the better the quality of the reconstructed image as it tries to minimize MSE between the images with respect to the maximum pixel value of the input image. When L is fixed, PSNR is only related to the pixel-wise distances between two images represented by MSE. The ability of MSE, and consequently PSNR, to capture perceptually relevant differences, such as high texture detail, is very limited meaning that PSNR does not care about human visual perception and photo-realistic characteristics of the image. This often leads to poor performance of PSNR when used to assess the quality of super-resolved images in natural scenes. However, due to the lack of an efficient and comprehensive IQM that considers image quality from all perspectives, PSNR remains the most widely used metric for evaluating image quality in SR tasks.

5.2. Structural Similarity (SSIM) Index

Similar to the human visual system, which is highly adapted for extracting structural information from the viewing scene, SSIM index provides a perceptual metric that quantifies image quality degradation based on perceived image quality [68]. Made up of three relatively independent terms, luminance, contrast, and structure, SSIM index estimates the visual impact of those factors when they are modified in the reconstructed image. Those modifications may comprise shifts in image luminance, alterations in image contrast, and any other remaining deviations collectively identified as structural changes [60].

For an original high quality image I and its reconstructed counterpart $\hat{I}$, the SSIM index is defined as follows [69]:

$$SSIM(I,\hat{I}) = [C_l(I,\hat{I})]^{\alpha}[C_c(I,\hat{I})]^{\beta}[C_s(I,\hat{I})]^{\gamma} \tag{16}$$

where $\alpha > 0$, $\beta > 0$, and $\gamma > 0$ control the relative significance of each of the three terms of the index. In some implementations, $\alpha = \beta = \gamma = 1$ [60]. The luminance, C_l, contrast, C_c, and structural, C_s, components of the SSIM index are defined as follows [69]:

$$C_l(I,\hat{I}) = \frac{2\mu_I\mu_{\hat{I}} + C_1}{\mu_I^2 + \mu_{\hat{I}}^2 + C_1} \tag{17}$$

$$C_c(I,\hat{I}) = \frac{2\sigma_I\sigma_{\hat{I}} + C_2}{\sigma_I^2 + \sigma_{\hat{I}}^2 + C_2} \tag{18}$$

$$C_s(I,\hat{I}) = \frac{\sigma_{I\hat{I}} + C_3}{\sigma_I\sigma_{\hat{I}} + C_3} \tag{19}$$

where μ_I, σ_I and $\mu_{\hat{I}}, \sigma_{\hat{I}}$ represent the means and standard deviations of the original high quality image and the corresponding reconstructed image, respectively, and $\sigma_{I\hat{I}}$ is the covariance of the two images. The constants C_1, C_2, and C_3 in Equations (17)–(19) help to avoid instability when the denominators are close to zero. The formulation given in Equation (16) guarantees *symmetry*, where $SSIM(I,\hat{I}) = SSIM(\hat{I},I)$. Moreover, the index ensures a *bounded* $SSIM(I,\hat{I}) \leq 1$. Furthermore, there is a *unique maximum*, where $SSIM(I,\hat{I}) = 1$ if and only if $I = \hat{I}$. For an 8-bit grayscale image containing $L = 2^8 = 256$ gray-levels, $C_1 = (k_1.L)^2$, $C_2 = (k_2.L)^2$, and $C_3 = C_2/2$, where $k_1 \ll 1$ and $k_2 \ll 1$ are very small constants for avoiding instability. According to the above formulas, SSIM can be represented as follows [69]:

$$SSIM(I,\hat{I}) = \frac{(2\mu_I\mu_{\hat{I}} + C_1)(\sigma_{I\hat{I}} + C_2)}{(\mu_I^2 + \mu_{\hat{I}}^2 + C_1)(\sigma_I^2 + \sigma_{\hat{I}}^2 + C_1)} \tag{20}$$

In addition, to deal with uneven distribution of image statistical features or distortions, it is more reliable to perform image quality assessment locally rather than globally. Thus, mean structural similarity (mSSIM) [60] is proposed for locally assessing SSIM. This technique splits the images into multiple windows in which the SSIM of each window is evaluated, and finally averages it over all windows across the image. Because it evaluates the image reconstruction quality from the perspective of the human visual system, SSIM index better meets the requirements of perceptual assessment. The efficiency of SSIM-based IQM outperforms those based on MSE and the related PSNR over natural images including a wide variety of image distortions [69]. Those properties make SSIM index a widely used IQM among others in most SR tasks [70,71]. However, in some cases, SSIM index may lead to similar results in evaluation of image performance with PSNR metric [60].

5.3. Task-Based Evaluation

Evaluating image reconstruction performance via other image analysis tasks is also an effective IQM [11–13,72]. Specifically, this technique feeds the original high quality image and the corresponding reconstructed image into a trained model for a specific vision task, and evaluates the reconstruction quality by comparing the relative impact of reconstructed images on the prediction performance with respect to that from high quality original HR images. The vision tasks used for this evaluation technique include face recognition [73,74], face alignment and parsing [65,75], and object recognition [12,76]. However, certain vision tasks may focus on some specific image attributes that are more favorable to the task, and may not be aware or care about the visual perceptual quality of the image. For example, most object recognition models mainly focus on the high-level semantics while ignoring the image contrast and noise. But on the other hand, in some domain-specific applications, such as super-resolving surveillance video for face recognition, task-based IQM may reflect the performance of the SR models.

6. Methods and Materials

6.1. Methodology

In this SISR experiment, enhanced SRGAN (ESRGAN) [32] model is employed which improves the original SRGAN model in three aspects. First, ESRGAN improves the network by designing a Residual-in-Residual Dense Block (RRDB), illustrated in Figure 4, which offers higher capacity and easier training. Second, the Relativistic average GAN (RaGAN) [77], which learns to distinguish a more realistic image from a corresponding less realistic image, replaces the original discriminator in SRGAN, which simply judges whether an image is real or fake. According to [77], this improvement allows the ESRGAN generator to recover more realistic texture details. Third, ESRGAN adjusts the perceptual loss in the original SRGAN model by using VGG features before activation, rather than features after activation. This empirically leads to sharper edges and more visually pleasing results. Some properties of ESRGAN model is discussed below in more details.

Network Architecture: ESRGAN employs the basic architecture of SRResNet [23] for feature learning in the LR feature space. ESRGAN introduces two modifications to the generator architecture of SRGAN to improve the quality of the super-resolved images, G: (1) it removes all batch normalization (BN) layers; (2) it replaces the original basic residual block (RB) in SRGAN with a more compact RRDB architecture. According to Figure 4, by optimally combining multi-level residual blocks, the RRDB design improves the perceptual quality of super-resolved images [32]. When the statistics of image batches for training and testing are significantly high, BN layers tend to introduce unpleasant artefacts limiting the generalization ability [32]. Removing BN layers, especially under the GAN framework which is more prone to artefact generation, leads to consistent higher performance, lower computational complexity, and better generalization in the network [32,59]. In addition to the architectural improvement, to facilitate training a very deep network, ESRGAN exploits residual scaling technique [55,59] to prevent instability in training by scaling down the residuals using a scaling factor between 0 and 1 before adding them to the main path. Moreover, ESRGAN employs a smarter initialization technique, which has empirically been shown to provide easier training when the initial parameter variance becomes smaller [32].

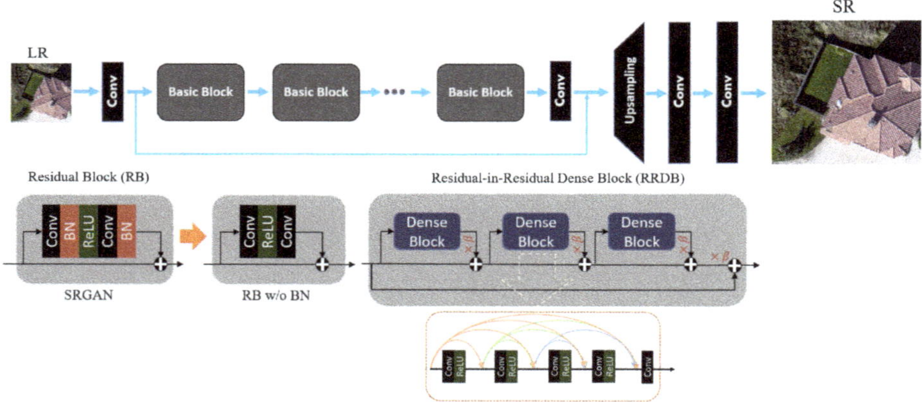

Figure 4. Basic architecture of SRResNet with different possible residual blocks.

Relativistic Discriminator: The original SRGAN model uses the standard discriminator expressed as $D(I) = \sigma(C(I))$, where σ is the sigmoid function and $C(I)$ is the discriminator output. This definition estimates the probability that the input image I is the original HR (real) image or the super-resolved (fake) image. In contrast, a relativistic discriminator predicts the probability that the original HR image I^{HR} is relatively more realistic than the super-resolved image I^{LR} as shown in Figure 5. The Relativistic average Discriminator (RaD) [77] is formulated as: $D_{Ra}(x_r, x_f) = \sigma(C(x_r) - \mathbb{E}_{x_f}[C(x_f)])$, where D_{Ra} is RaD function and x_r and x_f are the real (original HR) and fake (super-resolved) images, respectively. $\mathbb{E}_{x_f}[.]$ represents average over all generated or fake images in each individual mini-batch. The discriminator loss, $\mathcal{L}_D^{Ra}$, is defined as follows [32]

$$\mathcal{L}_D^{Ra} = -\mathbb{E}_{I^{HR}}\left[\log\left(D_{Ra}(I^{HR}, I^{SR})\right)\right] - \mathbb{E}_{I^{SR}}\left[\log\left(1 - D_{Ra}(I^{SR}, I^{HR})\right)\right] \quad (21)$$

The adversarial loss for generator, $\mathcal{L}_G^{Ra}$, is in a symmetrical form as [32]:

$$\mathcal{L}_G^{Ra} = -\mathbb{E}_{I^{HR}}\left[\log\left(1 - D_{Ra}(I^{HR}, I^{SR})\right)\right] - \mathbb{E}_{I^{SR}}\left[\log\left(D_{Ra}(I^{SR}, I^{HR})\right)\right] \tag{22}$$

where I^{LR} and $I^{SR} = G(I^{LR})$ stand for the input LR image and the predicted super-resolved image, respectively. In contrast to the adversarial loss for the generator in the original SRGAN model, $\mathcal{L}_{Gen}^{Ra}$ in Equation (7), in which only gradients from the generated images take part in adversarial training, the adversarial loss for the generator in ESRGAN, $\mathcal{L}_G^{Ra}$ in Equation (22), contains both I^{SR} and I^{HR}. This property causes the gradients from both real images and generated images to participate in adversarial training [32].

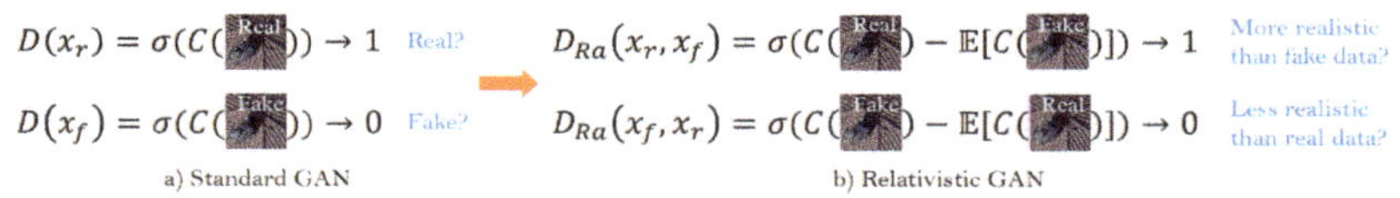

$D(x_r) = \sigma(C(\text{Real})) \rightarrow 1$ Real?

$D(x_f) = \sigma(C(\text{Fake})) \rightarrow 0$ Fake?

a) Standard GAN

$D_{Ra}(x_r, x_f) = \sigma(C(\text{Real}) - \mathbb{E}[C(\text{Fake})]) \rightarrow 1$ More realistic than fake data?

$D_{Ra}(x_f, x_r) = \sigma(C(\text{Fake}) - \mathbb{E}[C(\text{Real})]) \rightarrow 0$ Less realistic than real data?

b) Relativistic GAN

Figure 5. The standard and relativistic discriminators employed in the standard and relativistic GAN architectures, respectively [32].

Perceptual Loss: ESRGAN suggests a more effective perceptual loss $\mathcal{L}_{percep}$ by computing distances between corresponding feature maps before activation rather than after activation, as practiced in the original SRGAN model. Employing features before the activation layers overcomes two drawbacks in the original design including extreme sparsity in the activated feature maps, and inconsistent brightness reconstruction compared with the original HR image. Specially within a very deep network, sparsity within feature maps leads to weak supervision and inferior performance. The loss function for the generator in ESRGAN model is as follows [32]:

$$\mathcal{L}_G = \mathcal{L}_{percep} + \lambda\mathcal{L}_G^{Ra} + \eta\mathcal{L}_1 \tag{23}$$

where $\mathcal{L}_1 = \mathbb{E}_{I^{LR}}\|G(I^{LR}) - I^{HR}\|_1$ is the content loss that evaluates the L_1 distance between super-resolved image $G(I^{LR})$ and the original HR image I^{HR}, and λ and η are coefficients to balance different loss terms.

6.2. IQMs for SR Images

In this experiment, a comprehensive quantitative and qualitative assessment is performed on the resulting SR images by exploiting some standard IQMs that are frequently used for assessing the performance of different SISR models. Furthermore, a task-based IQM based on the SfM photogrammetry [78] procedure is carried out. Applying any type of image processing algorithm on a raw aerial image set can dramatically affect the precision and accuracy of retrieving the interior and exterior geometry of a camera at image acquisition time. That, consequently, may lead to a significant decrease in the quality and final accuracy of the main SfM photogrammetry products, such as point clouds, DSMs, and orthoimages. The authors believe that the chosen task-based IQM can more accurately exhibit the effectiveness and performance of DCNN-based SISR to enhance the spatial resolution of LR imagery in RS applications. More specifically, where highly accurate spatial products from processing RS images are required.

6.2.1. Standard IQM methods

PSNR and SSIM index are evaluated as standard IQMs for quantitative assessment of predicted SR images. Choosing those two IQMs enables performance comparison in DCNN-based SISR applications when it is applied on two different categories of images (general images and aerial RS images).

6.2.2. SfM Photogrammetry for Task-Based IQM

SfM photogrammetry procedure, as illustrated in Figure 6, is employed on all available image sets including HR ground truth, LR, and predicted SR image sets. SfM photogrammetry is a low-cost method, based on stereoscopic photogrammetry, for highly accurate topographic reconstruction using a series of overlapping images acquired from multiple viewpoints [78]. In contrast to traditional photogrammetry, in SfM photogrammetry, interior geometry of the camera, usually referred to as interior orientation (*IO*) parameters, position and orientation of each camera station with respect to the scene's global coordinate system, commonly called exterior orientation (*EO*) parameters, and the geometry of the scene, i.e., the 3D coordinate of each point of the 3D scene, are resolved automatically. All required parameters are calculated simultaneously based on the highly redundant and iterative bundle adjustment (BA) computations using a rich database of corresponding image features automatically extracted from a set of multiple overlapping images [79]. SfM photogrammetry addresses the key problem of determining the 3D locations of a large number of corresponding features extracted from multiple overlapping images, taken from different positions and angles with respect to the 3D scene.

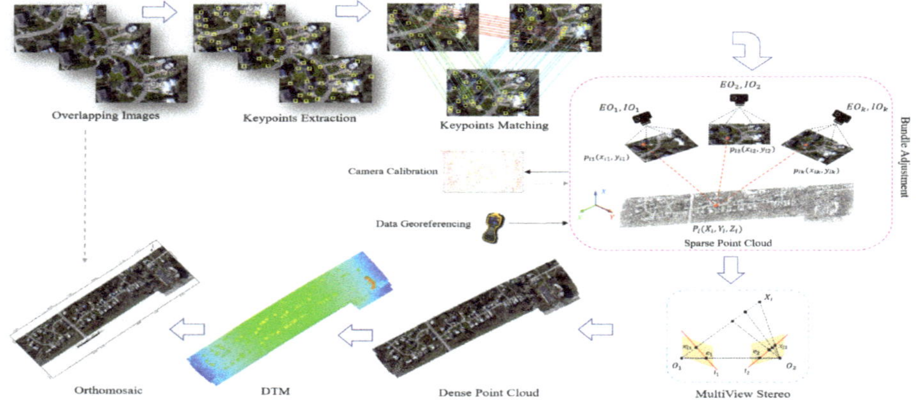

Figure 6. Steps of SfM photogrammetry.

Most image-based 3D reconstruction software that work based on the SfM photogrammetry principle, first solve for camera *IO* and *EO* parameters followed by a multi-view stereo (MVS) algorithm to escalate the density of the sparse point cloud generated by the SfM algorithm [78]. In the first step, several overlapping images are imported into the software, and a keypoints detection algorithm, usually the popular scale invariant feature transform (SIFT) algorithm [80], is applied to detect keypoints and keypoint correspondences across and between all images using a keypoint descriptor. In the SIFT algorithm, for example, the keypoint descriptor is determined by computing local image gradients and transforming them into a representation substantially insensitive to some image feature variations, including illumination, orientation, and scale [80]. These descriptors are unique enough to allow features to be matched in large image datasets. The BA technique is performed to minimize the errors in the phase of finding point correspondences [78].

In addition to solving for *IO* and *EO* parameters, which indicate camera calibration and pose parameters, respectively, the SfM algorithm generates a sparse point cloud using the image coordinates of all corresponding keypoints, *IO*, and *EO* parameters of the camera in all imaging stations. The coordinate system related to the generated point cloud is arbitrary. In order to transform the point cloud coordinate system to any local or global coordinate system, a georeferencing phase should be adopted. In that phase, a few ground control points (GCPs) with known 3D coordinates in a local or global coordinate reference frame using land surveying or initial camera positions, e.g., using global navigation satellite system (GNSS), is typically required. In this experiment, it is not necessary to perform the georeferencing step since all images are processed in the same reference frame. The *IO* and *EO* parameters for each camera are used as the input to the MVS algorithm. Leveraging the known *IO* and *EO* parameters for each individual camera, MVS initiates an intense search algorithm to find more correspondences along all existing epipolar lines in all overlapping images. The accuracy of the MVS algorithm and the quality of the dense point cloud generated by the MVS algorithm is highly dependent on the reliability of the *IO* and *EO* parameters calculated from the initial BA computations [81].

Images captured at high spatial resolutions, in general, return the most keypoints and keypoints correspondences in overlapping images. In addition to the major contribution of the natural texture in the 3*D* scene, the quality of the generated point cloud highly depends on several other factors including the density, sharpness, contrast, and resolution of the image content within the image set [78]. Moreover, decreasing the image acquisition distance, or flight height above ground, leads to an increase in the image spatial resolution or a finer GSD. This will further enhance the spatial density and spatial resolution of the resulting point cloud [78]. However, the uncertainty in keypoints extraction and matching, which is a typical issue in all low quality LR images, may result in poor estimation of a camera's *IO* and *EO* parameters leading to a very inaccurate and erroneous 3D point cloud.

6.3. Study Site and Dataset

Port Aransas is a town located on Mustang Island along the southern Texas Gulf of Mexico coastline, USA Figure 7. In 2017, Hurricane Harvey, a category 4 hurricane, made landfall to the north of Port Aransas along San Jose Island on the night of 25 August 2017. The southern portion of the eye wall passed within close proximity to Port Aransas causing extensive damage, primarily due to extreme winds but also surge coming from the bay side of the island.

Figure 7. Port Aransas study site located along the southern Texas Gulf of Mexico coastline. The square box (top figure) shows the UAS flight area, which has been illustrated with more details in the UAS-derived ortho-image (bottom figure).

A few days after the landfall of Harvey, a small UAS photogrammetric survey was conducted over a section of the town directly bordering the Gulf-facing shoreline Figure 7. The purpose was to inspect and evaluate structural damages to residential and commercial properties caused by the catastrophic storm. The flight mission covers almost 0.275 km^2 of Port Aransas. Phantom 4 Pro multi-rotor UAS (SZ DJI Technology C.o., Ltd., Shenzhen, China) was employed to conduct the survey. The platform was equipped with a 1 inch CMOS RGB sensor to capture 20 megapixel imagery at a resolution of 5472 × 3648 pixels. The flight altitude was designed to achieve a GSD of 2.5 cm, resulting in a flying height above ground level of about 90 m with forward lap and side lap around 80% and 70%, respectively. A total of 450 HR images were acquired over the study site. These images are used for the purposes of this study.

6.4. Data Preparation and Model Training

In order to fine-tune pre-trained ESRGAN parameters with the existing dataset, 50 non-overlapping images were chosen from the original HR dataset as ground truth for fine-tuning ESRGAN during training phase. Scaling factor of ×4 was set between LR and HR images. LR training images were obtained by down-sampling corresponding HR images. MATLAB bicubic kernel function was employed for image down-sampling, where its scale factor was set to 0.25. To make the SISR problem more complicated and realistic, additive white Gaussian noise with mean 0 and standard deviation of one-tenth of the standard deviation of each channel in RGB image was later added to the LR image set. Due to the high resolution of the original imagery, feeding the full-size images into the DCNN model rapidly exhausts the whole GPU's memory. However, in training phase, large image patches help very deep convolutional networks with wider receptive fields to capture more semantic information from the training samples. Therefore, this experiment was performed by extracting 1500 random image patches of resolution 1000 × 1000 pixels from the original HR images. Figure 8 illustrates a LR image and corresponding ground truth HR image for a training sample. The model is trained in the RGB channels, and data augmentation with random horizontal flips and 90 degree rotations is employed on the training image set. Testing and evaluation of model performance is then done on 1000 image patches randomly extracted from the remaining 400 images in the original HR and corresponding LR image sets.

It should be emphasized here that due to the large overlap between the employed UAS images, objects are sometimes captured by multiple images resulting in the appearance of the same object in the training and testing image sets. However, it should also be noted that such objects are captured from different viewing angles, causing different perspective and radiometric distortions for each specific object, or portion of the object, appearing in multiple images. Furthermore, the presence of such similar scenes within the training image set is necessary for performing transfer learning effectively, in which the weight parameters from a pre-trained DCNN model trained over a large dataset is applied to leverage complex mappings learned by very deep CNN models for performing a downstream task [82]. The weight parameters taken from the pre-trained model are, then, fine-tuned by training the model using a new dataset specific to the prediction task. In fact, one of the main reasons behind the transfer learning technique is to help the DCNN model to effectively capture a priori information related to the new task by fine-tuning the parameters of the underlying model using a new dataset for a different but related task. In the SISR technique, such a priori information can be provided to the SISR model by introducing information related to objects that are present in the acquired scene. Furthermore, the main goal of this study is to show the effectiveness of the SISR technique for recovering degraded or lost image details in the LR UAS images by fine-tuning a DCNN-based SISR model on a very limited set of HR UAS images.

The original ESRGAN model, before fine-tuning, is also employed to investigate the capability of the pre-trained ESRGAN, to enhance the image content and downgrade the inherent noise in the original HR images. The idea is that such a pre-trained model, trained on some standard datasets, may be capable of capturing the behavior of some types of noise that might be common in many imaging systems. To do this experiment, the original HR image set is fed to the original pre-trained ESRGAN with scaling factor of ×1.

Figure 8. LR and corresponding HR image patches.

The pytorch [83] implementation of ESRGAN model was chosen for training over the UAS dataset. The training process starts by initializing the ESRGAN model with weights from the pre-trained network trained on some of the well-known benchmarks in SISR such as the DIV2K dataset [84], the Flickr2K dataset [85], and the OutdoorSceneTraining (OST) dataset [66], which include thousands of high quality HR images with a broad diversity in texture and contextual information. The performance of the trained model has already been tested on widely used SR benchmarks such as Set5 [47], Set14 [49], BSD100 [86], Urban100 [87], and the PIRM self-validation dataset [88]. Table 1 summarizes the information related to the ESRGAN model setup and optimization settings for training the model on the UAS image set. According to the table, dense block architecture for generator was set to $64 \times 5 \times 5$, which includes 64 kernels of size 5×5. The generator is comprised of 23 residual-in-residual dense blocks (RRDBs). The learning rate α was set to 0.0001, and Adam optimizer was chosen for updating weights during training. Two exponential decay rate parameters in Adam optimizer β_1 and β_2, were set to 0.9, and 0.999, respectively. ϵ parameter in the optimization algorithm was set to 1×10^{-7} to avoid any division by zero. The experiment was carried out with 100 epochs on Google Colab, *Google's free cloud service*, with one Intel(R) Xeon(R) CPU 2.30GHz and one high-performance Tesla K80 GPU, having 2496 CUDA cores and 12GB GDDR5 VRAM. Fine-tuning the network took around 48 hours and inference time for predicting the super-resolved image was 10 sec/image.

Table 1. ESRGAN model and training parameters setup .

Dense Block	RRDB	Learning Rate	Adam Optimization Parameters
$64 \times 5 \times 5$	23	$\alpha = 0.0001$	$\beta_1 = 0.9, \beta_2 = 0.999, \epsilon = 1 \times 10^{-7}$

7. Results

This section provides comprehensive qualitative and quantitative experimental results on predicted super-resolved, SR_{pre}, images from LR images, virtually downsampled form original (ground truth) HR, HR_{gt}, UAS image set with additive white Gaussian noise. Also, the result of applying ESRGAN model on HR_{gt} with scale factor $\times 1$, as an image enhancement network, to generate enhanced HR images, HR_{enh}, is investigated. Furthermore, the results of the task-based IQM using the SfM photogrammetry procedure implemented with the original and super-resolved imagery is reported.

7.1. Qualitative Assessment

Figure 9 illustrates the qualitative assessment of the SISR performance using ESRGAN model on two different test samples. According to the visual inspection, and as observed in Figure 9, the ESRGAN model is able to upscale the LR images by factor 4 and predict SR images with high similarity in perceptual and visual quality when they are compared with the corresponding HR counterparts. A closer look at the qualitative results in this experiment reveals some noise removal properties learned within the SISR model trained on a sufficient number of LR and corresponding HR images.

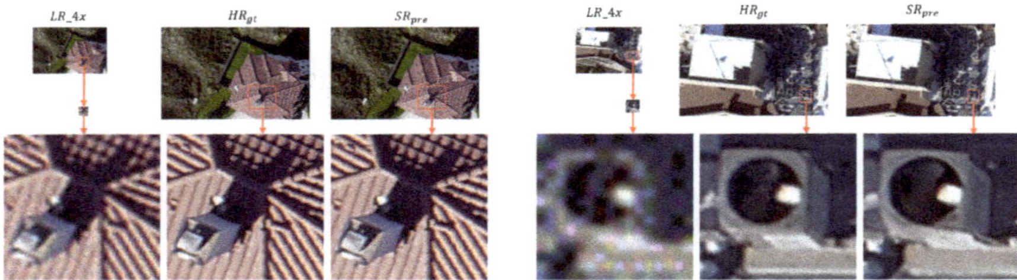

Figure 9. Illustration of the qualitative comparison between the predicted SR image and corresponding LR and ground truth HR images for two test images.

7.2. Quantitative Results

For quantitative evaluation of the SISR performance, in this experiment with ESRGAN model, PSNR value and SSIM index were calculated for the test image set and enhanced HR (HR_{enh}) image set. Table 2 illustrate the lowest, highest, and average PSNR values and SSIM indices for both image sets. The range of values for both PSNR and SSIM index in Table 2, resulting from evaluating ESRGAN performance on SR_{pre} image set, is comparable in values reported for those IQMs when ESRGAN, or any other high-performance DCNN-based SISR model, is applied on standard SISR image sets [23,25,32]. The values of the standard IQMs represented in Table 2 confirm that SISR can be effectively applied for recovering lost or degraded details in LR UAS imagery, and hopefully on a wide range of imagery in RS applications, including aerial and satellite imagery, with a comparable performance.

Table 2. PSNR and SSIM index claculated on image sets.

Image Set	Lowest PSNR/SSIM	Highest PSNR/SSIM	Mean PSNR/SSIM
SR_{pre}	25/0.6675	32/0.9011	28/0.8550
HR_{enh}	43/0.9145	49/0.9940	82/0.9601

7.3. Task-Based IQM and Related Results

Further investigation of ESRGAN model performance in a task-based image quality evaluation using SfM photogrammetry reveals more about the impact of image super-resolving on the internal and external camera imaging geometry and the geometry of the reconstructed 3D scene. All available UAS image sets including the downsampled noisy LR image set (LR), the original ground truth HR image set (HR_{gt}), the predicted super-resolved image set (SR_{pre}), and enhanced HR image set (HR_{enh}) were separately imported to Agisoft Metashape software [89] for SfM photogrammetric processing. Each image set was processed using the exact same settings and workflow procedure to ensure a fair comparative evaluation could be made on the impact of SR imagery to the BA computations and 3D reconstruction (i.e., point cloud).

BA computations, using keypoints extracted from each individual image in each image set, also result in an accurate estimation of camera calibration (IO) parameters in a self-calibration procedure using a pre-defined camera calibration model. Camera parameters evaluated within BA computations include the focal distance f, principal point coordinates (C_x, C_y), radial distortion coefficients (K_1, K_2, K_3, K_4), decentering distortion coefficients (P_1, P_2, P_3, P_4), and affinity and skew transformation coefficients (B_1, B_2), which represent a specific distortion in digital imaging sensors accounting for scale distortion and non-orthogonality of pixel elements in the x, and y directions of the digital sensor [90]. Table 3 illustrates the camera calibration results for LR, HR_{gt}, SR_{pre}, and HR_{enh} UAS image sets. According to Table 3, the evaluated values of IO parameters for SR_{pre} image set, especially, the sensor element (or pixel) size, focal distance, f, principal point offset C_x, C_y, and the first coefficient of radial lens distortion, K_1, which are among the most critical camera calibration parameters, closely approximate the real values derived from HR_{gt} image set. Referring to Table 3, the calibrated IO parameters for LR image set are different from IO parameters for HR_{gt}, SR_{pre}, and HR_{enh}, meaning that the parameters defining the internal imaging geometry in LR UAS image set is different than those in the other HR UAS image sets. It should be emphasized here that the number of selected keypoints and the level of certainty in finding their correspondences in multiple images within an image set can have a significant impact on the stability of BA computations and the accuracy of the estimated IO and EO parameters.

Table 3. Camera calibration results.

Parameters	LR	HR_{gt}	SR_{pre}	HR_{enh}
$Pixelsize(mm)$	0.00964	0.00241	0.00241	0.00241
$f(pix)$	911.785	3689.370	3701.798	3681.261
$C_x(pix)$	−0.9885	−49.8694	−57.7129	−40.4323
$C_y(pix)$	0.7271	−13.8803	−16.2507	−15.3213
K_1	0.00726	0.00512	0.00656	0.00402
K_2	−0.04381	−0.00924	−0.01842	−0.01004
K_3	0.07859	0.01028	0.02948	0.01011
K_4	−0.04655	−0.00124	−0.01439	−0.00140
P_1	0.00187	-1.7070×10^{-5}	-2.8148×10^{-5}	-1.6030×10^{-5}
P_2	0.00068	-1.0218×10^{-5}	-1.4783×10^{-5}	-1.0199×10^{-5}
P_3	0.28067	−11.0844	−3.01011	−10.7841
P_4	−0.06669	4.86345	−0.51856	4.00345
B_1	0.19185	0.00048	0.12109	0.00078
B_2	0.69768	0.62977	0.63074	0.60117

Figure 10 displays plots representing the average reprojection error vectors from BA computations across the image space for LR, SR_{pre}, HR_{enh}, and HR_{gt} UAS image sets. This error quantifies the distance between a certain keypoint location on an image and the location of the corresponding 3D point reprojected on that image. The magnitude of reprojection error in the image space depends on the quality of estimated camera calibration parameters and pose parameters, as well as on the quality of the extracted keypoints on each individual image [89]. Maximum and RMS of reprojection errors across the image space, and the average camera location errors with respect to the 3D scene have been depicted in Table 4 for LR, HR_{gt}, SR_{pre}, and HR_{enh} image sets. According to the table, both the maximum and RMS of the reprojection errors in SR_{pre} image space are closely comparable with those derived from HR_{gt} image set. The errors related to the quality of the 3D space, reconstructed by SR_{pre} image set, confirm the same quality in scene reconstruction when HR_{gt} image set is employed. In addition, Figure 11 illustrates a graphical view of the camera locations and their errors represented by the error ellipsoids for all UAS image sets.

The process of point cloud densification was carried out on each individual UAS image set after BA computations and digital surface models (DSMs) were later generated from the 3D point cloud data by the post-processing within the SfM photogrammetry software. Figure 12 displays the dense point cloud over a small area of the study site for all UAS image sets. Moreover, Table 5 summarizes the processing report from SfM photogrammetry for each individual image set. According to Figure 12 and Table 5, visual and quantitative inspections on the density of the resulting dense point cloud, which is the average number of points per square meter, demonstrate that the dense point cloud generated from HR_{gt}, SR_{pre}, and HR_{enh} are about ×17 denser than the dense point cloud generated from the LR image set.

To investigate how closely the DSM generated based on the SR_{pre} image set approximates the corresponding DSM generated from HR_{gt} image set, DSM from SR_{pre} was subtracted from the DSM generated from HR_{gt} image set. Figure 13 displays the resulting differential surface. Referring to Figure 13, the average height difference between the two DSMs is about −0.5 cm. However, there are some areas showing large height differences. These areas are mostly related to the edges of tall man-made and natural objects. Areas with lack of texture, such as water bodies, also contribute to the large height differences observed in Figure 13. The histogram in Figure 14 displays a statistical representation of the pixel-wise height differences based on the frequency of occurrence for pixel values in differential DSMs after filtering blunders.

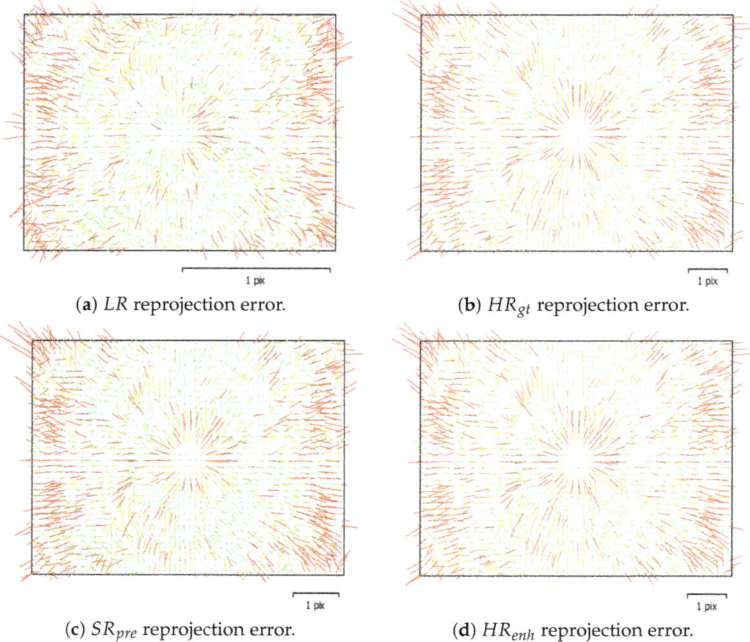

(**a**) LR reprojection error. (**b**) HR_{gt} reprojection error. (**c**) SR_{pre} reprojection error. (**d**) HR_{enh} reprojection error.

Figure 10. Average reprojection error vectors plotted on image space. Colors of the error vectors represent increasing magnitudes of the reprojection error progressing from blue to red respectively. The scale bar at bottom shows the magnitude of the error vector in pixel units.

Table 4. Bundle adjustment results for reprojection and camera location errors.

Image Set	LR	HR_{gt}	SR_{pre}	HR_{enh}
Max reprojection error (pix)	15.90	56.96	57.21	55.05
Reprojection error (pix)	0.4984	0.7868	0.9932	0.6348
X error (m)	1.7702	2.4005	2.4174	2.3241
Y error (m)	2.3225	2.6635	2.6691	2.3993
Z error (m)	0.5504	4.3415	4.1831	3.9901
XY error (m)	2.9202	3.5856	3.6012	3.503
Total error (m)	2.9716	5.6307	5.5197	5.4201

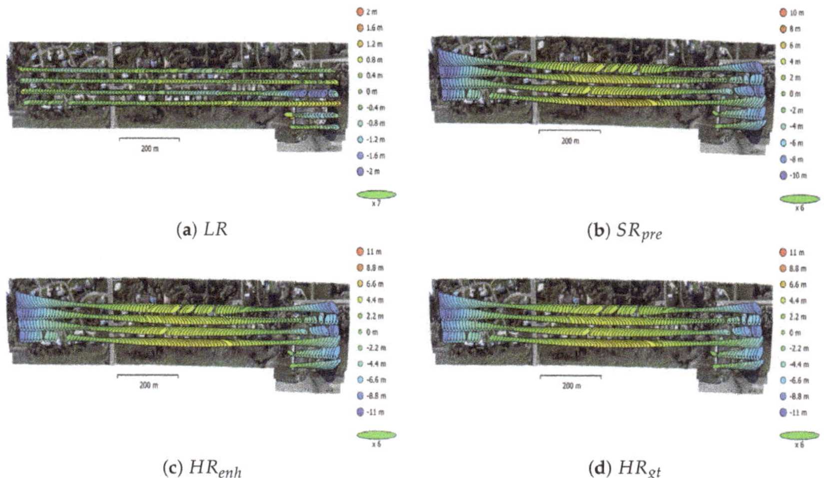

(**a**) LR (**b**) SR_{pre} (**c**) HR_{enh} (**d**) HR_{gt}

Figure 11. Camera locations and related uncertainties for image data sets. Ellipse color represents Z error. Errors in X and Y directions are represented by ellipse shape. Black dot within each individual ellipse represents estimated camera locations.

Table 5. SFM photogrammetry report summary for different image sets.

Parameters	LR	HR_{gt}	SR_{pre}	HR_{enh}	LR **to** SR_{pre}	HR_{gt} **to** HR_{enh}
Num. of images	440	440	440	440	0.0%	0.0%
Flying altitude (m)	106	106	107	106	0.9%	0.0%
Tie points (points)	1,398,877	11,051,665	8,268,475	11,630,227	490.0%	5.2%
Dense cloud (points)	1,805,966	31,041604	31,052,606	31,940,817	1619.4%	2.8%
Point density (points/m^2)	5.82	94.5	94.4	94.9	1521.9%	0.4%
DSM resolution (cm/pix)	41.40	10.30	10.30	10.30	75.1%	0.0%

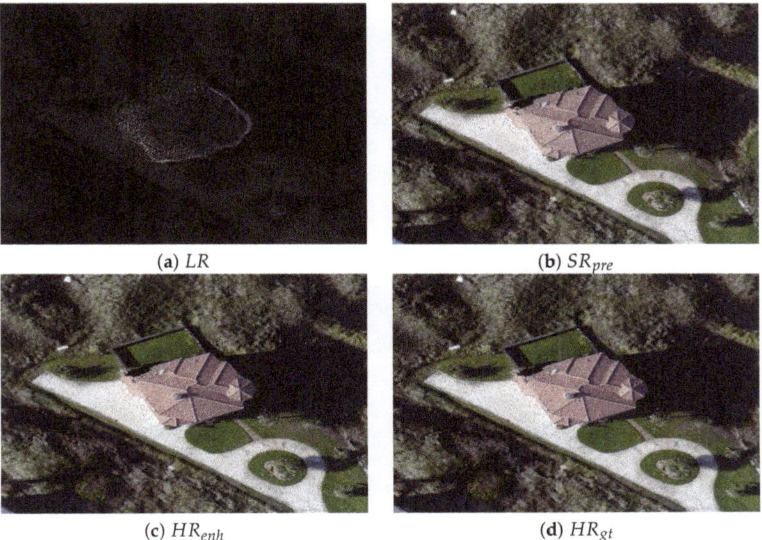

(**a**) LR (**b**) SR_{pre} (**c**) HR_{enh} (**d**) HR_{gt}

Figure 12. Resulting dense RGB point cloud computed within the SfM photogrammetry process using different image sets.

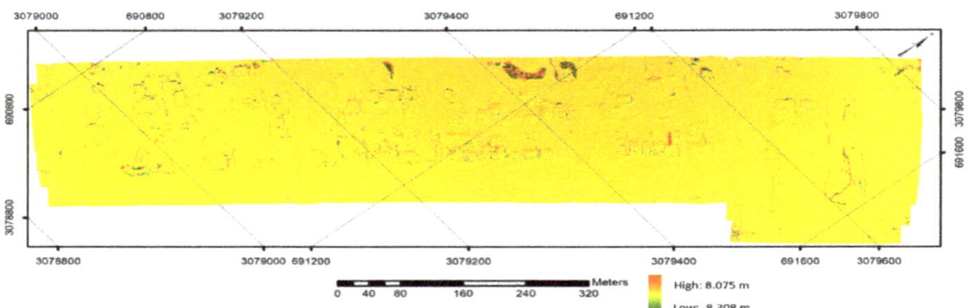

Figure 13. Illustration of DSM difference between HR_{gt} and SR_{pre} image set.

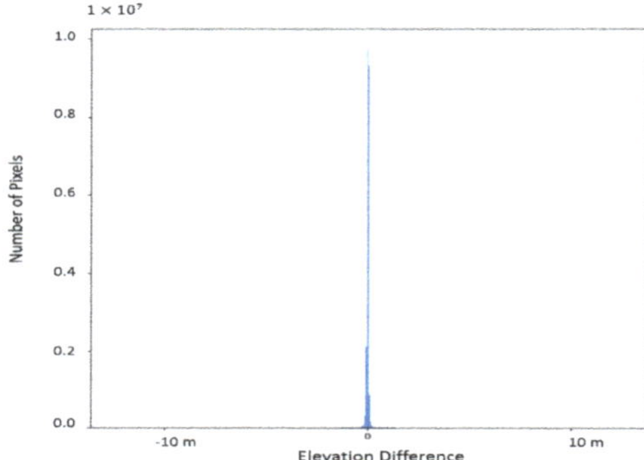

Figure 14. Height-difference histogram between DSMs from HR and SR.

8. Discussion

Visual inspection of image samples in SR_{pre} and corresponding HR_{gt} image sets confirms that the ESRGAN model performs much better over man-made objects and natural objects with definite boundaries than other targets, as shown in Figure 9. One reason may be due to the fact that natural objects usually comprise extremely intricate structures and severely random patterns with very fine details. In addition, natural objects, such as vegetation, may be moving due to the wind during image acquisition in an outdoor environment, inducing dynamic image motions in the recorded images. More accurate visual inspection on SR_{pre} images demonstrates that the model is able to predict super-resolved images with lower level of noise and blur when they are visually compared with the corresponding HR_{gt} images. This noise reduction property of the model, however, may result in removing unpleasing pseudo-noise patterns within some natural targets, such as vegetated areas. This noise reduction capability of the ESRGAN model is more evident over man-made structures and surfaces as illustrated in the right example of Figure 9.

Such image enhancement and noise removal characteristics can also be observed on both natural and man-made objects that appear in HR_{enh} image set, where the HR_{gt} images were used as input and the naive pre-trained SISR model, with scale factor $\times 1$, was used as an image restoration network. This observation demonstrates that pre-trained ESRGAN, on several standard image sets for SISR, has been able to capture, to some extent, the behavior of some types of noise that are common in almost all digital imaging systems. Considering the fact that this model has already been trained to predict SR images with scale factor $\times 2$ and $\times 4$, the observations with scale factor $\times 1$ divulges that there might be some types of noise that may commonly appear in different image scales where the pre-trained network has been able to differentiate them from the real signal.

The high IQM values reported for the HR_{enh} image set in Table 2 is due to the high degree of similarity in image content and quality between corresponding images in HR_{enh} and HR_{gt} image sets. This observation demonstrates that pre-trained ESRGAN can be used as an image restoration network when it is employed with scale factor $\times 1$.

It is worth mentioning that employing pre-trained ESRGAN, without fine-tuning the parameters using LR and corresponding HR_{gt} UAS image sets for predicting the super-resolved images (SR_{pre}), decreases the model performance around 15% for both PSNR and SSIM index in this experiment. The relatively high values for those standard image quality metrics on SR_{pre} UAS image set, whose contents are intrinsically different from those on which the vanilla ESRGAN model has been trained, verifies that the transfer learning technique and fine-tuning of the pre-trained parameters significantly helps the DCNN-SISR model to extract more related semantic information from the UAS images. This information is optimally encoded as abstract information within multiple layers of a DCNN-SISR model. Interestingly, according to Table 2, the vanilla ESRGAN model trained on standard image sets, resulted in high values for PSNR and SSIM index when it was employed on the HR_{gt} image set as an image restoration network. This is regardless of the fact that the model did not previously see the UAS images for which it has been employed to predict on in this experiment.

Results of the task-based IQM using SfM photogrammetry adds more to the previous findings. Referring to Table 3, calibrated sensor element size, or image pixel size, for LR images is about 4 times bigger than that for images in other image sets, which is compatible with our experiment. The calibrated focal lengths in SR_{pre} and HR_{enh} image sets closely approximate the real focal length evaluated in HR_{gt} ground truth image set. The difference in calibrated focal length for LR, SR_{pre}, and HR_{enh} image sets from the calibrated focal length for HR_{gt} image set are -0.010 mm, -0.030 mm and 0.020 mm, respectively. Furthermore, calibrated C_x and C_y values shows an accurate estimation of the principal point location in SR_{pre} images with respect to the HR_{gt} images. For LR images, however, those calibrated parameters show a very different location for the principal point in LR image space.

Referring again to Table 3, the remaining calibration parameters, including radial and decentering lens distortion coefficients, affinity, and skew transformation parameters in SR_{pre} and HR_{enh} image sets show a high degree of compatibility with HR_{gt} parameters confirming that lens distortion parameters and other sensor related distortions can be accurately estimated in both super-resolved SR_{pre} images and restored HR_{enh} images. However, interpreting the values of those coefficients, especially between LR and HR_{gt} images, is not very meaningful because some of them are usually highly correlated with other parameters, especially the focal length, principal point location, and the first coefficient of radial lens distortion [90,91].

Referring to Figure 10, the behavior of the average reprojection error in SR_{pre} image space accurately approximates that in the original HR_{gt} image space. This finding can be supported further by our above findings when referring to the calibrated camera parameters, where results showed that the internal geometry of the sensor can be accurately recovered in the SR_{pre} images. The plot related to the average reprojection error in LR image space represents less similarity with the error behavior in HR_{gt} and SR_{pre} image space, especially in the center of the image space. On the other hand, the average reprojection error plot for HR_{enh} image space (Figure 10d) is very similar to the reprojection error plot for the HR_{gt} image space (Figure 10b). This observation demonstrates that image restoration processing carried out on the HR_{gt} images within the pre-trained ESRGAN has not meaningfully changed the IO parameters of the camera derived from the SfM analytical self-calibration procedure.

According to Table 4, investigation on maximum reprojection error and its RMS in the SR_{pre} and $HRenh$ image spaces shows that they closely approximate those values in the HR_{gt} image space with sub-pixel magnitudes. However, RMS of reprojection error in HR_{enh} image space is about 20% less than it is in HR_{gt} image space. Part of this decrease in reprojection error might be due to the noise reduction process in HR_{enh} image space with respect to the original HR_{gt} image space. Referring to the average camera location errors in Table 4, SR_{pred} and HR_{enh} image sets closely approximate those in the original HR_{gt} image set. This suggests that the SISR process employed with factor $\times 4$ on the LR

image set, and employed with the image restoration process on HR_{gt}, preserves the external imaging geometry with respect to the 3D scene. As depicted in Table 4, pre-trained ESRGAN model with scaling factor ×1, as image restoration network, resulted in 3% improvement on total error in camera positions for HR_{enh} image set. There is also 2% improvement in that error for SR_{pre} dataset. Figure 11 shows that camera locations and their positional errors in the HR UAS imagery can be accurately retrieved in the predicted SR image set. Furthermore, it shows that image enhancement performed with the employed pre-trained ESRGAN model does not dramatically change the external imaging geometry.

Carefully exploring the differential DSM in Figure 13 reveals that large differential offsets are occurring in areas that include natural and man-made water bodies with lack of texture and along the edges of tall natural and man-made structures. Filtering out those areas from the original differential DSM and calculating some statistics over them shows that the minimum, maximum, and standard deviation (SD) of height difference in those areas are −8.308 m, 8.075 m, and 30 cm respectively. The height-difference histogram in Figure 14, for filtered differential DSM, confirms that the geometry of the reconstructed 3D scene, as reflected by the DSM, can be accurately retrieved with a SD around 2.50 cm. The minimum, maximum, and mean of height-differences within the filtered differential DSM are about −4.85 cm, 5.73 cm, and −0.02 cm, respectively.

It is worth mentioning that there are numerous environmental and sensor-related factors as well as flight design parameters which contribute to the quality and the spatial resolution of images captured by the UAS. Texture quality, related to each individual object in the scene, can highly affect the training and inference phases of the DCNN-based SISR model, which subsequently affects the results of the SfM process. Ambient environmental conditions, such as lighting or any instability of the platform during image capturing, such as due to the wind, can impact the above results. Similarly, flight design including altitude above ground and camera perspective (e.g., oblique versus nadir) will impact the GSD and appearance of land cover features. As a result, the visual representation of the same target may deviate from one exposure to another in a single UAS flight mission and across repeat data acquisitions. Thus, the authors emphasize that the results shown here, are valid for the specific data set acquired at a certain time over the specific study site. The results presented here, in terms of reconstruction accuracy, cannot be necessarily generalized to other sites with very different targets and textures, or the same area imaged at a different time and during different environmental conditions, without further experimentation. However, we believe that the high capacity of deep CNN models to efficiently extract informative contextual features from the raw UAS images in an end-to-end manner have the potential to be extended further by training DCNN-based SISR models using a time-series of UAS images acquired over the same area, or UAS images captured from the same area under different weather conditions. Also, training and evaluating the performance of a certain DCNN-based SISR model on multiple UAS image sets including images from different areas with a wider range of targets and varying textures may be considered for further analyses.

9. Conclusions

SISR seeks to obtain HR images from corresponding LR images, which is a notoriously arduous and ill-posed problem. Investigating different IQMs evaluated on SR images predicted from corresponding LR images in a DCNN-based SISR network revealed two important findings with respect to this study's experiment on UAS imagery. First, the quantitative measures of image quality, including PSNR and SSIM index, applied to the super-resolved UAS imagery, confirm that the DCNN-based super-resolution technique employed here (ERSGAN architecture) can achieve the same level of performance for spatial-resolution and pictorial information enhancement relative to the original HR ground truth image set. Both quantitative and qualitative assessment of SR images showed that the level of additive white noise to the LR image remarkably decreases in the SR image. Furthermore, visual comparison of SR images with corresponding HR images in some areas showed that the SR image may exhibit less amount of noise.

The second important finding relates to the task-based IQM performed using SfM photogrammetry. Results confirmed that the geometry of UAS image acquisition can be recovered in SR images with high accuracy. Camera interior and exterior parameters, evaluated by processing SR images in auto-calibration module within the SfM photogrammetry procedure, closely approximate the original results derived from the same procedure on the ground truth HR images. Preserving the geometry of imagery can significantly increase the reliability of using super-resolution techniques in many different RS applications, specifically where extracting spatial information from RS images is required. The densified point cloud generated by SfM photogrammetry on the SR UAS images is about 15 times richer than the point cloud generated from the artificially degraded LR UAS images, which provides more details about the underlying terrain. Furthermore, the differential DSM and related height-difference histogram show the STD around 2.5 cm, which confirms the closeness of the two reconstructed surfaces generated from the SR and HR image sets.

Overall, results from this study's experiment on UAS imagery show that DCNN-based SISR enhancement techniques can exploit spatial and non-spatial information in LR and HR imagery for effectively discriminating the signal from noise in image space resulting in high performance in recovering image details and more visually appealing images for different RS applications. For example, one practical application of the SR technique for UAS mapping is that it can potentially enable flights at higher altitudes and lower GSDs to cover more area in a certain time duration, thereby leading to more flight efficiency. Then, a DCNN-based SISR technique, such as presented in this study, could be applied to super-resolve the imagery to a specific resolution and generate a dense point cloud from SfM photogrammetry, and subsequently DSM or orthoimage, as though the data were acquired from a UAS flight conducted at a lower altitude and with similar quality.

Future work will seek to investigate the real scenario of employing SISR to reduce UAS image acquisition flight time for aerial surveying operations when mapping of a relatively large area at high resolution is demanded. This will be investigated by employing two UAS image sets acquired at two different altitudes over the same area. Performance of the DCNN-based SISR model to super-resolve the LR (high altitude) images can then be assessed by comparing SfM processing results with the super-resolved LR images and original HR (low altitude) images in terms of 3D reconstruction fidelity and image quality. The effect of different lighting and environmental conditions, and the impact of different study sites with different objects of varying textures, on model performance may also be explored. Finally, examining the most optimized DCNN-based SISR techniques, with the lowest time-complexity in training and inference phases, might be a topic of great interest where it can help pave the path for integration of SISR into real-time remote sensing application scenarios.

Author Contributions: M.P. and M.J.S. conceived the overall study concept and approach; M.P. formulated experimental design; J.B. carried out the field operation for UAS imagery. M.P. and H.K. prepared training and validation image sets. M.P. developed computational code, performed the experiments. M.P. and J.B. designed and performed the SfM photogrammetry experiment on all image sets. M.P. and J.B. analyzed the results; M.J.S. and H.K. helped with results interpretation; H.K. designed figures for the paper. M.P. and M.J.S. wrote the paper. All authors have read and agreed to the published version of the manuscript.

Funding: This publication was prepared by Texas A&M University-Corpus Christi using Federal funds under award NA18NOS4000198 from the National Oceanic and Atmospheric Administration, U.S. Department of Commerce. The statements, findings, conclusions, and recommendations are those of the author(s) and do not necessarily reflect the views of the National Oceanic and Atmospheric Administration or the U.S. Department of Commerce.

Acknowledgments: The authors gratefully acknowledge James Rizzo of the Conrad Blucher Institute for Surveying and Science for his support and encouragement.

Conflicts of Interest: The authors declare no conflict of interest.

References

1. Stumpf, R.P.; Holderied, K.; Sinclair, M. Determination of water depth with high-resolution satellite imagery over variable bottom types. *Limnol. Oceanogr.* **2003**, *48*, 547–556. [CrossRef]

2. Aizawa, K.; Komatsu, T.; Saito, T. Acquisition of very high resolution images using stereo cameras. In Proceedings of the Visual Communications and Image Processing'91: Visual Communication, Boston, MA, USA, 10 November 1991; International Society for Optics and Photonics; Volume 1605, pp. 318–328.
3. Dare, P.M. Shadow analysis in high-resolution satellite imagery of urban areas. *Photogramm. Eng. Remote Sens.* **2005**, *71*, 169–177. [CrossRef]
4. Yang, J.; Wright, J.; Huang, T.S.; Ma, Y. Image super-resolution via sparse representation. *IEEE Trans. Image Process.* **2010**, *19*, 2861–2873. [CrossRef] [PubMed]
5. Chaudhuri, S. *Super-Resolution Imaging*; Springer Science & Business Media: Berlin/Heidelberg, Germany, 2001; Volume 632.
6. Al-falluji, R.A.A.; Youssif, A.A.H.; Guirguis, S.K. Single Image Super Resolution Algorithms: A Survey and Evaluation. *Int. Adv. Res. Comput. Eng. Technol.* **2017**, *6*, 1445–1451.
7. Vega, M.; Mateos, J.; Molina, R.; Katsaggelos, A.K. Super-resolution of multispectral images. *Comput. J.* **2009**, *52*, 153–167.
8. Zhang, H.; Zhang, L.; Shen, H. A super-resolution reconstruction algorithm for hyperspectral images. *Signal Process.* **2012**, *92*, 2082–2096.
9. Zhang, H.; Yang, Z.; Zhang, L.; Shen, H. Super-resolution reconstruction for multi-angle remote sensing images considering resolution differences. *Remote Sens.* **2014**, *6*, 637–657.
10. Greenspan, H. Super-resolution in medical imaging. *Comput. J.* **2009**, *52*, 43–63. [CrossRef]
11. Haris, M.; Shakhnarovich, G.; Ukita, N. Task-driven super resolution: Object detection in low-resolution images. *arXiv* **2018**, arXiv:1803.11316.
12. Sajjadi, M.S.; Scholkopf, B.; Hirsch, M. Enhancenet: Single image super-resolution through automated texture synthesis. In Proceedings of the IEEE International Conference on Computer Vision, Venice, Italy, 29 October 2017; pp. 4491–4500.
13. Bai, Y.; Zhang, Y.; Ding, M.; Ghanem, B. Sod-mtgan: Small object detection via multi-task generative adversarial network. In Proceedings of the European Conference on Computer Vision (ECCV), Munich, Germany, 8–14 September 2018; pp. 206–221.
14. Tuna, C.; Unal, G.; Sertel, E. Single-frame super resolution of remote-sensing images by convolutional neural networks. *Int. J. Remote Sens.* **2018**, *39*, 2463–2479.
15. Yue, L.; Shen, H.; Li, J.; Yuan, Q.; Zhang, H.; Zhang, L. Image super-resolution: The techniques, applications, and future. *Signal Process.* **2016**, *128*, 389–408.
16. Borman, S.; Stevenson, R.L. Super-resolution from image sequences-a review. In Proceedings of the 1998 Midwest Symposium on Circuits and Systems, Notre Dame, IN, USA, 12 August 1998; pp. 374–378.
17. Hardie, R.C.; Barnard, K.J.; Armstrong, E.E. Joint MAP registration and high-resolution image estimation using a sequence of undersampled images. *IEEE Trans. Image Process.* **1997**, *6*, 1621–1633. [PubMed]
18. Tipping, M.E.; Bishop, C.M. Bayesian image super-resolution. *Adv. Neural Inf. Process. Syst.* **2003**, *15*, 1303–1310.
19. He, K.; Zhang, X.; Ren, S.; Sun, J. Deep residual learning for image recognition. In Proceedings of the IEEE Conference on Computer Vision and Pattern Recognition, Las Vegas, NV, USA, 27 June 2016; pp. 770–778.
20. Pashaei, M.; Starek, M.J. Fully Convolutional Neural Network for Land Cover Mapping In A Coastal Wetland with Hyperspatial UAS Imagery. In Proceedings of the 2019 IEEE International Geoscience and Remote Sensing Symposium (IGARSS), Yokohama, Japan, 28 July–2 August 2019; pp. 6106–6109.
21. Pashaei, M.; Kamangir, H.; Starek, M.J.; Tissot, P. Review and Evaluation of Deep Learning Architectures for Efficient Land Cover Mapping with UAS Hyper-Spatial Imagery: A Case Study Over a Wetland. *Remote Sens.* **2020**, *12*, 959. [CrossRef]
22. Kamangir, H.; Rahnemoonfar, M.; Dobbs, D.; Paden, J.; Fox, G. Deep hybrid wavelet network for ice boundary detection in radra imagery. In Proceedings of the 2018 the IEEE International Geoscience and Remote Sensing Symposium (IGARSS), Valencia, Spain, 22–27 July 2018; pp. 3449–3452.
23. Ledig, C.; Theis, L.; Huszár, F.; Caballero, J.; Cunningham, A.; Acosta, A.; Aitken, A.; Tejani, A.; Totz, J.; Wang, Z.; et al. Photo-realistic single image super-resolution using a generative adversarial network. In Proceedings of the IEEE Conference on Computer Vision and Pattern Recognition, Honolulu, HI, USA, 21–26 July 2017; pp. 4681–4690.
24. Dong, C.; Loy, C.C.; He, K.; Tang, X. Image super-resolution using deep convolutional networks. *IEEE Trans. Pattern Anal. Mach. Intell.* **2015**, *38*, 295–307.

25. Yang, W.; Zhang, X.; Tian, Y.; Wang, W.; Xue, J.H.; Liao, Q. Deep learning for single image super-resolution: A brief review. *IEEE Trans. Multimed.* **2019**, *21*, 3106–3121. [CrossRef]
26. Liebel, L.; Körner, M. Single-image super resolution for multispectral remote sensing data using convolutional neural networks. *ISPRS-Int. Arch. Photogramm. Remote Sens. Spat. Inf. Sci.* **2016**, *41*, 883–890.
27. Wang, C.; Liu, Y.; Bai, X.; Tang, W.; Lei, P.; Zhou, J. Deep residual convolutional neural network for hyperspectral image super-resolution. In Proceedings of the International Conference on Image and Graphics, Shanghai, China, 2–4 July 2017; pp. 370–380.
28. Haut, J.M.; Fernandez-Beltran, R.; Paoletti, M.E.; Plaza, J.; Plaza, A.; Pla, F. A new deep generative network for unsupervised remote sensing single-image super-resolution. *IEEE Trans. Geosci. Remote Sens.* **2018**, *56*, 6792–6810. [CrossRef]
29. Arun, P.V.; Herrmann, I.; Budhiraju, K.M.; Karnieli, A. Convolutional network architectures for super-resolution/sub-pixel mapping of drone-derived images. *Pattern Recognit.* **2019**, *88*, 431–446. [CrossRef]
30. Burdziakowski, P. Increasing the Geometrical and Interpretation Quality of Unmanned Aerial Vehicle Photogrammetry Products using Super-Resolution Algorithms. *Remote Sens.* **2020**, *12*, 810. [CrossRef]
31. Goodfellow, I.; Pouget-Abadie, J.; Mirza, M.; Xu, B.; Warde-Farley, D.; Ozair, S.; Courville, A.; Bengio, Y. Generative adversarial nets. *Adv. Neural Inf. Process. Syst.* **2014**, *2*, 2672–2680.
32. Wang, X.; Yu, K.; Wu, S.; Gu, J.; Liu, Y.; Dong, C.; Qiao, Y.; Change Loy, C. Esrgan: Enhanced super-resolution generative adversarial networks. In Proceedings of the European Conference on Computer Vision (ECCV), Munich, Germany, 8–14 September 2018.
33. Dougherty, G. *Digital Image Processing for Medical Applications*; Cambridge University Press: Cambridge, MA, USA, 2009.
34. Park, S.C.; Park, M.K.; Kang, M.G. Super-resolution image reconstruction: A technical overview. *IEEE Signal Process. Mag.* **2003**, *20*, 21–36. [CrossRef]
35. Chang, H.; Yeung, D.Y.; Xiong, Y. Super-resolution through neighbor embedding. In Proceedings of the 2004 IEEE Computer Society Conference on Computer Vision and Pattern Recognition, Washington, DC, USA, 27 June–2 July 2004; Volume 1.
36. Goto, T.; Fukuoka, T.; Nagashima, F.; Hirano, S.; Sakurai, M. Super-resolution System for 4K-HDTV. In Proceedings of the 22nd International Conference on Pattern Recognition, Stockholm, Sweden, 24–28 August 2014; pp. 4453–4458.
37. Peled, S.; Yeshurun, Y. Superresolution in MRI: Application to human white matter fiber tract visualization by diffusion tensor imaging. *Magn. Reson. Med. Off. J. Int. Soc. Magn. Reson. Med.* **2001**, *45*, 29–35.
38. Shi, W.; Caballero, J.; Ledig, C.; Zhuang, X.; Bai, W.; Bhatia, K.; de Marvao, A.M.S.M.; Dawes, T.; O'Regan, D.; Rueckert, D. Cardiac image super-resolution with global correspondence using multi-atlas patchmatch. In Proceedings of the International Conference on Medical Image Computing and Computer-Assisted Intervention, Nagoya, Japan, 22–26 September 2013; pp. 9–16.
39. Thornton, M.W.; Atkinson, P.M.; Holland, D. Sub-pixel mapping of rural land cover objects from fine spatial resolution satellite sensor imagery using super-resolution pixel-swapping. *Int. J. Remote Sens.* **2006**, *27*, 473–491. [CrossRef]
40. Gunturk, B.K.; Batur, A.U.; Altunbasak, Y.; Hayes, M.H.; Mersereau, R.M. Eigenface-domain super-resolution for face recognition. *IEEE Trans. Image Process.* **2003**, *12*, 597–606. [PubMed]
41. Zhang, L.; Zhang, H.; Shen, H.; Li, P. A super-resolution reconstruction algorithm for surveillance images. *Signal Process.* **2010**, *90*, 848–859.
42. Yang, C.Y.; Huang, J.B.; Yang, M.H. Exploiting self-similarities for single frame super-resolution. In Proceedings of the Asian Conference on Computer Vision, Queenstown, New Zealand, 8–12 November 2010; pp. 497–510.
43. Shi, W.; Caballero, J.; Huszár, F.; Totz, J.; Aitken, A.P.; Bishop, R.; Rueckert, D.; Wang, Z. Real-time single image and video super-resolution using an efficient sub-pixel convolutional neural network. In Proceedings of the IEEE Conference on Computer Vision and Pattern Recognition, Las Vegas, NV, USA, 26 June–1 July 2016; pp. 1874–1883.
44. Wang, Z.; Chen, J.; Hoi, S.C. Deep learning for image super-resolution: A survey. *arXiv* **2019**, arXiv:1902.06068.
45. Yang, C.Y.; Ma, C.; Yang, M.H. Single-image super-resolution: A benchmark. In Proceedings of the European Conference on Computer Vision, Zurich, Switzerland, 6–12 September 2014; pp. 372–386.

46. Baker, S.; Kanade, T. Limits on super-resolution and how to break them. *IEEE Trans. Pattern Anal. Mach. Intell.* **2002**, *24*, 1167–1183. [CrossRef]
47. Bevilacqua, M.; Roumy, A.; Guillemot, C.; Alberi-Morel, M.L. Low-Complexity Single-Image Super-Resolution Based on Nonnegative Neighbor Embedding. 2012. Available online: http://people.rennes.inria.fr/Aline.Roumy/publi/12bmvc_Bevilacqua_lowComplexitySR.pdf (accessed on 24 April 2020).
48. Yang, J.; Wang, Z.; Lin, Z.; Cohen, S.; Huang, T. Coupled dictionary training for image super-resolution. *IEEE Trans. Image Process.* **2012**, *21*, 3467–3478. [CrossRef]
49. Zeyde, R.; Elad, M.; Protter, M. On single image scale-up using sparse-representations. In Proceedings of the International Conference on Curves and Surfaces, Avignon, France, 24–30 June 2010; pp. 711–730.
50. Dong, C.; Loy, C.C.; He, K.; Tang, X. Learning a deep convolutional network for image super-resolution. In Proceedings of the European conference on computer vision, Zurich, Switzerland, 6–12 September 2014; pp. 184–199.
51. Dong, C.; Loy, C.C.; Tang, X. Accelerating the super-resolution convolutional neural network. In Proceedings of the European conference on computer vision, Amsterdam, The Netherlands, 8–16 October 2016; pp. 391–407.
52. Tong, T.; Li, G.; Liu, X.; Gao, Q. Image super-resolution using dense skip connections. In Proceedings of the IEEE International Conference on Computer Vision, Venice, Italy, 22–29 October 2017; pp. 4799–4807.
53. Simonyan, K.; Zisserman, A. Very deep convolutional networks for large-scale image recognition. *arXiv* **2014**, arXiv:1409.1556.
54. Szegedy, C.; Liu, W.; Jia, Y.; Sermanet, P.; Reed, S.; Anguelov, D.; Erhan, D.; Vanhoucke, V.; Rabinovich, A. Going deeper with convolutions. In Proceedings of the IEEE Conference on Computer Vision and Pattern Recognition, Boston, MA, USA, 7–12 June 2015; pp. 1–9.
55. Szegedy, C.; Ioffe, S.; Vanhoucke, V.; Alemi, A.A. Inception-v4, inception-resnet and the impact of residual connections on learning. In Proceedings of the Thirty-First AAAI Conference on Artificial Intelligence, San Francisco, CA, USA, 4–10 February 2017.
56. Huang, G.; Liu, Z.; Van Der Maaten, L.; Weinberger, K.Q. Densely connected convolutional networks. In Proceedings of the IEEE Conference on Computer Vision and Pattern Recognition, Honolulu, HI, USA, 21–26 July 2017; pp. 4700–4708.
57. Lai, W.S.; Huang, J.B.; Ahuja, N.; Yang, M.H. Deep laplacian pyramid networks for fast and accurate super-resolution. In Proceedings of the IEEE Conference on Computer Vision and Pattern Recognition, Honolulu, HI, USA, 21–26 July 2017; pp. 624–632.
58. Bruhn, A.; Weickert, J.; Schnörr, C. Lucas/Kanade meets Horn/Schunck: Combining local and global optic flow methods. *Int. J. Comput. Vis.* **2005**, *61*, 211–231.
59. Lim, B.; Son, S.; Kim, H.; Nah, S.; Mu Lee, K. Enhanced deep residual networks for single image super-resolution. In Proceedings of the IEEE Conference on Computer Vision and Pattern Recognition Workshops, Honolulu, HI, USA, 21–26 July 2017; pp. 136–144.
60. Wang, Z.; Bovik, A.C.; Sheikh, H.R.; Simoncelli, E.P. Image quality assessment: From error visibility to structural similarity. *IEEE Trans. Image Process.* **2004**, *13*, 600–612.
61. Johnson, J.; Alahi, A.; Fei-Fei, L. Perceptual losses for real-time style transfer and super-resolution. In Proceedings of the European Conference on Computer Vision, Amsterdam, The Netherlands, 8–16 October 2016; pp. 694–711.
62. Bruna, J.; Sprechmann, P.; LeCun, Y. Super-resolution with deep convolutional sufficient statistics. *arXiv* **2015**, arXiv:1511.05666.
63. Gatys, L.; Ecker, A.S.; Bethge, M. Texture synthesis using convolutional neural networks. *Adv. Neural Inf. Process. Syst.* **2015**, *1*, 262–270.
64. Gatys, L.A.; Ecker, A.S.; Bethge, M. Image style transfer using convolutional neural networks. In Proceedings of the IEEE Conference on Computer Vision and Pattern Recognition, Las Vegas, NV, USA, 26 June–1 July 2016; pp. 2414–2423.
65. Bulat, A.; Tzimiropoulos, G. Super-FAN: Integrated facial landmark localization and super-resolution of real-world low resolution faces in arbitrary poses with GANs. In Proceedings of the IEEE Conference on Computer Vision and Pattern Recognition, Salt Lake City, UT, USA, 18–22 June 2018; pp. 109–117.

66. Wang, X.; Yu, K.; Dong, C.; Change Loy, C. Recovering realistic texture in image super-resolution by deep spatial feature transform. In Proceedings of the IEEE Conference on Computer Vision and Pattern Recognition, Salt Lake City, UT, USA, 18–22 June 2018; pp. 606–615.
67. Yuan, Y.; Liu, S.; Zhang, J.; Zhang, Y.; Dong, C.; Lin, L. Unsupervised image super-resolution using cycle-in-cycle generative adversarial networks. In Proceedings of the IEEE Conference on Computer Vision and Pattern Recognition Workshops, Salt Lake City, UT, USA, 18–22 June 2018; pp. 701–710.
68. Wang, Z.; Simoncelli, E.P.; Bovik, A.C. Multiscale structural similarity for image quality assessment. In Proceedings of the Thrity-Seventh Asilomar Conference on Signals, Systems & Computers, Pacific Grove, CA, USA, 9–12 November 2003; Volume 2, pp. 1398–1402.
69. Channappayya, S.S.; Bovik, A.C.; Caramanis, C.; Heath, R.W. SSIM-optimal linear image restoration. In Proceedings of the 2008 IEEE International Conference on Acoustics, Speech and Signal Processing, Las Vegas, NV, USA, 31 March–4 April 2008; pp. 765–768.
70. Sheikh, H.R.; Sabir, M.F.; Bovik, A.C. A statistical evaluation of recent full reference image quality assessment algorithms. *IEEE Trans. Image Process.* **2006**, *15*, 3440–3451. [CrossRef] [PubMed]
71. Wang, Z.; Bovik, A.C. Mean squared error: Love it or leave it? A new look at signal fidelity measures. *IEEE Signal Process. Mag.* **2009**, *26*, 98–117.
72. Dai, D.; Wang, Y.; Chen, Y.; Van Gool, L. Is image super-resolution helpful for other vision tasks? In Proceedings of the 2016 IEEE Winter Conference on Applications of Computer Vision (WACV), Lake Placid, NY, USA, 7–9 March 2016; pp. 1–9.
73. Fookes, C.; Lin, F.; Chandran, V.; Sridharan, S. Evaluation of image resolution and super-resolution on face recognition performance. *J. Vis. Commun. Image Represent.* **2012**, *23*, 75–93. [CrossRef]
74. Zhang, K.; Zhang, Z.; Cheng, C.W.; Hsu, W.H.; Qiao, Y.; Liu, W.; Zhang, T. Super-identity convolutional neural network for face hallucination. In Proceedings of the European Conference on Computer Vision (ECCV), Munich, Germany, 8–14 September 2018; pp. 183–198.
75. Chen, Y.; Tai, Y.; Liu, X.; Shen, C.; Yang, J. Fsrnet: End-to-end learning face super-resolution with facial priors. In Proceedings of the IEEE Conference on Computer Vision and Pattern Recognition, Salt Lake City, UT, USA, 18–22 June 2018; pp. 2492–2501.
76. Zhang, Y.; Li, K.; Li, K.; Wang, L.; Zhong, B.; Fu, Y. Image super-resolution using very deep residual channel attention networks. In Proceedings of the European Conference on Computer Vision (ECCV), Munich, Germany, 8–14 September 2018; pp. 286–301.
77. Jolicoeur-Martineau, A. The relativistic discriminator: A key element missing from standard GAN. *arXiv* **2018**, arXiv:1807.00734.
78. Westoby, M.J.; Brasington, J.; Glasser, N.F.; Hambrey, M.J.; Reynolds, J.M. 'Structure-from-Motion'photogrammetry: A low-cost, effective tool for geoscience applications. *Geomorphology* **2012**, *179*, 300–314.
79. Snavely, N. Scene reconstruction and visualization from internet photo collections: A survey. *IPSJ Trans. Comput. Vis. Appl.* **2011**, *3*, 44–66. [CrossRef]
80. Lowe, D.G. Distinctive image features from scale-invariant keypoints. *Int. J. Comput. Vis.* **2004**, *60*, 91–110. [CrossRef]
81. Furukawa, Y.; Hernández, C. Multi-view stereo: A tutorial. *Found. Trends Comput. Graph. Vis.* **2015**, *9*, 1–148. [CrossRef]
82. Yosinski, J.; Clune, J.; Bengio, Y.; Lipson, H. How transferable are features in deep neural networks? *Adv. Neural Inf. Process. Syst.* **2014**, *2*, 3320–3328.
83. Paszke, A.; Gross, S.; Chintala, S.; Chanan, G.; Yang, E.; DeVito, Z.; Lin, Z.; Desmaison, A.; Antiga, L.; Lerer, A. Automatic Differentiation in PyTorch. 2017. Available online: https://openreview.net/pdf?id=BJJsrmfCZ (accessed on 24 April 2020).
84. Agustsson, E.; Timofte, R. Ntire 2017 challenge on single image super-resolution: Dataset and study. In Proceedings of the IEEE Conference on Computer Vision and Pattern Recognition Workshops, Honolulu, HI, USA, 21–26 July 2017; pp. 126–135.
85. Timofte, R.; Agustsson, E.; Van Gool, L.; Yang, M.H.; Zhang, L. Ntire 2017 challenge on single image super-resolution: Methods and results. In Proceedings of the IEEE Conference on Computer Vision and Pattern Recognition Workshops, Honolulu, HI, USA, 21–26 July 2017; pp. 114–125.

86. Martin, D.; Fowlkes, C.; Tal, D.; Malik, J.; others. A database of human segmented natural images and its application to evaluating segmentation algorithms and measuring ecological statistics. In Proceedings of the Iccv Vancouver, Vancouver, BC, Canada, 7–14 July 2001.
87. Huang, J.B.; Singh, A.; Ahuja, N. Single image super-resolution from transformed self-exemplars. In Proceedings of the IEEE Conference on Computer Vision and Pattern Recognition, Boston, MA, USA, 7–12 June 2015; pp. 5197–5206.
88. Blau, Y.; Mechrez, R.; Timofte, R.; Michaeli, T.; Zelnik-Manor, L. The 2018 PIRM challenge on perceptual image super-resolution. In Proceedings of the European Conference on Computer Vision (ECCV), Munich, Germany, 8–14 September2018.
89. Agisoft. Metashape—Photogrammetric Processing of Digital Images and 3D Spatial Data Generation. 2019. Available online: https://www.agisoft.com/ (accessed on 24 April 2020).
90. Fraser, C.S. Digital camera self-calibration. *ISPRS J. Photogramm. Remote Sens.* **1997**, *52*, 149–159. [CrossRef]
91. Remondino, F.; Fraser, C. Digital camera calibration methods: Considerations and comparisons. *Int. Arch. Photogramm. Remote Sens.* **2006**, *36*, 266–272.

Article

Mapping Heterogeneous Urban Landscapes from the Fusion of Digital Surface Model and Unmanned Aerial Vehicle-Based Images Using Adaptive Multiscale Image Segmentation and Classification

Mohamed Barakat A. Gibril [1,2], Bahareh Kalantar [3,*], Rami Al-Ruzouq [1,4], Naonori Ueda [3], Vahideh Saeidi [5], Abdallah Shanableh [1,4], Shattri Mansor [2] and Helmi Z. M. Shafri [2]

1. GIS & Remote Sensing Center, Research Institute of Sciences and Engineering, University of Sharjah, 27272 Sharjah, UAE; mbgibril@sharjah.ac.ae (M.B.A.G.); ralruzouq@sharjah.ac.ae (R.A.-R.); shanableh@sharjah.ac.ae (A.S.)
2. Department of Civil Engineering, Faculty of Engineering, Universiti Putra Malaysia, Serdang, 43400 Selangor, Malaysia; shattri@upm.edu.my (S.M.); helmi@upm.edu.my (H.Z.M.S.)
3. RIKEN Center for Advanced Intelligence Project, Goal-Oriented Technology Research Group, Disaster Resilience Science Team, Tokyo 103-0027, Japan; naonori.ueda@riken.jp
4. Department of Civil and Environmental Engineering, University of Sharjah, 27272 Sharjah, UAE
5. Department of Mapping and Surveying, Darya Tarsim Consulting Engineers Co. Ltd., 1457843993 Tehran, Iran; saeidi@daryatarsim.com

* Correspondence: bahareh.kalantar@riken.jp; Tel.: +81-362-252-482

Received: 3 March 2020; Accepted: 24 March 2020; Published: 27 March 2020

Abstract: Considering the high-level details in an ultrahigh-spatial-resolution (UHSR) unmanned aerial vehicle (UAV) dataset, detailed mapping of heterogeneous urban landscapes is extremely challenging because of the spectral similarity between classes. In this study, adaptive hierarchical image segmentation optimization, multilevel feature selection, and multiscale (MS) supervised machine learning (ML) models were integrated to accurately generate detailed maps for heterogeneous urban areas from the fusion of the UHSR orthomosaic and digital surface model (DSM). The integrated approach commenced through a preliminary MS image segmentation parameter selection, followed by the application of three supervised ML models, namely, random forest (RF), support vector machine (SVM), and decision tree (DT). These models were implemented at the optimal MS levels to identify preliminary information, such as the optimal segmentation level(s) and relevant features, for extracting 12 land use/land cover (LULC) urban classes from the fused datasets. Using the information obtained from the first phase of the analysis, detailed MS classification was iteratively conducted to improve the classification accuracy and derive the final urban LULC maps. Two UAV-based datasets were used to develop and assess the effectiveness of the proposed framework. The hierarchical classification of the pilot study area showed that the RF was superior with an overall accuracy (OA) of 94.40% and a kappa coefficient (K) of 0.938, followed by SVM (OA = 92.50% and K = 0.917) and DT (OA = 91.60% and K = 0.908). The classification results of the second dataset revealed that SVM was superior with an OA of 94.45% and K of 0.938, followed by RF (OA = 92.46% and K = 0.916) and DT (OA = 90.46% and K = 0.893). The proposed framework exhibited an excellent potential for the detailed mapping of heterogeneous urban landscapes from the fusion of UHSR orthophoto and DSM images using various ML models.

Keywords: unmanned aerial vehicle; urban LULC; GEOBIA; multiscale classification

---

1. Introduction

Land use/land cover (LULC) maps play an indispensable part in gaining comprehensive insights into coupled human–environment systems, socioecological concerns, resource inventories, ecosystem management, planning activities, change monitoring, emergency response, and decision making. For instance, high-quality thematic LULC information is an essential input for versatile local and regional applications, such as natural disasters [1,2], agriculture [3], sustainable development [4], and land use suitability and management [5]. Therefore, producing accurate, up-to-date, and cost-efficient detailed LULC maps is crucial for resource managers, scientists, decision makers, and city planners.

Remote sensing technologies have been extensively used to retrieve LULC information using comprehensive options of platforms and sensors with versatile spatial, spectral, and temporal resolutions. Satellite and airborne remotely sensed data are usually expensive and constrained by the inability to deliver adequate spatial and temporal resolutions compared to drone-based data. Nowadays, unmanned aerial vehicles (UAVs) are used to collect remotely sensed data in a cost-effective manner at low altitudes below the cloud cover with ultrahigh spatial (UHSR) spectral and temporal resolutions. These advantages make the UAV system a powerful tool that can be used to fulfil the rapid monitoring and assessment during a natural disaster and real-time monitoring applications [6,7]. A plethora of studies have successfully used UAV platforms to acquire remotely sensed data for LULC applications [7–13].

Geographic object-based image analysis (GEOBIA), a paradigm that imitates the human visual perception of real-world targets by addressing the spectral variability amongst classes, has been a preferable classification approach because of its advantages over pixel-based classification [14,15]. The limitations of pixel-based approaches, such as misclassification and salt-and-pepper effects, are addressed through the hierarchical/multiscale (MS) exemplification of image objects, representation of image objects across single/multiple images at MS levels, and incorporation of spatial, spectral, textural, geometrical, elevation, backscattering, and contextual information in LULC classification [16,17]. GEOBIA has been widely used along with advanced machine learning (ML) algorithms for analyzing and classifying drone-based images in various applications. De Castro et al. [18] suggested an automatic GEOBIA approach using a random forest (RF) classifier for site-specific weed management with UAV-based images. Their results helped the farmers with timely decision making for crop optimization and management. Komárek et al. [19] utilized a three-level GEOBIA system with a support vector machine (SVM) algorithm to identify individual plant species from multispectral and thermal drone-based images. Kamal et al. [20] introduced a GEOBIA approach for mangrove canopy delineation using UAV-based data. The results showed that the UAV red, green, and blue (RGB) images are valuable inputs for GEOBIA regardless of the limit of spectral information. Mishra et al. [21] presented the potential for achieving species-level mapping from multispectral UAV data through GEOBIA. White et al. [22] proposed a GEOBIA approach to identify sapling Jak pine forests after wildfire. A MicaSense RedEdge 3 multispectral camera onboard a quadcopter UAV platform was used for data acquisition. The highest classification accuracy was achieved by including the red and near-infrared spectra. The following section reviews the various elements affecting the GEOBIA's overall quality, including image optimization of image segmentation parameters, several feature selection (FS) approaches, and machine learning (ML) algorithms.

Related Studies

GEOBIA is constructed on the basis of the conception of creating a meaningful representation of real-world targets (i.e., LULC types, such as buildings, roads, and vegetation) by generating homogenous regions from image pixels, and this procedure is referred to as image segmentation. This procedure groups the raw pixels into homogeneous segments that are jointly exhaustive and mutually disjointed, which are then used as the primary element for interpretation, classification, and modelling [23,24]. The overall performance of the GEOBIA's succeeding phases (i.e., feature computation, extraction, and classification) is immensely influenced by image segmentation quality [25].

Several image segmentation algorithms have been adopted to segment remotely sensed data in the GEOBIA domain. The four commonly used image segmentation algorithms are watershed [26,27], region-based [28], mean-shift [29], and hybrid segmentation [30]. Amongst them, the region-based and multiresolution segmentation algorithms have been widely adopted in various remote sensing applications because of their competency to produce meaningful image objects [31]. The scale parameter (SP), which explains the degree or the density level where a specific phenomenon can be presented, is the main parameter in all segmentation algorithms that require fine-tuning depending on the application [28]. The image segmentation on UHSR images with a single-scale (SS) value results in creating image objects that are either small (oversegmented) or large (undersegmented). Different issues should be considered when selecting image segmentation parameters [32]. Firstly, different urban LULC classes can differ in terms of size and structure and may require several optimum segmentation levels. Secondly, image objects that belong to the same class might correspond to different optimal scales because of their different surrounding contrast. Finally, different components within an object might require analysis at multiple scales. Therefore, finding the optimal SP(s) prior to the analysis is a crucial step in the GEOBIA framework because different objects can only be analyzed accurately on the basis of the scale(s) corresponding to their granularity [33].

Various supervised and unsupervised image segmentation quality evaluation techniques have been proposed to determine the optimal single or MS segmentation. Image segmentation results are usually assessed through the supervised image segmentation quality measures by evaluating the disparity between the manually digitized objects and the generated image objects from an image segmentation algorithm, whereas the unsupervised image segmentation quality measures evaluate MS segmentation results using various statistical-based image segmentation quality measures [34]. Considerable attention has been given to unsupervised segmentation quality measures [35,36]. The vast majority of the unsupervised optimization methods determine the optimum segmentation parameters to evaluate the segmentation outputs by computing the between-object heterogeneity and the within-object homogeneity metrics [37–44]. Moreover, other unsupervised techniques have been adopted in various applications to determine the optimum single or MS segmentation parameters. Xiao et al. [33] proposed a MS segmentation optimization using the unsupervised optimization technique to determine the optimum scales suitable for the extraction of urban green cover from the high-spatial-resolution dataset. Kamal et al. [45] mapped mangrove species using the MS GEOBIA approach from multiple images with a varied spatial resolution (i.e., WorldView 2, LiDAR, ALOS AVNIR-2, and Landsat TM).

FS, which is regarded as a crucial task that influences the GEOBIA classification accuracy, specifies the most relevant features to increase the effectiveness of the adopted classification approach and expedites the processing time by minimizing irrelevant or redundant features [46]. Various FS algorithms have been incorporated with the GEOBIA approach in various applications, and these methods include RF [47,48], SVM [47], ant colony optimization (ACO) [49,50], artificial bee colony [51], hybrid particle swarm optimization [52], correlation-based FS (CFS) [49,53], and chi-square [54]. Ridha and Pradhan [49] applied three FS methods, namely, CFS, RF, and ACO, to discriminate several types of landslides from LiDAR data. The results showed that CFS performs the best with 89.28% accuracy, followed by RF with 85.59% accuracy and ACO with 86.74% accuracy. Al-Ruzouq et al. [50] adopted ACO for feature reduction and identification of the most crucial features for date palm mapping from very-high-resolution aerial imageries. The results showed that ACO and CFS are superior to other algorithms, including principal component analysis, SVM, information gain, gain ratio, and chi-square. The effects of various feature importance evaluation methods, including gain ratio, chi-square, SVM-recursive feature elimination, CFS, Relief-F, SVM, and RF were investigated by Ma et al. [47] within the GEOBIA environment to map agricultural areas from UAV data. The results showed that CFS dominates other feature importance evaluation methods.

Recently, GEOBIA has been integrated with various ML methods to classify UAV-based images for various LULC applications. Ma et al. [47] used two ML classification approaches, namely, SVM and RF, to classify the UAV data into six categories, namely, building, crop, bare land, road, water, and woodland in Deyang, China. RF exhibits a higher classification accuracy compared to SVM. Cao et al. [55] applied two classification algorithms, namely, SVM and k-nearest neighbors (KNN), in the GEOBIA domain to map mangrove species from a UAV hyperspectral image. The result showed that SVM performs better with 89.55% accuracy compared to KNN with 81.70% accuracy. Akar [56] compared various ML algorithms to perform LULC classification using the UAV images collected from urban and rural areas. The results showed that rotation forest (92.52%) outperforms RF (90.52) and gentle AdaBoost (87.52%). Liu et al. [57] proposed a SVM-deep belief network (restricted Boltzmann machine) method to extract eight land cover classes, namely, tree, building 1, road, grass, river, building 2, building 3, and bare-land, using the fusion of LiDAR data and UAV images. Their proposed technique shows an overall accuracy (OA) of 92.16% and a kappa (K) value of 0.904%.

In this study, an adaptive MS segmentation and classification approach was adopted to classify heterogeneous urban areas through the fusion of the UHSR orthophoto and digital surface model (DSM). The main objectives of the current study are to (1) develop an adaptive MS-optimized image object approach for detailed urban LULC mapping from UAV-based data, (2) investigate the effects of MS segmentation on FS computation (CFS and SVM) and its impact on classification accuracy, (3) compare the performance of three mature ML classification algorithms, namely, RF, SVM, and decision tree (DT), at MS levels, and (4) assess the transferability of the adopted framework. The remainder of this paper is organized as follows. Section 2 outlines the geographical location of the study area and describes the ground truth (GT) data. Section 3 presents a generic overview of the methodological framework and detailed information about image processing, image segmentation optimization, FS, MS classification, and evaluation metrics. Section 4 describes the results, and Section 5 discusses the experimental findings. Section 6 provides the conclusions.

2. Study Area and Materials

2.1. Study Area

The location of the study area is geographically positioned at the University of Science, Malaysia (USM) campus, Penang, Malaysia. The study area represents an urban area of Penang island with different LULCs, including vegetation, water bodies, buildings, roads, and bare soil. The RGB images were acquired on February, 2018, (Figure 1) using a Canon PowerShot SX230 HS (4000 × 3000 resolution) boarding on a UAV from an altitude of 353 m. The ground resolution of the orthomosaic is approximately 10 cm, with an 8-bit radiometric resolution. The first dataset was a subset of 2.24 km^2 from the produced orthomosaic photos, located between 100°18′7.43″E, 5°21′51.574″N and 100°19′2″E, 5°21′8.143″N. A DSM with 0.8 m resolution was generated from 3500 points using Agisoft PhotoScan Professional (version 1.3.4, http://www.agisoft.com). The second subset (with the coordinates of 100°17′29.341″E, 5°21′38.6″N and 100°18′9.622″E, 5°21′5.345″N), covering an area of 1.27 km^2, was selected for investigating the transferability of the methodology.

2.2. GT Data

In the first study area, a total of 1177 GT samples for various urban LULC classes were prepared through field surveys with the aid of Google Earth images. Twelve different classes were identified, including water bodies, bare soil, grass, trees, clay tiles type 1, clay tiles type 2, metallic roofs type 1, metallic roofs type 2, concrete, dark concrete roofs, asbestos cement roofs, and asphalt. The training and testing GT samples were prepared as vector points and meticulously selected to ensure that all the available urban LULC classes are well represented. Table 1 presents several types of LULC classes available in the UAV-based images. Amongst the collected GT samples, 70% of each class was utilized in the training of ML models, and 30% was dedicated for testing them. For the second study area, sample statistics derived from the training samples of the first study area were used in the classification models, and 305 GT testing samples were used to evaluate the classification results.

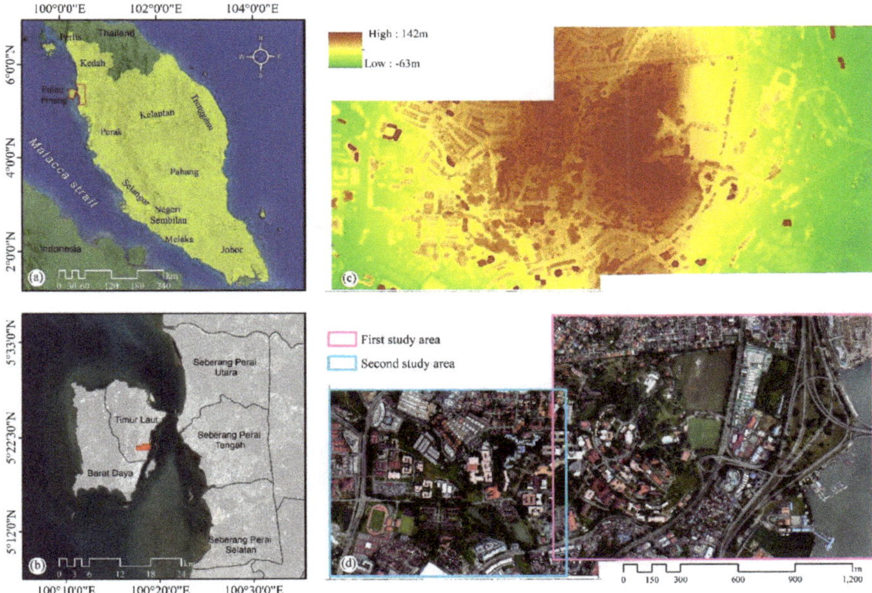

Figure 1. General location of the study sites: (**a**) Malaysian states; (**b**) location map; (**c**) digital surface model (DSM) of the study sites; (**d**) unmanned aerial vehicle (UAV) images of the study sites.

Table 1. Detailed description of different land use/land cover (LULC) classes available in the UAV-based images.

LULC Type	Images	Description
Water bodies		Water bodies with light blue and green colors
Trees and grass		Various tree species and grass thickness
Bare soil		Exposed soil with different colors
Concrete roofs		Concrete slab with bright white color

Table 1. *Cont.*

LULC Type	Images	Description
Dark concrete roofs	 	Residential and industrial buildings with dark brown color
Clay tiles type 1	 	Roofing material with different structural shapes and red color
Clay tiles type 2	 	Roofing material with different structures and bright peach color
Asbestos cement roofs	 	Roofs with regular shape and grey color
Metallic roofs type 1	 	Metal deck with blue color
Metallic roofs type 2	 	Rooftops with turquoise color
Roads	 	Urban roads with grey color

3. Methodology

3.1. Overview

In this study, MS image segmentation optimization, MS feature computation and evaluations, and supervised hierarchical ML models were conducted for accurate detailed mapping of a heterogeneous urban landscape from UAV-based images. As depicted in Figure 2, the adopted methodology comprises five main phases. Firstly, drone-based images were acquired and preprocessed to generate the orthophoto and the DSM. Secondly, the optimum MS segmentation parameters were identified using unsupervised segmentation quality metrics. Thirdly, the most significant features were selected at MS levels on the basis of CFS and SVM wrapper approaches. Fourthly, adaptive MS segmentation optimization and classification were conducted for detailed urban LULC mapping using the RF, SVM, and DT algorithms. Finally, the transferability of the proposed methodology to a different study area was investigated.

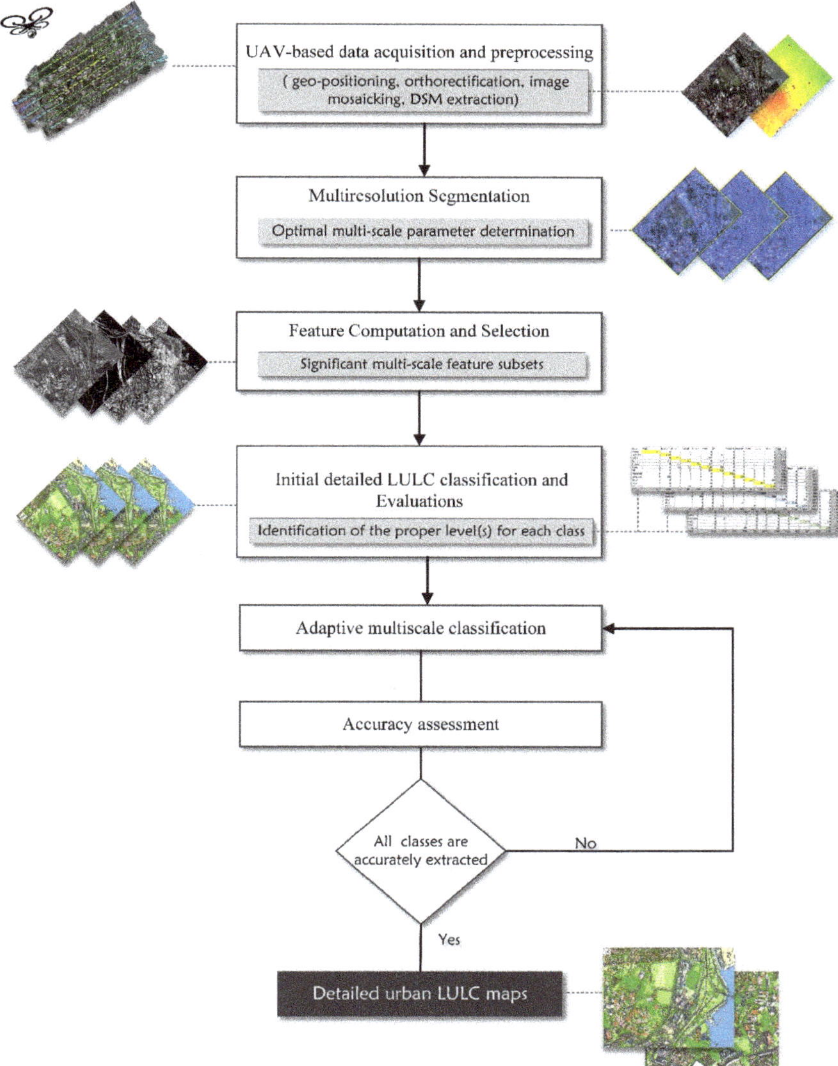

Figure 2. Framework of the proposed methodology.

3.2. Image Preprocessing

Various photogrammetric steps, such as interior, relative, and absolute orientations, have been conducted to establish the mathematical relationship between the image and the ground and subsequently generate the digital elevation model and the orthophoto (an image with the same characteristics of the map, where distortions caused by relief displacement are removed and the image has a uniform scale). Throughout this process, image matching, automatic aerial triangulation, geopositioning, orthorectification, and image mosaicking were performed to create the orthomosaic image and the DSM from the UAV data using Agisoft PhotoScan and ArcGIS 10.4.1. The process commenced by estimating the exterior and interior orientation parameters that estimate the positions of the camera in each image and the camera calibration parameters. The RGB images were geometrically corrected and geotagged to the WGS1984 (world geodetic reference system) using the files extracted from the Global Positioning System units in the drone and the ground reference station. The images were projected to a Universal Transverse Mercator coordinate system (zone 38 North) and converted from JPEG to GeoTiff format. The following steps, such as aligning images, building field geometry, and orthophoto generation, were implemented to create a DSM (a 3D polygon mesh representing surface ground) and an orthomosaic. The DSM was generated with the nearest-neighbor interpolation method and resampled to the same resolution of the orthomosaic. The spatial resolution of the final orthomosaic for the two study areas was 10 cm, and the spatial resolution of the DSM was 80 cm.

3.3. MS Image Segmentation Optimization

The optimal segmentation level is defined in most of the unsupervised methods as the level that maximizes the between-object heterogeneity (i.e., adjacent objects can be distinguished from their surroundings) and the within-object homogeneity (i.e., pixels belonging to the same objects are similar) [40,41]. The likeness between each image and its neighbors, known as the undersegmentation metric, is determined through spatial autocorrelation (Moran's I (MI)) [58], whereas the internal homogeneity of an image object, known as the oversegmentation metric, is determined through the area-weighted variance (WV) [41].

An adaptive segmentation optimization approach that integrates unsupervised quality measures, namely, the *F*-measure, accompanied with a machine learning classification model was adopted in this study to identify the optimal scale(s) for each urban LULC class. The *F*-measure quality measure [39] was utilized to determine the hierarchical scale values from a set of given segmentation outputs. The *F*-measure value for estimating the optimum MS of an application can be computed using Equation (1).

$$F\text{-measure} = \left(1+\varphi^2\right)\frac{MI_{norm} \times WV_{norm}}{\varphi^2.\, MI_{norm} + WV_{norm}}, \tag{1}$$

where WV_{norm} and MI_{norm} represent the normalized area-WV (oversegmentation metric) and the normalized Moran's I (undersegmentation metric), respectively. The relative weights of WV_{norm} and MI_{norm} are controlled through a scene-independent factor (φ). The φ values are selected to ensure that the generated segmentation levels vary considerably in terms of the within-object homogeneity and between-object heterogeneities. For instance, $\varphi = 3$ signifies that triple weighting is assigned to WV_{norm}, $\varphi = 0.5$ indicates half weighting for WV_{norm}, and $\varphi = 1$ denotes that equal weighting is considered for WV_{norm} and MI_{norm}. Additional details about the unsupervised parameter optimization can be found in [39,59]. The levels defined by the *F*-measure are used in the second phase to perform a single scale (SS) classification of each defined segmentation scale. The class-specific accuracy (*F*-measure) is used to evaluate the accuracy of each class at multiple levels, as shown in Section 4.3. Then, the optimal scale(s) for extracting each class is determined and used for subsequent analysis.

3.4. Feature Computation and Selection

Considering the spectral similarity between the various urban LULC classes in the UHSR RGB images, various features, including spectral values, color invariants, and geometrical textural features, were computed and assessed at multiple scales, as shown in Table 2. Selecting the significant features prior to classification is necessary to minimize the computational time by excluding the redundant attributes and enhance the accuracy of an *ML* classifier [47]. In this study, CFS and SVM as wrapper FS techniques were utilized to identify the most relevant MS features of image objects from UAV datasets.

Table 2. Detailed description of the evaluated attributes (features).

Feature Type	Tested Feature Name	Description	Reference
Spectral	Mean	The mean intensity values computed for an image segment of the RGB channels and the DSM	[60]
	Standard deviation	The standard deviation values computed for an image segment of the RGM channels and the DSM.	[60]
	Max_ difference	The maximum difference between the RGB channels.	[60]
	Brightness	The average of means of the RGB channels.	[60]
	NDRG	$\frac{Red-Green}{Red+Green}$	[61]
	NDGB	$\frac{Green-Blue}{Green+Blue}$	[61]
	NDBG	$\frac{Blue-Green}{Blue+Green}$	[61]
	NDRB	$\frac{Red-Blue}{Red+Blue}$	[61]
	NDBR	$\frac{Blue-Red}{Blue+Red}$	[61]
	NDGR	$\frac{Green-Red}{Green+Red}$	[61]
	RB	$\frac{Red}{Blue}$	[61]
	Ratio-R	$\frac{Red}{Green+Blue+Red}$	[61]
	Ratio-G	$\frac{Green}{Green+Blue+Red}$	[61]
	Ratio-B	$\frac{Blue}{Green+Blue+Red}$	[61]
	V	$\frac{4}{\pi}\ .\ \arctan\left(\frac{Green-Blue}{Green+Blue}\right)$	[62]
	S	$\frac{4}{\pi}\ .\ \arctan\left(\frac{1-\sqrt{Red^2+Green^2+Blue^2}}{1+\sqrt{Red^2+Green^2+Blue^2}}\right)$	[63]
Texture	Mean	The grey level co-occurrence matrix (GLCM) mean sum of all directions determined for each band from the RGB channels and the DSM.	[64]
	Homogeneity	The GLCM homogeneity sum of all directions determined for each band from the RGB channels, and the DSM.	[64]
	Contrast	The GLCM contrast sum of all directions determined for each band from the RGB channels, and the DSM. The grey level difference vector (GLDV) matrix contrast sum of all directions determined for each band from the RGB channels, and the DSM.	[64]
	Entropy	The GLCM and GLDV entropy sum of all directions determined for each band from the RGB channels, and the DSM.	[64]
	Correlation	The GLCM correlation sum of all directions determined for each band from the RGB channels, and the DSM.	[64]
	Standard deviation	The GLCM standard deviation sum of all directions determined for each band from the RGB channels, and the DSM.	[64]
	Dissimilarity	The GLCM dissimilarity sum of all directions determined for each band from the RGB channels, and the DSM.	[64]
	Angular second moment	The GLCM angular second-moment sum of all directions determined for each band from the RGB channels, and the DSM.	[64]
Geometric	Length\Width	The ratio between the length and width.	[60]
	Rectangular Fit	A ratio that is based on how well an image object fits into a rectangle.	[60]
	Shape_index	A ratio that defines border smoothness of image objects and can be computed by dividing the border length of an image object by four times the square root of its area.	[60]
	Density	It can be computed by dividing the area covered by an image object by its radius.	[60]
	Elliptic_fit	A ratio based on how well an image object can fit into an ellipse.	[60]
	Compactness	It is expressed as the ratio of the area of an image object to the area of a circle with a similar perimeter.	[60]

The seventy features listed above were computed for three optimized MS image objects, and two efficient FS methods, namely, CFS and SVM, were used to find the relevant feature subset for each optimized image object level.

3.4.1. CFS

CFS performs fast processing to appropriately select the optimal feature subset [53,65,66]. It uses a search algorithm that heuristically assesses each attribute's predictive capability and the degree of intercorrelation between the attributes [67]. In other words, this evaluating mechanism calculates the correlations between the features and classes to classify highly correlated features to the target class whilst considering the low correlations and low level of redundancy amongst the features [68]. The estimations of the correlation between the subset of attributes and target classes are performed using Equation (2).

$$R_s = \frac{s\bar{r}_{ci}}{\sqrt{s + s(s-1)\bar{r}_{ii}}} \tag{2}$$

where s denotes the number of features, $\bar{r}_{ci}$ represents the correlation average between the subset features and the class variable, and $\bar{r}_{ii}$ denotes the intercorrelation average between the subset features. Accordingly, the high correlation coefficients between the feature attributes and the target labels are considered to be relevant to the respective class characterization with a high level of association, whilst lower intercorrelation ($\bar{r}_{ii}$) is desired [68].

3.4.2. SVM

SVM is a widely applied regression algorithm with a nonparametric supervised statistical learning task and is highly suitable for GEOBIA FS and classification tasks [51,69]. This algorithm seeks an optimal separating hyperplane using the training dataset of so-called support vectors that can effectively separate the input features (datasets) into target classes with a minimum misclassification and a maximum margin amongst the target classes [70–72]. When the task is linearly separable, the hyperplane can be represented using Equation (3):

$$y_i(w.x_i + b) \geq 1 - \delta_i, \tag{3}$$

where w indicates the coefficient vector that determines the orientation of the hyperplane in the feature space. The offsets of the hyperplane from the original and positive slack variables are represented by b and δ_i, respectively [73]. Equation (4) determines the optimized hyperplane, where many hyperplanes can be designed to distinguish between classes.

$$\text{Minimise} \sum_{i=1}^{n} a_i - \frac{1}{2} \sum_{i=1}^{n} \sum_{j=1}^{n} a_i a_j y_i y_j \left(x_i x_j\right), \tag{4}$$

$$\text{Subject to} \sum_{i=1}^{n} a_i y_i = 0,\ 0 \leq a_i \leq C, \tag{5}$$

where a_i denotes the Lagrange multipliers and C is the penalty.

3.5. Supervised MS Image Object Classification

Image classification is the final phase in GEOBIA, and the common classification methods used in this phase are supervised ML models or rule-based methods. In this study, the MS image object classification was implemented using three supervised classification algorithms, namely, RF, SVM, and DT. The classification models were trained using the sample statistics derived from the GT dataset of the first study area. The object-based classification outcomes at different scales were used to quantitatively evaluate the MS segmentation results and select the optimum scale for each urban LULC class. Then, the classification scheme started with a single classification of each optimized image-level using the selected feature subsets for each level. After acquiring the proper information about the optimal scale(s) for each class, ML models were used to initially classify large objects at large SP. The classification results were then copied to a new level, where the unclassified objects were only resegmented to a fine segmentation level, and the ML models were then used to classify the resegmented objects on the basis of the selected significant features at that level. The process iteratively continued until all classes were accurately classified or no improvement was detected in the OA and class-specific accuracy (*F*-measure). The two of the aforementioned ML algorithms are briefly described in the following paragraphs.

A DT is a supervised and nonparametric ML technique that is operable without prior knowledge on data distribution, with easy interpretation and capability to model and handle the data complexity reduction and the relationships between variables [74–79]. It is a flexible, fast, and robust algorithm that can be used to control the nonlinearity between the input features and discrete classes [75]. DT hierarchically utilizes IF-THEN rules to label the variables of each class, where the tree structures, leaves, and end nodes represent the discrete class labels (decision), and the branches assist in assigning the labels on the basis of the attributes and majority voting [76]. A heuristic DT recursively partitions a dataset into homogenous subsets in conjunction with the attribute values at each branch or node in the single tree [77].

The RF algorithm is an ensemble of DT classifiers that improves the classification of variables with high accuracy, and its robustness against overfitting the training dataset along with insensitivity to nonnormal and noisy data makes it suitable for LULC classification [51,78,79]. RF is an ensemble method that exploits many DTs as a forest generated from bootstrap and utilizes each tree's vote to assign the most frequent class label to the input variables [78,80]. Each tree then randomly selects the predictors and object features from the input vector of every tree node to increase the generalization error [78,81]. The prediction of the samples is calculated on the basis of the majority votes amongst the trees [80,81]. The discrimination assignment is calculated using Equation (6):

$$\mathrm{H(x)} = \mathrm{argmax}_{\mathrm{Y}} \sum_{\mathrm{i=1}}^{\mathrm{k}} \mathrm{I}(\mathrm{h_i}\ (\mathrm{X},\ \theta_{\mathrm{k}}) = \mathrm{Y}), \tag{6}$$

where θ_k is a random vector for the *kth* tree, *X* is an input vector, $I(\cdot)$ is an indicator function, $h(\cdot)$ is a single DT, *Y* is an output variable, and $argmax_y$ denotes the *Y* value in the maximization of $\sum_{i=1}^{k} I(h_i\ (X,\ \theta_k) = Y)$.

3.6. Evaluation Metrics

The evaluation metrics of the classified images were generated through the frequently applied confusion matrix and its derivatives, including the OA, *K*, precision, recall, and *F*-measure. The error matrix (confusion matrix) evaluates the classification results versus the reference data in two dimensions as actual classes in rows and predicted classes in columns.

3.6.1. OA

The OA, which is a percentage indicator of the classification performance, can be defined as the sum of the correctly classified variables into discrete classes (true positives plus true negatives) to the total tested variables. OA can be computed from the confusion matrix by dividing the total number of correctly classified objects/pixels ($\sum \mathrm{D_{ij}}$ or the sum of the major diagonal) with the total number of objects/pixels (N):

$$\mathrm{OA} = \frac{\sum \mathrm{D_{ij}}}{\mathrm{N}}. \tag{7}$$

3.6.2. K Statistics

The *K* statistic is another statistical measure that defines the observed level of agreement or accuracy between a detailed map and reference data. The *K* value approaches +1 when the contribution of the chance of agreement diminishes and becomes negative when the effects of chance agreements increases. Conversely, a *K* value equaling 0 indicates no agreement, indicating that the classification is entirely conducted by chance or random assignment. A negative *K* value signifies that the agreement is worse than occurring by chance. The *K* statistic is computed using Equation (8):

$$K = \frac{N\sum_{i,j=1}^{m} D_{ij} - \sum_{i,j=1}^{m} R_i \times C_j}{N^2 - \sum_{i,j=1}^{m} R_i \times C_j}, \tag{8}$$

where m denotes the number of urban LULC classes in the confusion matrix, $\mathrm{D_{ij}}$ denotes the number of observations (objects/pixels) that are correctly classified in row i and column j, $\mathrm{R_i}$ denotes the total number of objects/pixels in row i, $\mathrm{C_j}$ denotes the total number of observations in column j, and N denotes the total number of objects/pixels.

3.6.3. Precision, Recall, and F-measure

The *F*-measure is the weighted average or harmonic mean of two ratios known as precision (p). Recall (r) metric is another performance measure used to assess the class-specific accuracy from retrieved information [82,83]. It can be computed using Equation (9) on the basis of the average of p and r. The *F*-measure value ranges from 0 (lowest) to 1 (highest).

$$\mathrm{F_{measure}} = 2 \times \frac{\mathrm{p} \times \mathrm{r}}{\mathrm{p} + \mathrm{r}}. \tag{9}$$

The p or the confidence of a LULC class is determined by dividing the number of true positives (number of objects\pixels correctly belonging to the actual class) by the total number of objects categorized as the positive class (i.e., the sum of true positives and false positives, which are objects/pixels incorrectly categorized as belonging to the class). The r or the sensitivity shows the proportion of true positive objects/pixels that are correctly predicted and identified and can be defined as the number of true positives divided by the total number of objects/pixels that are members of the positive class (i.e., the sum of true positives and false negatives). p and r can be calculated using Equations (10) and (11), respectively. A perfect predictor's value for p and r would be described as 1.

$$p = \frac{\text{true positives}}{\text{true positives} + \text{false positives}}, \tag{10}$$

$$\mathrm{r} = \frac{\text{true positives}}{\text{true positives} + \text{false negatives}}. \tag{11}$$

4. Results

This section summarizes the various outcomes of this study, including the MS image segmentation optimization and parameter selection, FS, and classification results.

4.1. Results of MS Image Segmentation

In this study, the quantitative evaluation of image segmentation results at MS levels through unsupervised segmentation quality measures aims to determine the optimal SP that allows excellent delineation and extraction of urban LULC classes that may share a similar spectral response with each other and vary in structure, size, and their surrounding contrast. The oversegmentation (WV) and undersegmentation (MI) metrics were computed from the three RGB channels, and their mean values were normalized and used to compute the *F*-measure (for selecting the three optimum SPs), as shown in Table 3. Three values, namely, 3, 1, and 0.33, of the scene-independent variables (φ) were selected to pinpoint the three SPs from the computation of Equation (1). These values were empirically selected and supported by the study of Johnson et al. [62] to ensure that the adopted segmentation levels vary remarkably from each other in terms of the between-object homogeneity and within-object heterogeneity. The highest values on the last three columns in Table 2 correspond to the optimal MS levels, and these scales are 200, 100, and 50. Figure 3a,b depict the image segmentation results of a small subset at the scale of 200, where large homogenous objects, such as water bodies, grass, bare soil, and some clay tiles, are well delineated. Figure 3c,d show the image segmentation results of a small subset at the scale of 100, where medium objects, such as some types of roofing materials, are well identified. Figure 3e,f display the image segmentation results of a small subset at the scale of 50, where large and medium objects are oversegmented but small roofing materials and trees are well distinguished.

Table 3. Results of the applied unsupervised segmentation quality measures at multiscale (MS) levels.

Scale	No of Objects	WV_{mean}	MI_{mean}	WV_{norm}	MI_{norm}	*F*-Measure		
						$\varphi = 3$	$\varphi = 1$	$\varphi = 0.33$
25	340731	78.606	0.548	1.000	0.000	0	0	0
50	104840	132.924	0.452	0.858	0.242	0.684	0.377	0.260
75	53011	178.661	0.395	0.739	0.385	0.677	0.506	0.404
100	33217	217.177	0.344	0.639	0.514	0.624	0.570	0.524
125	22978	253.258	0.314	0.545	0.588	0.549	0.566	0.584
150	16878	288.418	0.286	0.453	0.658	0.468	0.537	0.630
175	12887	322.978	0.250	0.363	0.748	0.383	0.489	0.674
200	10181	354.309	0.229	0.281	0.801	0.301	0.416	0.678
225	8216	384.925	0.217	0.202	0.833	0.218	0.325	0.637
250	6841	412.203	0.195	0.130	0.888	0.143	0.227	0.566
275	5738	439.515	0.169	0.059	0.953	0.065	0.112	0.384
300	4961	462.250	0.150	0.000	1.000	0	0	0

Figure 3. Optimized image objects of a subset UAV image mosaic: (**a**,**b**) Scale 200; (**c**,**d**) scale 100; (**e**,**f**) scale 50.

4.2. Results of FS

Following the optimization of segmentation SPs, several spectral, geometrical, and textural features were computed at MS levels for FS, as shown in Table 1. Two wrapper approaches, namely,

CFS and SVM, combined with the KNN algorithm, were used to assess all features as a part of classification. Table 4 compares the OA, *K*, and other relevant features selected by SVM and CFS at scales of 50, 100, and 200. The results of CFS and SVM exhibited significant differences in terms of the number and type of selected features in each scale. However, the two methods eliminated 60% from the total number of features, whereas less than 40% of the features contributed to achieving high accuracy. CFS attained a slight improvement in terms of the OA and number of selected features, as presented in Table 2, and was selected for subsequent processing.

4.3. Classification Results

The detailed mapping of impervious surfaces in a heterogeneous urban area from UAV-based images is particularly challenging when only three spectral channels, RGB, are used because of the spectral similarity of various urban LULC classes. In such a case, a successful extraction of urban objects should consider the information of the variation in size and the surroundings of the different types of LULC that exist in the image. For instance, asbestos cement and dark concrete roofs or cemented pavements may share similar spectral responses because of the presence of cement in their contents. To minimize the confusion between different LULC classes, the information of the suitable scale(s) that provides the best accuracy and ensures the strong differentiation between classes is necessary to obtain a holistic view and to perform hierarchical classification.

The initial stage of classification in this study is to find the optimum level for extracting each class, which can be achieved using ML models, followed by a class-specific accuracy measure. Three standard classification algorithms, namely, RF, SVM, and DT, were used to classify the first study area at the selected optimal scales (SP 200, SP 100, and SP 50). The accuracy of each classification level was evaluated on the basis of OA, *K*, and *F*-measure. Figure 4a–c show the SS classification results of RF, Figure 4d–f display the SS classification results of SVM, and Figure 4g–i show the SS classification results of DT. Table 5 shows the SS classification results for the first study area. The highest SS classification results were obtained by SS-RF at scale 50, with an OA of 92.2 and a *K* of 9.14, followed by SS-SVM at scale 100 with an OA of 90.5 and a *K* of 0.896 and SS-DT at scale 50 with an OA of 88.1% and a *K* of 0.87. Finding the optimum scale for extracting the LULC in heterogeneous urban areas can vary on the basis of the adopted classification algorithm by comparing the class-specific accuracy measures of SS-RF, SS-SVM, and SS-DT classification results. For instance, the SS-RF classification results showed that the SP 50 exhibited the highest OA for the extraction of water bodies, trees, grass, dark concrete, type 2 clay tiles, and type 2 metallic roofs, whereas the SP 200 showed enhanced extraction of bare soil, asphalt, type 1 metallic roofs, concrete, type 1 clay tiles, and asbestos cement roofs. The classification results of SS-SVM showed better extraction for clay tiles (types 1 and 2) at the largest optimized SP, whereas the smallest optimized SP was optimal for extracting water bodies, bare soil, trees, and metallic roofs (types 1 and 2). The previous step was adopted prior to the hierarchical classification approach to provide a diagnostic result where SP is suitable for extracting 12 urban LULC classes and for ensuring reasonable discrimination between the classes.

Table 4. Results of correlation-based feature selection (CFS) and support vector machine (SVM) at MS levels.

Scale	Feature Type	CFS				SVM			
		Selected Features	No	OA	K	Selected Features	No	OA	K
50	Spectral	Red, Blue, DSM, SD_DSM, Vegetation, Ratio_G, Ratio_B, NDRG, NDGR NDBR, Max_diff	16	91.63	0.93	Red, Green, Blue, DSM, Vegetation, Ratio_G, Ratio_B, NDGR, NDBR, NDRB, NDGB, NDBG, Shadow, RB, Max_diff,	21	91.61	0.91
	Textural	GLCM_Mean_DSM, GLCM_Entropy_Green, GLCM_ Dissimilarity _Blue, GLCM_SD_Blue				GLCM_ Entropy _R, GLCM_ Entropy _Blue, GLCM_ Entropy _Green, GLCM_ Homogeneity _Red, GLDV_Mean_Blue			
	Geometrical	Shape_index				Length/Width			
100	Spectral	Red, Green, Blue, DSM, SD_DSM, SD_R, Vegetation, Ratio_R, Ratio_G, NDRG, NDBR, NDBG, Shadow, RB, Max_diff,	22	93.3	0.92	Green, Red, Blue, DSM, Vegetation, Ratio_G, Ratio_Blue, NDRG, NDGR, NDGB, RB, NDBR, NDRB, Shadow, NDBG, Max_diff, Brightness	28	92.11	0.914
	Textural	GLCM_Mean_Red, GLCM_Mean_DSM, GLCM_SD_Green, GLCM_Correlation_DSM, GLCM_ Dissimilarity _Blue, GLCM_Ang 2nd moment_Green				GLCM_Mean_DSM, GLCM_Mean_Blue, GLCM_ Entropy _Blue, GLCM_ Entropy _Red, GLCM_ Entropy _Green, GLCM_ Homogeneity _Red, GLDV_Mean_Blue, GLDV_Conrast_Blue, GLDV_Entropy_Red.			
	Geometrical	Shape_index				Shape_index, Length/Width			
200	Spectral	Red, Green, Blue, DSM, SD_DSM, Vegetation, Ratio_R, Ratio_G, Ratio_B, NDRG, NDGR, NDBR, NDGB, NDRB, Shadow Max_diff,	27	93.78	0.93	Red, Green, Blue, DSM, SD_DSM, Vegetation, Ratio_G, Ratio_B, NDGR, NDBG, Shadow, RB, Max_diff	21	92.3	0.92
	Textural	GLCM_Mean_Red, GLCM_Mean_Blue, GLCM_SD_DSM, GLCM_SD_Green, GLCM_ Homogeneity _DSM, GLCM_Ang 2nd moment _Blue, GLCM_ Dissimilarity _Blue, GLCM_ Correlation _Rede, GLCM_ Homogeneity _Red				GLCM_Mean_Blue, GLCM_ Homogeneity _Green, GLCM_Entropy_R, GLCM_ Correlation _Blue, GLDV_Mean_Blue, GLDV_Entropy_DSM			
	Geometrical	Shape_index, Length/Width				Length/Width, Compactness			

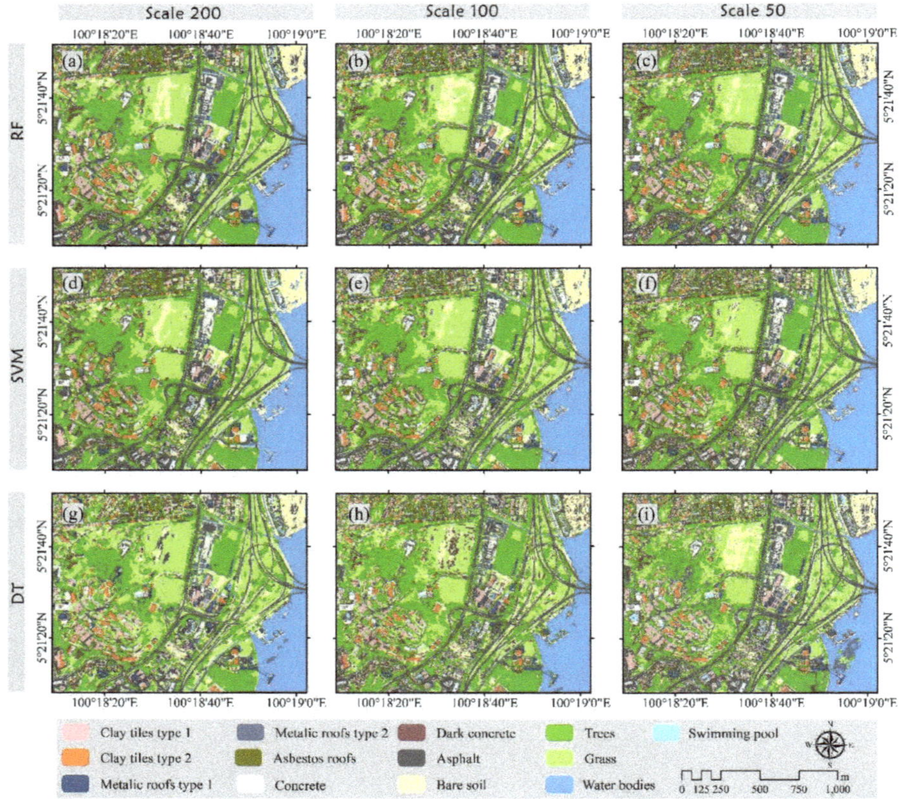

Figure 4. Single-scale (SS) classification results: (**a**–**c**) Random forest (RF) classification at scales of 200, 100, and 50, respectively; (**d**–**f**) SVM classification at scales of 200, 100, and 50, respectively; and (**g**–**i**) decision tree (DT) classification at scales of 200, 100, and 50, respectively.

Table 5. Performance of the extraction of LULC classes using RF, SVM, and DT at single scales.

	SS-RF								
Class	**SP 200**			**SP 100**			**SP 50**		
	Precision	**Recall**	***F*-Measure**	**Precision**	**Recall**	***F*-Measure**	**Precision**	**Recall**	***F*-Measure**
Water bodies	1.000	0.980	0.990	1.000	0.997	0.999	1.000	0.997	0.999
Bare soil	0.879	0.918	0.898	0.653	0.404	0.499	0.850	0.942	0.893
Grass	0.790	0.352	0.487	0.871	0.878	0.874	0.871	0.994	0.928
Asphalt	0.855	0.757	0.803	0.674	0.626	0.649	0.771	0.569	0.655
Metallic roofs 2	0.992	0.862	0.922	1.000	0.924	0.961	1.000	1.000	1.000
Trees	0.580	0.830	0.683	0.790	0.999	0.883	0.999	0.973	0.986
Dark concrete	0.741	0.961	0.836	1.000	0.899	0.947	0.994	0.969	0.981
Metallic roofs 1	1.000	1.000	1.000	1.000	1.000	1.000	1.000	1.000	1.000
Concrete	1.000	0.925	0.961	1.000	0.906	0.951	0.983	0.914	0.947
Clay tiles type 2	1.000	0.964	0.981	0.980	0.874	0.924	0.985	0.983	0.984
Clay tiles type 1	1.000	0.996	0.998	0.474	0.902	0.621	0.977	1.000	0.989
Asbestos	0.913	0.951	0.932	0.743	0.838	0.788	0.624	0.849	0.719
OA		88.4%			84.25%			92.2%	
Kappa		0.873			0.827			0.914	

Table 5. *Cont.*

Class	SS-SVM								
	SP 200			SP 100			SP 50		
	Precision	Recall	*F*-Measure	Precision	Recall	*F*-Measure	Precision	Recall	*F*-Measure
Water bodies	0.964	1.000	0.982	1.000	0.998	0.999	1.000	1.000	1.000
Bare soil	0.861	0.823	0.842	0.940	0.974	0.956	0.936	1.000	0.967
Grass	0.790	0.742	0.765	0.787	0.866	0.825	0.835	0.874	0.854
Asphalt	0.861	0.836	0.849	0.906	0.854	0.879	0.878	0.745	0.806
Metallic roofs 2	0.854	1.000	0.921	0.891	1.000	0.942	0.998	1.000	0.999
Trees	0.943	0.939	0.941	0.941	0.953	0.947	0.974	0.963	0.969
Dark concrete	0.835	0.624	0.714	0.999	0.651	0.788	0.977	0.633	0.768
Metallic roofs 1	0.969	0.808	0.881	1.000	0.848	0.918	1.000	1.000	1.000
Concrete	1.000	0.907	0.951	1.000	0.993	0.996	1.000	0.942	0.970
Clay tiles type 2	1.000	1.000	1.000	1.000	1.000	1.000	0.928	0.982	0.954
Clay tiles type 1	0.485	0.992	0.651	0.485	0.911	0.633	0.474	0.873	0.614
Asbestos	0.983	0.933	0.958	0.889	0.967	0.926	0.750	0.952	0.839
OA		88%			90.5%			89.7%	
Kappa		0.868			0.896			0.886	
	SS-DT								
Water bodies	1.000	0.788	0.881	1.000	1.000	1.000	0.915	1.000	0.956
Bare soil	0.889	0.853	0.871	0.758	0.886	0.817	0.729	0.924	0.815
Grass	0.790	0.246	0.375	0.871	0.244	0.381	0.864	0.687	0.765
Asphalt	0.752	0.812	0.781	0.707	0.677	0.692	0.710	0.541	0.614
Metallic roofs 2	0.957	0.858	0.905	0.998	1.000	0.999	0.998	0.846	0.916
Trees	0.529	0.895	0.665	0.806	0.945	0.870	0.904	0.965	0.934
Dark concrete	0.732	0.966	0.833	0.946	0.843	0.891	0.943	0.925	0.934
Metallic roofs 1	1.000	0.964	0.982	1.000	1.000	1.000	1.000	1.000	1.000
Concrete	1.000	0.925	0.961	0.983	0.911	0.946	0.975	1.000	0.987
Clay tiles type 2	0.948	0.521	0.672	0.886	0.902	0.894	1.000	0.823	0.903
Clay tiles type 1	0.474	0.948	0.632	0.385	0.882	0.536	1.000	1.000	1.000
Asbestos	0.676	1.000	0.806	0.756	0.833	0.793	0.639	0.852	0.730
OA		79%			83.4%			88.1%	
K		0.771			0.819			0.869	

Utilizing the preliminary information acquired from the SS classification of the RF, SVM, and DT algorithms, the hierarchical classification was conducted for the first study area. The results are shown in Figure 5. Table 6 illustrates the OA, *K*, and class-specific accuracies of the first study area using the hierarchical RF, SVM, and DT classification algorithms. Similar to SS classification, the MS-RF classification was superior with an OA of 94.40% and a *K* of 0.938, followed by MS-SVM with an OA of 92.50% and a *K* of 0.917 and MS-DT with an OA of 91.60% and a *K* of 0.908.

Table 6. Class-specific accuracy measures for the MS RF, SVM, and DT of the first study area.

Class	MS-RF			MS-SVM			MS-DT		
	Precision	Recall	*F*-Measure	Precision	Recall	*F*-Measure	Precision	Recall	*F*-Measure
Water bodies	1.000	0.967	0.983	1.000	0.998	0.999	1.000	1.000	1.000
Bare soil	0.761	0.754	0.757	0.919	1.000	0.958	0.899	0.810	0.852
Grass	0.871	0.933	0.901	0.996	0.892	0.941	0.869	0.698	0.774
Asphalt	0.863	0.916	0.889	0.851	0.944	0.895	0.765	0.705	0.734
Metallic roofs 2	1.000	0.999	1.000	0.998	1.000	0.999	0.998	0.873	0.932
Trees	0.972	0.972	0.972	0.941	0.999	0.969	0.888	0.999	0.940
Dark concrete	0.832	0.963	0.893	0.996	0.646	0.783	0.768	0.950	0.849
Metallic roofs 1	1.000	1.000	1.000	1.000	1.000	1.000	1.000	1.000	1.000
Concrete	1.000	0.925	0.961	1.000	0.993	0.996	0.975	1.000	0.987
Clay tiles type 2	1.000	0.964	0.981	1.000	1.000	1.000	0.931	0.878	0.904
Clay tiles type 1	1.000	0.923	0.960	0.485	0.904	0.631	1.000	1.000	1.000
Asbestos	0.962	0.969	0.966	0.983	0.927	0.955	0.935	0.975	0.954
OA		94.40%			92.50%			91.60%	
K		0.938			0.917			0.908	

Compared to SS classification, the hierarchical classification results noticeably improved the extraction of urban LULC classes. For instance, an improvement of 2.24% in the OA was observed in the MS-RF algorithm, along with a significant improvement in the differentiation and extraction of asbestos cement, concrete, and asphalt roofs. Similarly, the MS-SVM classification exhibited an enhancement in the class-specific accuracies, OA, and K of trees, grass, and asphalt classes. The OA accuracy of MS-DT showed an improvement with 3.57%, which achieved an overall improvement in the extraction of trees, grass, and asbestos cement roofs.

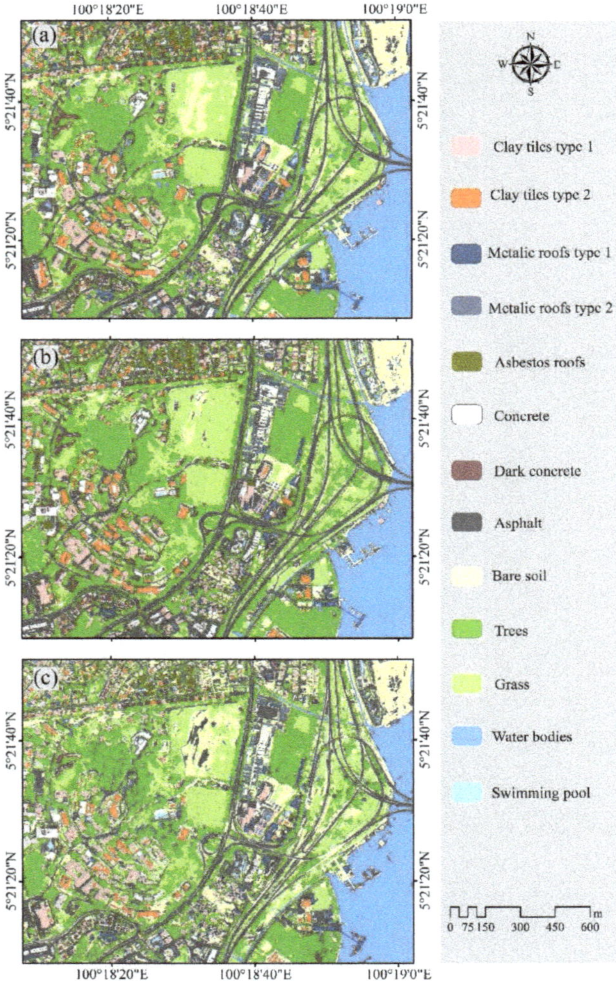

Figure 5. Results of MS classification using the integrated approach of the first dataset: (**a**) RF, (**b**) SVM, and (**c**) DT.

To validate the transferability of the hierarchical classification approach, the MS-RF, MS-SVM, and MS-DT classifications were applied in the second study area using the sample statistic file derived from the image of the first study area. Figure 6 and Table 7 show the classification results for the second study area. The results of the second dataset showed that the MS-SVM classification was superior with an OA of 94.45% and a *K* of 0.938, followed by MS-RF with an OA of 92.46% and a *K* of 0.916 and MS-DT with an OA of 90.46% and a *K* of 0.893. The proposed hierarchical classification approach demonstrates excellent potential for the detailed mapping of heterogenous urban areas from RGB-UAV images and DSM. The proposed methodology can be adopted for various areas with different LULCs.

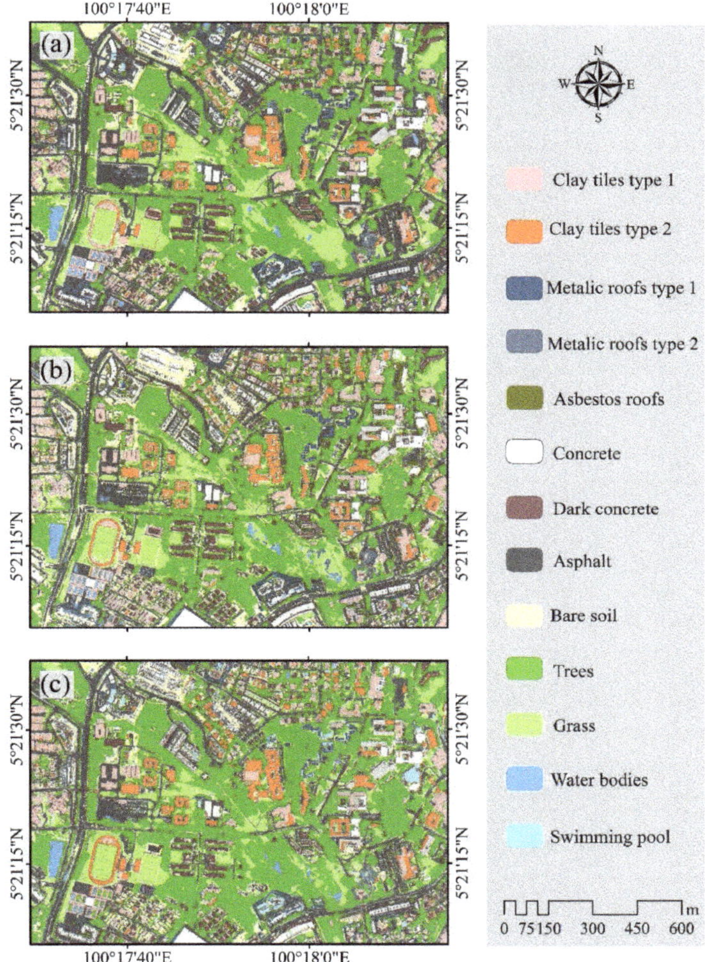

Figure 6. Results of MS classification using the integrated approach of the second study area: (**a**) RF, (**b**) SVM, and (**c**) DT.

Table 7. Class-specific accuracy measures for the MS RF, SVM, and DT of the second study area.

	RF			SVM			DT		
Class	**Precision**	**Recall**	***F*-Measure**	**Precision**	**Recall**	***F*-Measure**	**Precision**	**Recall**	***F*-Measure**
Water bodies	0.531	0.989	0.691	0.594	0.969	0.737	0.585	0.974	0.731
Bare soil	0.960	0.884	0.921	0.982	0.960	0.971	0.953	0.788	0.862
Grass	0.972	0.650	0.779	0.984	0.786	0.874	0.912	1.000	0.954
Asphalt	0.925	1.000	0.961	0.976	1.000	0.988	0.973	1.000	0.986
Metallic roofs 2	1.000	1.000	1.000	0.969	0.680	0.800	0.745	1.000	0.854
Trees	1.000	0.959	0.979	1.000	0.984	0.992	1.000	0.618	0.764
Dark concrete	0.971	0.975	0.973	0.992	1.000	0.996	0.901	0.990	0.944
Metallic roofs 1	0.992	1.000	0.996	0.980	1.000	0.990	0.986	1.000	0.993
Concrete	0.953	1.000	0.976	1.000	1.000	1.000	0.953	1.000	0.976
Clay tiles type 2	1.000	0.961	0.980	1.000	0.983	0.991	1.000	0.856	0.923
Clay tiles type 1	1.000	1.000	1.000	1.000	1.000	1.000	1.000	1.000	1.000
OA		92.46%			94.45%			90.46%	
K		0.916			0.938			0.893	

5. Discussion

Considering that segmenting UHSR UAV-based images of a heterogeneous and complex urban landscape is a challenging task in GEOBIA, the selection of the optimum SP(s) is an imperative step to ensure that different landscapes are well delineated at different scales. This study conducted a detailed mapping of a heterogeneous urban area, an area covered with various natural and impervious surfaces that vary in size and structure, from the fusion of the UAV-based orthophoto and DSM by improving the GEOBIA frameworks with different solutions to some of the issues stated in related studies section. An adaptive MS segmentation that assimilates an unsupervised image segmentation evaluation metric (i.e., *F*-measure) and ML algorithms were proposed to identify the optimal MS parameters for extracting the detailed urban LULC classes.

Although GEOBIA can leverage the computation and use various features in the classification process, adding many features can reduce the classification accuracy and increase the computational time. CFS and SVM were used in this study to select the most significant features computed for each level from the optimized three-scale levels. An object's spectral, geometrical, and textural feature values are different because the size of the generated image objects (i.e., roofing material and roads) varies on the basis of the selected scale level. CFS obtained a maximum OA of 93.78% ($K = 0.93$) at the scale level of 200 by selecting 27 significant features, whereas SVM obtained a minimum accuracy at the scale level of 50, with the value of OA = 91.61 ($K = 0.91$) by selecting 21 features. CFS and SVM selected a set of features that vary in terms of the number and type in each segmentation level. However, various spectral features, such as R, B, DSM, Ratio-G, Ratio-B, Vegetation, the normalized difference between the red and green channels (NDRG), the standard deviation of image objects derived from the DSM (SD-DSM), and the normalized difference between the blue and red channels (NDBR), were commonly selected at all levels. The selection and incorporation of DSM-derived features, along with other selected features, remarkably contributed in the differentiation of spectrally similar classes, such as asbestos cement roofs, dark concrete roofs, old pavements, and asphalt. In a complex landscape without height information, an ML model might erroneously categorize bare soil as a roofing material or the opposite in accordance with the parallel spectral and textural characteristics. Al-Najjar et al. [7] utilized the fusion of DSM and optical images to generate generic automatic LULC classes for a complex urban area.

As stated in Section 3, the SS RF, SVM, and DT classification models were initially examined in the first study area at the optimal scales, identified through the *F*-measure along with the significant features, and selected by CFS at each optimal scale level. The SS classification results varied from one level to another when each SS model was applied, and the OA classification accuracies ranged from 79% to 92.2%. The comparison of SS classification maps showed that clear misclassifications were obtained by the DT algorithm, especially on clay tiles, asphalt, grass, and trees.

Following the FS and SS classification, iterative adaptive MS classification models were developed in the first study area and were tested in the second study area. The adopted hierarchical classification scheme of the first study area using the RF algorithm showed outstanding performance (OA = 94.4%) on the extraction of urban LULC classes compared to SVM (OA = 92.5%) and DT (OA = 91.6%). However, slight confusion was observed between different classes, as shown in Figure 7. For instance, MS-SVM showed a minor confusion between metal type1, asphalt, bare soil, and grass, as demonstrated in Figure 7b,k. Similarly, MS-DT showed a great confusion amongst the asphalt, grass, and metal type 1, as shown in Figure 7c. In addition, a remarkable confusion was found between some asbestos cement roofs, dark concrete roofs, asphalt, clay tiles type 2, and bare soil classes in some areas when MS-DT was used, as shown in Figure 7f,i,l. MS-RF showed an outstanding performance but exhibited a minor confusion between some asphalt objects that mixed with shadows as water bodies, as depicted in Figure 7d,j. The comparison of SS and MS approaches showed that the accuracy of some classes (i.e., trees, type 1 clay tiles, and asbestos classes) clearly improved with the use of the proposed approach.

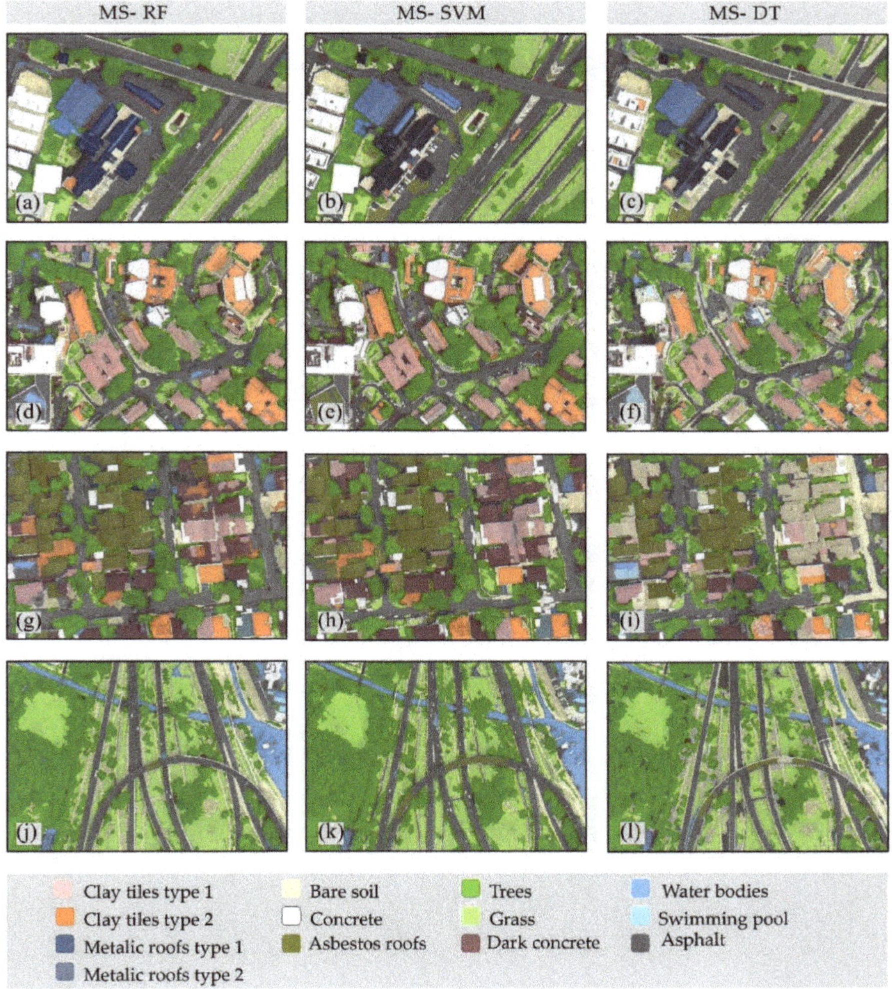

Figure 7. MS classification results for four different regions of the first study area using RF, SVM, and DT: (**a–c**) first region, (**d–f**) second region, (**g–i**) third region, and (**j–l**) fourth region.

The applicability of the adopted scheme in the second study area indicated that MS-RF and MS-SVM exhibited relatively similar classification results. However, the MS-SVM algorithm (OA = 94.45% and $K = 0.938$) was superior to RF (OA = 92.46% and $K = 0.816$), with a slight improvement in the OA and K values. All MS algorithms showed some degrees of confusion between some objects with grass and water bodies, as shown in Figure 8j– l, which may be attributed to the existence of new water objects that vary in spectral characteristics as the second study area was classified on the basis of the sample statistics derived from the first study area. As represented in Figure 8j, the water body was poorly classified using MS-RF and showed confusion with the grass class. In this scenario, the present water body in the second study area was a pond with extremely different reflectance from the training samples obtained in the first study area and was more obvious in the MS-RF classified map compared to other algorithms because of the RF algorithm sensitivity to training. RF is sensitive to the size of training samples and the selection of an accurate representative of each class for classification [84].

Moreover, utilizing MS-DT resulted in misclassification between the tree and dark concrete classes, and between the grass and tree classes (Figure 8c). DT demonstrated a minor confusion between the bare soil, type 2 clay tiles, dark concrete, and asphalt, as shown in Figure 8c,f,i,l. As shown in Figure 8d,e, most of the roof types were categorized in an extremely similar manner by utilizing RF and SVM. MS-SVM showed a minor confusion in some areas between the asphalt and dark concrete, as shown in Figure 8h, whereas MS-RF showed a relatively better differentiation between asphalt and dark concrete for the same area, as shown in Figure 8g.

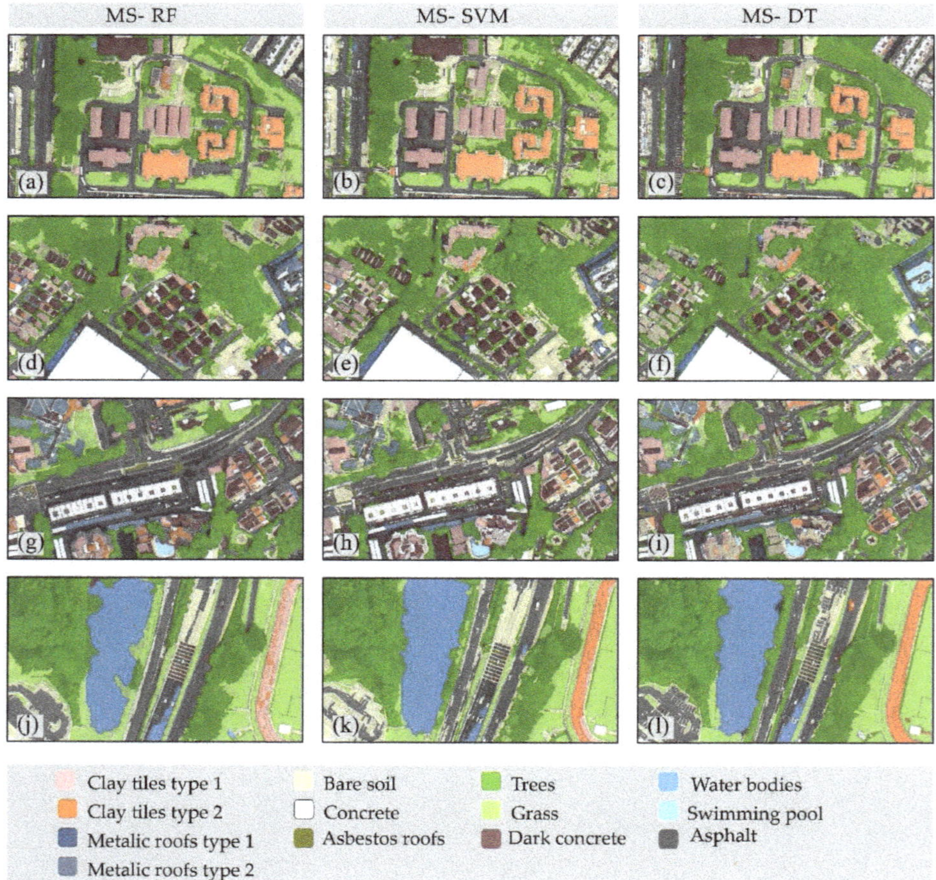

Figure 8. MS classification results of four different subsets of the second study area using RF, SVM, and DT: (**a–c**) first subset, (**d–f**) second subset, (**g–i**) third subset, and (**j–l**) fourth subset.

6. Conclusions

Accurate and up-to-date urban LULC information is crucial for urban planning, management and environmental applications. UAVs allow the acquisition of remotely sensed data with UHSR, as high as 1 cm, in a flexible and inexpensive manner, significantly contributing to the initiation of a wide spectrum of applications. This study aimed to achieve an accurate and detailed urban LULC classification in a heterogeneous landscape using GEOBIA and ML models from UHSR drone-based images. Given the high-level details of UAV images and the limited amount of spectral information, a MS GEOBIA approach that integrates MS image segmentation evaluation, MS FS, and hierarchical

ML classification algorithms was used to generate detailed LULC urban maps from the fusion of orthophotos and DSMs. Two UAV-based images were used to implement and evaluate the efficiency of the proposed method. Three commonly used supervised ML models, namely, RF, SVM, and DT, were compared within the MS/hierarchical segmentation and classification approach. The MS-RF classification achieved the highest accuracy, with an OA of 94.40% and a K of 0.938, followed by MS-SVM with an OA of 92.50% and a K of 0.917 and MS-DT with an OA of 91.60% and a K of 0.908. The applicability of the proposed approach to the dataset of the second study area showed excellent performance when MS-SVM and MS-RF were used. The proposed framework exhibited enormous potential for the detailed mapping of heterogeneous urban areas from UHSR RGB and DSM images. The results obtained from this approach can serve as vital information and input for scientists, decision makers, and city planners.

Author Contributions: B.K. and S.M. acquired the UAV data; M.B.A.G., B.K. and R.A.-R. conceptualized and performed the analysis; M.B.A.G., B.K., R.A.-R. and V.S. wrote the manuscript; N.U. supervised the funding acquisition; M.B.A.G., R.A.-R., A.S. and H.Z.M.S. edited, restructured, and professionally optimized the manuscript. All authors have read and agreed to the published version of the manuscript.

Funding: The APC is supported by the RIKEN Centre for Advanced Intelligence Project (AIP), Tokyo, Japan.

Acknowledgments: The authors would like to thank the University of Sharjah, RIKEN Centre for AIP, Japan, and Universiti Putra Malaysia for providing all facilities during the research.

Conflicts of Interest: The authors declare no conflict of interest.

References

1. Glade, T. Landslide occurrence as a response to land use change: A review of evidence from New Zealand. *Catena* **2003**, *51*, 297–314. [CrossRef]
2. Scanlon, B.R.; Reedy, R.C.; Stonestrom, D.A.; Prudic, D.E.; Dennehy, K.F. Impact of land use and land cover change on groundwater recharge and quality in the southwestern US. *Glob. Chang. Biol.* **2005**, *11*, 1577–1593. [CrossRef]
3. Bonato, M.; Cian, F.; Giupponi, C. Combining LULC data and agricultural statistics for A better identification and mapping of High nature value farmland: A case study in the veneto Plain, Italy. *Land Use Policy* **2019**, *83*, 488–504. [CrossRef]
4. Acheampong, M.; Yu, Q.; Enomah, L.D.; Anchang, J.; Eduful, M. Land use/cover change in Ghana's oil city: Assessing the impact of neoliberal economic policies and implications for sustainable development goal number one—A remote sensing and GIS approach. *Land Use Policy* **2018**, *73*, 373–384. [CrossRef]
5. McDowell, R.W.; Snelder, T.; Harris, S.; Lilburne, L.; Larned, S.T.; Scarsbrook, M.; Curtis, A.; Holgate, B.; Phillips, J.; Taylor, K. The land use suitability concept: Introduction and an application of the concept to inform sustainable productivity within environmental constraints. *Ecol. Indic.* **2018**, *91*, 212–219. [CrossRef]
6. Lin, Y.; Jiang, M.; Yao, Y.; Zhang, L.; Lin, J. Use of UAV oblique imaging for the detection of individual trees in residential environments. *Urban For. Urban Green.* **2015**, *14*, 404–412. [CrossRef]
7. Al-najjar, H.A.H.; Kalantar, B.; Pradhan, B.; Saeidi, V. Land cover classification from fused DSM and UAV images using convolutional neural networks. *Remote Sens.* **2019**, *11*, 1461. [CrossRef]
8. Ahmed, O.S.; Shemrock, A.; Chabot, D.; Dillon, C.; Williams, G.; Wasson, R.; Franklin, S.E. Hierarchical land cover and vegetation classification using multispectral data acquired from an unmanned aerial vehicle. *Int. J. Remote Sens.* **2017**, *38*, 2037–2052. [CrossRef]
9. Kalantar, B.; Mansor, S.B.; Sameen, M.I.; Pradhan, B.; Shafri, H.Z.M. Drone-Based land-cover mapping using a fuzzy unordered rule induction algorithm integrated into object-based image analysis. *Int. J. Remote Sens.* **2017**, *38*, 2535–2556. [CrossRef]
10. Zhang, X.; Chen, G.; Wang, W.; Wang, Q.; Dai, F. Object-Based land-cover supervised classification for very-high-resolution UAV images using stacked denoising autoencoders. *IEEE J. Sel. Top. Appl. Earth Obs. Remote Sens.* **2017**, *10*, 3373–3385. [CrossRef]
11. Akar, Ö. Mapping land use with using Rotation Forest algorithm from UAV images. *Eur. J. Remote Sens.* **2017**, *50*, 269–279. [CrossRef]

12. Mancini, A.; Frontoni, E.; Zingaretti, P. A multi/hyper-spectral imaging system for land use/land cover using unmanned aerial systems. In Proceedings of the 2016 International Conference on Unmanned Aircraft Systems (ICUAS), Arlington, VA, USA, 7–10 June 2016; Institute of Electrical and Electronics Engineers (IEEE): Piscataway, NJ, USA, 2016; pp. 1148–1155.
13. Iizuka, K.; Itoh, M.; Shiodera, S.; Matsubara, T.; Dohar, M.; Watanabe, K. Advantages of unmanned aerial vehicle (UAV) photogrammetry for landscape analysis compared with satellite data: A case study of postmining sites in Indonesia. *Cogent Geosci.* **2018**, *4*, 1–15. [CrossRef]
14. Randall, M.; Fensholt, R.; Zhang, Y.; Jensen, M.B. Geographic object based image analysis of world view-3 imagery for urban hydrologic modelling at the catchment scale. *Water* **2019**, *11*, 1133. [CrossRef]
15. Mboga, N.; Georganos, S.; Grippa, T.; Lennert, M.; Vanhuysse, S.; Wolff, E. Fully convolutional networks and geographic object-based image analysis for the classification of VHR imagery. *Remote Sens.* **2019**, *11*, 597. [CrossRef]
16. Gu, H.; Han, Y.; Yang, Y.; Li, H.; Liu, Z.; Soergel, U.; Blaschke, T.; Cui, S. An efficient parallel multi-scale segmentation method for remote sensing imagery. *Remote Sens.* **2018**, *10*, 590. [CrossRef]
17. Gibril, M.B.A.; Idrees, M.O.; Yao, K.; Shafri, H.Z.M. Integrative image segmentation optimization and machine learning approach for high quality land-use and land-cover mapping using multisource remote sensing data. *J. Appl. Remote Sens.* **2018**, *12*, 1. [CrossRef]
18. De Castro, A.I.; Torres-Sánchez, J.; Peña, J.M.; Jiménez-Brenes, F.M.; Csillik, O.; López-Granados, F. An automatic random forest-OBIA algorithm for early weed mapping between and within crop rows using UAV imagery. *Remote Sens.* **2018**, *10*, 285. [CrossRef]
19. Komárek, J.; Klouček, T.; Prošek, J. The potential of unmanned aerial systems: A tool towards precision classification of hard-to-distinguish vegetation types. *Int. J. Appl. Earth Obs. Geoinf.* **2018**, *71*, 9–19. [CrossRef]
20. Kamal, M.; Kanekaputra, T.; Hermayani, R.; Juniansah, A. Geographic object based image analysis (GEOBIA) for mangrove canopy delineation using aerial photography. *IOP Conf. Ser. Earth Environ. Sci.* **2019**, *313*, 12048. [CrossRef]
21. Mishra, N.; Mainali, K.; Shrestha, B.; Radenz, J.; Karki, D. Species-Level vegetation mapping in a Himalayan treeline ecotone using unmanned aerial system (UAS) imagery. *ISPRS Int. J. Geo Inf.* **2018**, *7*, 445. [CrossRef]
22. White, R.; Bomber, M.; Hupy, J.; Shortridge, A. UAS-GEOBIA approach to sapling identification in jack pine barrens after fire. *Drones* **2018**, *2*, 40. [CrossRef]
23. Gonçalves, J.; Pôças, I.; Marcos, B.; Mücher, C.A.; Honrado, J.P. SegOptim—A new R package for optimizing object-based image analyses of high-spatial resolution remotely-sensed data. *Int. J. Appl. Earth Obs. Geoinf.* **2019**, *76*, 218–230. [CrossRef]
24. Blaschke, T.; Hay, G.J.; Kelly, M.; Lang, S.; Hofmann, P.; Addink, E.; Queiroz Feitosa, R.; Van der Meer, F.; Van der Werff, H.; Van Coillie, F.; et al. Geographic object-based image analysis—Towards a new paradigm. *ISPRS J. Photogramm. Remote Sens.* **2014**, *87*, 180–191. [CrossRef]
25. Ma, L.; Li, M.; Ma, X.; Cheng, L.; Du, P.; Liu, Y. A review of supervised object-based land-cover image classification. *ISPRS J. Photogramm. Remote Sens.* **2017**, *130*, 277–293. [CrossRef]
26. Vincent, L.; Soille, P. Watersheds in digital spaces: An efficient algorithm based on immersion simulations. *IEEE Trans. Pattern Anal. Mach. Intell.* **1991**, *13*, 583–598. [CrossRef]
27. Deren, L.; Zhang, G.; Wu, Z.; Yi, L. An edge embedded marker-based watershed algorithm for high spatial resolution remote sensing image segmentation. *IEEE Trans. Image Process.* **2010**, *19*, 2781–2787. [CrossRef]
28. Benz, U.C.; Hofmann, P.; Willhauck, G.; Lingenfelder, I.; Heynen, M. Multi-Resolution, object-oriented fuzzy analysis of remote sensing data for GIS-ready information. *ISPRS J. Photogramm. Remote Sens.* **2004**, *58*, 239–258. [CrossRef]
29. Comaniciu, D.; Meer, P. Mean shift: A robust approach toward feature space analysis. *IEEE Trans. Pattern Anal. Mach. Intell.* **2002**, *24*, 603–619. [CrossRef]
30. Wang, Y.; Meng, Q.; Qi, Q.; Yang, J.; Liu, Y. Region merging considering within-and between-segment heterogeneity: An improved hybrid remote-sensing image segmentation method. *Remote Sens.* **2018**, *10*, 781. [CrossRef]
31. Kavzoglu, T.; Tonbul, H. A comparative study of segmentation quality for multi-resolution segmentation and watershed transform. In Proceedings of the 8th International Conference on Recent Advances in Space Technologies (RAST), Istanbul, Turkey, 19–22 June 2017; pp. 113–117.

32. Zhang, X.; Du, S. Learning selfhood scales for urban land cover mapping with very-high-resolution satellite images. *Remote Sens. Environ.* **2016**, *178*, 172–190. [CrossRef]
33. Xiao, P.; Zhang, X.; Zhang, H.; Hu, R.; Feng, X. Multiscale optimized segmentation of urban green cover in high resolution remote sensing image. *Remote Sens.* **2018**, *10*, 1813. [CrossRef]
34. He, Y.; Weng, Q. *High Spatial Resolution Remote Sensing: Data, Analysis, and Applications*; CRC Press: Boca Raton, FL, USA, 2018; ISBN 9781498767682.
35. Shen, Y.; Chen, J.; Xiao, L.; Pan, D. Optimizing multiscale segmentation with local spectral heterogeneity measure for high resolution remote sensing images. *ISPRS J. Photogramm. Remote Sens.* **2019**, *157*, 13–25. [CrossRef]
36. Wang, Y.; Qi, Q.; Liu, Y.; Jiang, L.; Wang, J. Unsupervised segmentation parameter selection using the local spatial statistics for remote sensing image segmentation. *Int. J. Appl. Earth Obs. Geoinf.* **2019**, *81*, 98–109. [CrossRef]
37. Drăguţ, L.; Tiede, D.; Levick, S.R. ESP: A tool to estimate scale parameter for multiresolution image segmentation of remotely sensed data. *Int. J. Geogr. Inf. Sci.* **2010**, *24*, 859–871. [CrossRef]
38. Drăguţ, L.; Csillik, O.; Eisank, C.; Tiede, D. Automated parameterisation for multi-scale image segmentation on multiple layers. *ISPRS J. Photogramm. Remote Sens.* **2014**, *88*, 119–127. [CrossRef] [PubMed]
39. Johnson, B.; Bragais, M.; Endo, I.; Magcale-Macandog, D.; Macandog, P. Image segmentation parameter optimization considering within-and between-segment heterogeneity at multiple scale levels: Test case for mapping residential areas using landsat imagery. *ISPRS Int. J. Geo Inf.* **2015**, *4*, 2292–2305. [CrossRef]
40. Johnson, B.; Xie, Z. Unsupervised image segmentation evaluation and refinement using a multi-scale approach. *ISPRS J. Photogramm. Remote Sens.* **2011**, *66*, 473–483. [CrossRef]
41. Espindola, G.M.; Camara, G.; Reis, I.A.; Bins, L.S.; Monteiro, A.M. Parameter selection for region-growing image segmentation algorithms using spatial autocorrelation. *Int. J. Remote Sens.* **2006**, *27*, 3035–3040. [CrossRef]
42. Yang, J.; He, Y.; Weng, Q. An automated method to parameterize segmentation scale by enhancing intrasegment homogeneity and intersegment heterogeneity. *IEEE Geosci. Remote Sens. Lett.* **2015**, *12*, 1282–1286. [CrossRef]
43. Ming, D.; Li, J.; Wang, J.; Zhang, M. Scale parameter selection by spatial statistics for GeOBIA: Using mean-shift based multi-scale segmentation as an example. *ISPRS J. Photogramm. Remote Sens.* **2015**, *106*, 28–41. [CrossRef]
44. Yang, J.; Li, P.; He, Y. A multi-band approach to unsupervised scale parameter selection for multi-scale image segmentation. *ISPRS J. Photogramm. Remote Sens.* **2014**, *94*, 13–24. [CrossRef]
45. Kamal, M.; Phinn, S.; Johansen, K. Object-Based approach for multi-scale mangrove composition mapping using multi-resolution image datasets. *Remote Sens.* **2015**, *7*, 4753–4783. [CrossRef]
46. Pedergnana, M.; Marpu, P.R.; Mura, M.D.; Benediktsson, J.A.; Bruzzone, L. A novel technique for optimal feature selection in attribute profiles based on genetic algorithms. *IEEE Trans. Geosci. Remote Sens.* **2013**, *51*, 3514–3528. [CrossRef]
47. Ma, L.; Fu, T.; Blaschke, T.; Li, M.; Tiede, D.; Zhou, Z.; Ma, X.; Chen, D. Evaluation of feature selection methods for object-based land cover mapping of unmanned aerial vehicle imagery using random forest and support vector machine classifiers. *ISPRS Int. J. Geo Inf.* **2017**, *6*, 51. [CrossRef]
48. Martha, T.R.; Kerle, N.; Van Westen, C.J.; Jetten, V.; Kumar, K.V. Segment optimization and data-driven thresholding for knowledge-based landslide detection by object-based image analysis. *IEEE Trans. Geosci. Remote Sens.* **2011**, *49*, 4928–4943. [CrossRef]
49. Ridha, M.; Pradhan, B. An improved algorithm for identifying shallow and deep-seated landslides in dense tropical forest from airborne laser scanning data. *Catena* **2018**, *167*, 147–159. [CrossRef]
50. Al-Ruzouq, R.; Shanableh, A.; Barakat, A.; Gibril, M.; AL-Mansoori, S.; Al-Ruzouq, R.; Shanableh, A.; Barakat, A.; Gibril, M.; AL-Mansoori, S. Image segmentation parameter selection and ant colony optimization for date palm tree detection and mapping from very-high-spatial-resolution aerial imagery. *Remote Sens.* **2018**, *10*, 1413. [CrossRef]
51. Hamedianfar, A.; Gibril, M.B.A. Large-Scale urban mapping using integrated geographic object-based image analysis and artificial bee colony optimization from worldview-3 data. *Int. J. Remote Sens.* **2019**, *40*, 6796–6821. [CrossRef]

52. Hamedianfar, A.; Gibril, M.B.A.; Pellikka, P.K.E. Synergistic use of particle swarm optimization, artificial neural network, and extreme gradient boosting algorithms for urban LULC mapping from WorldView-3 images. *Geocart. Int.* **2020**, 1–19. [CrossRef]
53. Shanableh, A.; Al-ruzouq, R.; Gibril, M.B.A.; Flesia, C. Spatiotemporal mapping and monitoring of whiting in the semi-enclosed gulf using moderate resolution imaging spectroradiometer (MODIS) time series images and a generic ensemble. *Remote Sens.* **2019**, *11*, 1193. [CrossRef]
54. Shahi, K.; Shafri, H.Z.M.; Hamedianfar, A. Road condition assessment by OBIA and feature selection techniques using very high-resolution WorldView-2 imagery. *Geocart. Int.* **2017**, *32*, 1389–1406. [CrossRef]
55. Cao, J.; Leng, W.; Liu, K.; Liu, L.; He, Z.; Zhu, Y. Object-based mangrove species classification using unmanned aerial vehicle hyperspectral images and digital surface models. *Remote Sens.* **2018**, *10*, 89. [CrossRef]
56. Akar, Ö. The Rotation Forest algorithm and object-based classification method for land use mapping through UAV images. *Geocart. Int.* **2018**, *33*, 538–553. [CrossRef]
57. Liu, L.; Liu, X.; Shi, J.; Li, A. A land cover refined classification method based on the fusion of LiDAR data and UAV image. *Adv. Comput. Sci. Res.* **2019**, *88*, 154–162. [CrossRef]
58. Fotheringham, A.S.; Brunsdon, C.; Charlton, M. *Quantitative Geography: Perspectives on Spatial Data Analysis*; SAGE: Thousand Oaks, CA, USA, 2000.
59. Grybas, H.; Melendy, L.; Congalton, R.G. A comparison of unsupervised segmentation parameter optimization approaches using moderate-and high-resolution imagery. *GISci. Remote Sens.* **2017**, *54*, 515–533. [CrossRef]
60. Trimble, T. *ECognition Developer 8.7 Reference Book*; Trimble Germany GmbH: Munich, Germany, 2011; pp. 319–328.
61. Gevers, T.; Smeulders, A.W.M. PicToSeek: Combining color and shape invariant features for image retrieval. *IEEE Trans. Image Process.* **2000**, *9*, 102–119. [CrossRef]
62. Sirmacek, B.; Unsalan, C. Building detection from aerial images using invariant color features and shadow information. In Proceedings of the 23rd International Symposium on Computer and Information Sciences, Istanbul, Turkey, 27–29 October 2008; pp. 1–5.
63. Cretu, A.M.; Payeur, P. Building detection in aerial images based on watershed and visual attention feature descriptors. In Proceedings of the International Conference on Computer and Robot Vision CRV 2013, Regina, SK, Canada, 28–31 May 2013; pp. 265–272. [CrossRef]
64. Haralick, R.M.; Shanmugam, K.; Dinstein, I. Textural features for image classification. *IEEE Trans. Syst. Man Cybern.* **1973**, *3*, 610–621. [CrossRef]
65. Hall, M.A.; Holmes, G. Benchmarking attribute selection techniques for discrete class data mining. *IEEE Trans. Knowl. Data Eng.* **2003**, *15*, 1437–1447. [CrossRef]
66. Shang, M.; Wang, S.; Zhou, Y.; Du, C.; Liu, W. Object-based image analysis of suburban landscapes using Landsat-8 imagery. *Int. J. Digit. Earth* **2019**, *12*, 720–736. [CrossRef]
67. Al-ruzouq, R.; Shanableh, A.; Mohamed, B.; Kalantar, B. Multi-scale correlation-based feature selection and random forest classification for LULC mapping from the integration of SAR and optical Sentinel imagess. *Proc. SPIE* **2019**, *11157*. [CrossRef]
68. Hall, M.A.; Smith, L.A. Practical feature subset selection for machine learning. In Proceedings of the 21st Australasian Computer Science Conference ACSC'98, Perth, Australia, 4–6 February 1998; Volume 98, pp. 181–191.
69. Fong, S.; Zhuang, Y.; Tang, R.; Yang, X.; Deb, S. Selecting optimal feature set in high-dimensional data by swarm search. *J. Appl. Math.* **2013**, *2013*. [CrossRef]
70. Raczko, E.; Zagajewski, B. Comparison of support vector machine, random forest and neural network classifiers for tree species classification on airborne hyperspectral APEX images. *Eur. J. Remote Sens.* **2017**, *50*, 144–154. [CrossRef]
71. Tzotsos, A. Preface: A support vector machine for object-based image analysis. In *Approach Object Based Image Analysis*; Springer: Cham, Switzerland, 2008; Volume 5–8. [CrossRef]
72. Taner San, B. An evaluation of SVM using polygon-based random sampling inlandslide susceptibility mapping: The Candir catchment area(western Antalya, Turkey). *Int. J. Appl. Earth Obs. Geoinf.* **2014**, *26*, 399–412. [CrossRef]
73. Cortes, C.; Vapnik, V. Support-Vector networks. *Mach. Learn.* **1995**, *20*, 273–297. [CrossRef]

74. Pradhan, B.; Seeni, M.I.; Kalantar, B. Performance evaluation and sensitivity analysis of expert-based, statistical, machine learning, and hybrid models for producing landslide susceptibility maps. In *Laser Scanning Applications in Landslide Assessment*; Springer International Publishing: Cham, Switzerland, 2017; pp. 193–232.
75. Pal, M.; Mather, P.M. An assessment of the effectiveness of decision tree methods for land cover classification. *Remote Sens. Environ.* **2003**, *86*, 554–565. [CrossRef]
76. Hamedianfar, A.; Shafri, H.Z.M. Integrated approach using data mining-based decision tree and object-based image analysis for high-resolution urban mapping of WorldView-2 satellite sensor data. *J. Appl. Remote Sens.* **2016**, *10*, 025001. [CrossRef]
77. Pal, M.; Mather, P.M. Decision tree based classification of remotely sensed data. In Proceedings of the 22nd Asian Conference on Remote Sensing, Singapore, 5–9 November 2001; p. 9.
78. Eisavi, V.; Homayouni, S.; Yazdi, A.M.; Alimohammadi, A. Land cover mapping based on random forest classification of multitemporal spectral and thermal images. *Environ. Monit. Assess.* **2015**, *187*, 1–14. [CrossRef]
79. Berhane, T.M.; Lane, C.R.; Wu, Q.; Autrey, B.C.; Anenkhonov, O.A.; Chepinoga, V.V.; Liu, H. Decision-Tree, rule-based, and random forest classification of high-resolution multispectral imagery for wetland mapping and inventory. *Remote Sens.* **2018**, *10*, 580. [CrossRef]
80. Goetz, J.N.; Brenning, A.; Petschko, H.; Leopold, P. Computers & geosciences evaluating machine learning and statistical prediction techniques for landslide susceptibility modeling. *Comput. Geosci.* **2015**, *81*, 1–11. [CrossRef]
81. Tian, S.; Zhang, X.; Tian, J.; Sun, Q. Random forest classification of wetland landcovers from multi-sensor data in the arid region of Xinjiang, China. *Remote Sens.* **2016**, *8*, 954. [CrossRef]
82. Zerrouki, N.; Bouchaffra, D. Pixel-based or object-based: Which approach is more appropriate for remote sensing image classification? In Proceedings of the IEEE International Conference on Systems, Man, and Cybernetics (SMC), San Diego, CA, USA, 5–8 October 2014; Volume 2014, pp. 864–869.
83. Li, W.; Guo, Q. A new accuracy assessment method for one-class remote sensing classification. *IEEE Trans. Geosci. Remote Sens.* **2014**, *52*, 4621–4632. [CrossRef]
84. Millard, K.; Richardson, M. On the importance of training data sample selection in random forest image classification: A case study in peatland ecosystem mapping. *Remote Sens.* **2015**, *7*, 8489–8515. [CrossRef]

MDPI
St. Alban-Anlage 66
4052 Basel
Switzerland
Tel. +41 61 683 77 34
Fax +41 61 302 89 18
www.mdpi.com

Remote Sensing Editorial Office
E-mail: remotesensing@mdpi.com
www.mdpi.com/journal/remotesensing

www.ingramcontent.com/pod-product-compliance
Lightning Source LLC
LaVergne TN
LVHW070128170726
843515LV00007B/1766

* 9 7 8 3 0 3 6 5 1 4 5 4 3 *